AF778158

PROCEEDINGS OF THE INTERNATIONAL CONFERENCE
ON HYDROLOGY AND WATER RESOURCES,
NEW DELHI, INDIA, DECEMBER 1993

VOLUME 4

WATER RESOURCES PLANNING AND MANAGEMENT

Water Science and Technology Library

VOLUME 16/4

Editor-in-Chief
V. P. Singh, *Louisiana State University, Baton Rouge, U.S.A.*

Editorial Advisory Board

M. Anderson, *Bristol, U.K.*
L. Bengtsson, *Lund, Sweden*
A. G. Bobba, *Burlington, Ontario, Canada*
S. Chandra, *New Delhi, India*
M. Fiorentino, *Potenza, Italy*
W. H. Hager, *Zürich, Switzerland*
N. Harmancioglu, *Izmir, Turkey*
A. R. Rao, *West Lafayette, Indiana, U.S.A.*
M. M. Sherif, *Giza, Egypt*
Shan Xu Wang, *Wuhan, Hubei, P.R. China*
D. Stephenson, *Johannesburg, South Africa*

The titles published in this series are listed at the end of this volume.

PROCEEDINGS OF THE INTERNATIONAL CONFERENCE ON HYDROLOGY AND WATER RESOURCES, NEW DELHI, INDIA, DECEMBER 1993

Volume 4

WATER RESOURCES PLANNING AND MANAGEMENT

edited by

VIJAY P. SINGH

Department of Civil and Environmental Engineering, Louisiana State University, Baton Rouge, U.S.A.

and

BHISHM KUMAR

National Institute of Hydrology, Roorkee, India

KLUWER ACADEMIC PUBLISHERS

DORDRECHT / BOSTON / LONDON

A C.I.P. Catalogue record for this book is available from the Library of Congress

ISBN 0-7923-3653-4 (Vol. 4)
ISBN 0-7923-3654-2 (Set)

Published by Kluwer Academic Publishers,
P.O. Box 17, 3300 AA Dordrecht, The Netherlands.

Kluwer Academic Publishers incorporates
the publishing programmes of
D. Reidel, Martinus Nijhoff, Dr W. Junk and MTP Press.

Sold and distributed in the U.S.A. and Canada
by Kluwer Academic Publishers,
101 Philip Drive, Norwell, MA 02061, U.S.A.

In all other countries, sold and distributed
by Kluwer Academic Publishers Group,
P.O. Box 322, 3300 AH Dordrecht, The Netherlands.

Printed on acid-free paper

All Rights Reserved
© 1996 Kluwer Academic Publishers
No part of the material protected by this copyright notice may be reproduced or
utilized in any form or by any means, electronic or mechanical,
including photocopying, recording or by any information storage and
retrieval system, without written permission from the copyright owner.

Printed in the Netherlands

To

Mother India

TABLE OF CONTENTS

SECTION 3: WATER-RESOURCE SYSTEM OPERATION AND MANAGEMENT

SECTION 4: REMOTE SENSING IN WATER RESOURCES INVESTIGATIONS

PREFACE

Water is vital to life, maintenance of ecological balance, economic development, and sustenance of civilization. Planning and management of water resources and its optimal use are a matter of urgency for most countries of the world, and even more so for India with a huge population. Growing population and expanding economic activities exert increasing demands on water for varied needs--domestic, industrial, agricultural, power generation, navigation, recreation, etc. In India, agriculture is the highest user of water.

The past three decades have witnessed numerous advances as well as have presented intriguing challenges and exciting opportunities in hydrology and water resources. Compounding them has been the growing environmental consciousness. Nowhere are these challenges more apparent than in India. As we approach the twenty-first century, it is entirely fitting to take stock of what has been accomplished and what remains to be accomplished, and what accomplishments are relevant, with particular reference to Indian conditions.

India has had a long tradition of excellence spanning over several millennia in construction of water resources projects. She has produced civil engineers of highest quality, some of whom rank among the best in the world. In keeping with India's tradition of honoring distinguished engineers at their retirement, it was decided to hold an International Conference on Hydrology and Water Resources, December 20-22, 1993, in New Delhi, to celebrate the retirement of Professor Satish Chandra, Director of the National Institute of Hydrology at Roorkee, who has had a long and distinguished career spanning over three decades in hydrologic education, research, and development. Through this forum, it was hoped to bring together from all over the world researchers, educationists, and practitioners to New Delhi, India, to present the results of their latest work, to share their experiences, to interact with Indian water resources professionals, and to, hopefully, set the stage for future course of hydrology and water resources.

The objectives of this conference were therefore (1) to assess the current state of the art of hydrology and water resources, (2) to discuss applicability of hydrologic and water resources models, (3) to enhance interdisciplinary interaction, (4) to discuss research needs in hydrology and water resources, and (5) to determine the directions for further research.

We received an overwhelming response to our call for papers. The subject matter of the conference was divided into 36 major topics encompassing virtually all facets of hydrology and water resources. Each topic comprised an invited state-of-the-paper and a number of contributed papers. These contributions provided a natural blend to evolve a synthesized body of knowledge on that topic. Extended abstracts of all the invited and contributed papers were assembled in a pre-conference proceedings volume. This helped stimulate discussion and exchange of ideas during the conference.

The papers presented at the conference were reviewed by the Organizing Committee. As a result, some of the papers did not pass the review and were therefore eliminated from inclusion in the final proceedings. The papers contained in this book, WATER RESOURCES PLANNING AND MANAGEMENT, represent one part of the conference contributions. The other parts are embodied in three separate companion books, SURFACE-WATER HYDROLOGY, SUBSURFACE-WATER HYDROLOGY, and WATER-QUALITY HYDROLOGY, which are being published simultaneously. Arrangement of the contributions in these books under four different titles was a natural consequence of the diversity of topics presented in the papers. These books can be treated almost independently, although some overlap may exist between them.

This book contains six sections encompassing major aspects of water resources planning and management. Each section starts usually with invited state-of-the-art papers, followed by contributed papers. Beginning with water resources planning in the first section, the papers go on to discuss water resources development in Section 2, water resource system operation and management in Section 3, remote sensing in water resources investigation in Section 4, hydraulic hazards and their control in Section 5, and environmental consequences of water resources projects in the concluding Section 6.

The book will be of interest to researchers as well as those engaged in practice of Civil Engineering, Agricultural Engineering, Hydrology, Water Resources, Earth Resources, Forestry, and Environmental Sciences. The graduate students as well as those wishing to conduct research in water resources planning and management will find this book to be of particular value.

We wish to take this opportunity to express our sincere appreciation to all the members of the Organizing and Advisory Committees, the Louisiana State University administration, and the Indian Association of Hydrologists. Numerous other people contributed to the conference in one way or another, and a lack of space does not allow us to list all of them by name here. We would, nonetheless, like to single out Mr. S. S. Kanwar, who served as Assistant Conference Organizing Secretary, for his inordinate assistance and effort without which the conference would not have come to fruition. Numerous other people contributed to the conference to varying degrees. The authors, including the invited keynote speakers, contributed to the conference technically and made the conference what it was; and this book is largely a result of their efforts. The session chairmen/co-chairmen administered the sessions in a positive and professional manner. We owe our sincere gratitude to all of these individuals.

If the success of a conference is measured in terms of the quality of participants and presentations, then most people would agree that this was a very successful conference. A large number of internationally well-known people, who have long been recognized for their contributions and have long been at the forefront of hydrologic and water resources research, came to participate in the conference. More than 20 countries, covering five continents and most of the countries of the world active in research in hydrology and water resources, were represented. It is hoped that long and productive personal associations and friendships will develop as a result of this conference.

V. P. Singh
Conference Chairman
B. Kumar
Organizing Secretary

ACKNOWLEDGMENTS

The International Conference on Hydrology and Water Resources was sponsored and co-sponsored by a number of organizations. The sponsors supported the conference financially without which it might not have come to fruition. Their financial support is gratefully acknowledged. The co-sponsors extended their support by announcing the conference through their journals, periodicals, transactions, newsletters, or magazines. The publicity so gained helped with increased participation in the conference, and is sincerely appreciated. The following is a list of symposium sponsors and co-sponsors.

Sponsors

Wallac Oy
Finland

Pharmacia United Limited
New Delhi, India

Electrotek Instruments Inc.
Madras, India

Elron Instruments Company Pvt. Ltd.
New Delhi, India

Department of Power
Govt. of Arunachal Pradesh
Itanagar, India

Environmental Measurements & Control
Cochin, India

AIC Agro Instruments (P) Ltd.
Calcutta, India

Spectrum Associates
New Delhi, India

Anubhav Printers & Packers
Roorkee, India

Ajay Printers
Roorkee, India

Co-Sponsors

American Geophysical Union
USA

International Association for
Hydraulic Research
The Netherlands

International Water Resources
Association
USA

Soil and Water Conservation Society
USA

Association of American Geographers
USA

International Statistical Institute
The Netherlands

SECTION 1:

WATER RESOURCES PLANNING

INTEGRATED RIVER BASIN PLANNING AND DEVELOPMENT - SOME CASE STUDIES

S.K. SINHA
Deputy Director
Central Water Commission

Synopsis

Water has been recognised as a basic and crucial natural resources for economic development. However, the availability of water does not match with the deamnd in terms of quantity, quality, time and place. Further, it is replensible natural resources that is often scarce. The Conservation and optimum utilisation of this scarce resources is extremely important for national economic development. This essentially calls for integrated river basin planning and development for short and long range conservation, utilisation, development and management of basin's water and related land resources in optimum ways. In this paper, an attempt has been made to bring out the various aspects of integrated river basin planning and present some case studies.

2.0 National Water Policy (1987)

The National Water Policy adopted by the National Water Resources Council in September 1987 has highlighted the

V. P. Singh and B. Kumar (eds.), Water Resources Planning and Management, 3–15.
© 1996 *Kluwer Academic Publishers. Printed in the Netherlands.*

importance of planning of water resources with river basins as rational units because all the water related activities in a river basin are inter-connected and inter-dependent. Further, it also envisages that Appropriate Organisation should be established for the planned development and management of river basin as a whole. Special multidisciplinary units should be set up in each state to prepare comprehensive plans taking into account not only the needs of irrigation but, also harmonising various other water uses, so that the available water resources are determined and put to optimum use having regard to subsisting agreements or awards of Tribunals under relevant laws.

3.0 Integrated River Basin Planning - an overview

The ever increasing demand for water in adequate quantity and desired quality, properly distributed in time and space has forced engineers and planners to formulate comprehensive co-ordinated and integrated basin plan for short and long range conservation, utilisation, planning, development and management of basin's water and related land resources in optimum ways. Individual water projects, whether single or multipurpose, can neither be undertaken nor managed with optimum benefit to promote human welfare unless there is a broad plan for the entire basin.

Integrated river basin planning calls for multi disciplinary approach to the planning and management of the water related activities. It has to take into account multiple uses, multiple purposes and multiple objectives. It combines various disciplines, such as, economics, agriculture, agro-economic, hydrology, environment and ecology, navigation, pisciculture, water shed management, Hydropower generation, forestry, etc. There is an imperative need for interaction and co-ordination of the activities in these aspects while the projects is under formulation.

4.0 **Planning Process**

Planning is the orderly consideration of a project from the original statement of purpose through the evaluation of alternatives to find decision on course of action. It includes all of work associated with the design of project except detailed engineering of structures. Here, it will not be out of place to mention the importance of sound economic decision arrived through the evaluation of various alternatives in planning process. Major water rersources Central Structures can be altered with great difficulty or not at all. There are relatively few suitable dam sites, and once they are appropriated, the possibilities for developament are very limited. Once an irrigation project is developed, it cannot be moved because unfavourable soil or climate factors are discovered.

Generally, speaking the planning process comprises of four steps, namely, (i) Definition of objective; (ii) Project studies; (iii) Projection for planning and (iv) Project formulation.

4.1 **Definition of objectives:**

All possible uses of water i.e. Domestic and Municipal

supply, Irrigation, Industrial use, Hydropower development, Stock water, Recreation, Fisheries and wild life, Navigation, water storages for stream flow regulation are pertinent to the problem. One or more of these ends may obviously be inappropriate for any particular project and can be deleted in the beginning of the project formulation.

4.2 **Project Studies:**

The second step is the studies to be undertaken for arriving at sound economic decision which will meet the requirement of the project in optimum way. A brief outlines of the studies to be carried out are indicated below:

4.2.1 **Test of Feasibility:**

Studies should permit a sound analysis and conclusion with respect to the specific engineering environmental considerations. These are basically; (a) The project is responsive to an urgent present/projected social or economic need; (b) The project as planned will adequately serve the intended purpose; (c) The services proposed to be performed through the project and the benefits it will produce will justify the cost and (d) The project will cause nominal disturbances to the ecology and the environment of the area.

The study should determine that the difficulties inherent in sites which affect economy, safety of construction and quality of operation have been satisfactorily foreseen; and that the designs are technically sound and reasonably representative of actual structures that may be expected to be built after more detailed investigation.

4.2.2 **Esthetic Values:**

Esthetic value should be given appropriate consideration which should be recognised in the first studies made and carried out in later studies and in construction operation. However, esthetic values should not be allowed to outweigh the safety or adequacy of the structural design.

4.2.3 **Stages of investigation:**

Investigation if carried out to completion is an expansive and time consuming phase of project development. Moreover, it may indicate that project is not economically or technically sound. In view of the above, there is an imperative need that an investigation should be planned and executed in stages so that the probable soundness of project will be determined as early and as inexpensively as possible. To accomplish this objective, the investigation may be divided into three stages. (a) The first or reconnaissance stage is designed primarily to support a decision on whether to proceed with more detailed investigation on the basis of rough data and short-cut studies. (b) The second or feasibility stage, determines the scope, magnitude, essential plan and feature, and the approximate benefits and costs of project with sufficient dependability to suppoprt project authorization or approval for construction. (c) The third or the specification stage supplements the feasibility stage to the degree needed for preparation of final plans and specification after authorisation or approval has been obtained and construction is imminent.

4.2.4 **Related Projects and Studies of Need:**

The proposed project should be in accordance with any large range planning programme which may have been adopted for the vicinity. The entire area to be served by the proposed project should be studied to determine whether there will be conflicts with other projects of a similar nature for use of land or water resources, power drops, etc. or whether economics are possible by securing optimum development of resources through joint use.

4.2.5 **Development of the General Plan:**

The Project plan normally originates with the desire to satisfy sponsor's needs, objectives or purposes. As support for project construction develops, those needs may increase, the objectives may broaden and the purposes may multiply through process of project formulation until selections of the final magnitude and scopes are reached.

At the outset of the reconnaissance study, considerable basic data are usually available in forms of maps, aerial photographs, stream flow records, regional geological reports, Census statistics, crop yields, market statistics, power loads, previous investigation reports, etc. The investigator must evaluate these data and conceive a workable basic plan that utilises available resources to meet the needs. The basic plan may then be compared roughly with alternatives to accomplish the desired ends as progressively increased or diminished scope and scale. It will be feasible by judgement or cursary study to eliminate many alternatives so that an approximate plan emerges which includes purposes; approximate location and height of dam, capacity of reservoir, spillway, outlets, canal, power plans and other features and which minimizes the environmental impact of projects. This plan will be refined in the feasibility and specific design stages.

4.3.0 Projection for Planning:

This is a very important step in project planning as all projects are for the future and must be planned to meet future needs. The projection should be more than a simple extrapolation of past growth curves for social, economic and technological development may cause significant changes in trends. Every efforts to anticipate such changes must be undertaken.

Among possible technological advances are the possibilities of conserving or augmenting natural water supplies, development in nuclear power production (or other means of producing electricity), improved agricultural technique with increased crop production on existing farm lands, new system of transportation, changing pattern of living and new modes of recreation may all play a role in the economy of a Project. In most cases, it would be difficult if not impossible to foresee the exact impact of any future technological breakthrough, but it may be possible to plan with sufficient flexibility so that expansion of the project may he governed by advances achieved when the expansion is undertaken. Stage construction of all or part of the project feature is an example of planned flexibility. Studies to be undertaken under this category are briefly indicated in Table I.

TABLE 1 : Studies for Projection

STUDIES

Sl.No.

1. General Studies

a) Population - Place and Time

b) Land use - Place and Time

c) Economic - Market; Tourism, etc.

2. Specific Studies

a) Agricultural -Market, Crops, Technological development; Water demand.

b) Municipal - Domestic water demand; industrial water need, Technological changes.

c) Power - Market demand; Growth of alternative sources; technological improvement.

d) Flood Control: Zoning possibilities; projected flood damage (place and frequency); Possibilities of flood warning.

e) Navigation - Growth of demand; Growth of competing facilities; Transportation pattern; Technological advance (Navigation and competing facilities).

f) Recreation - Growth of demand; changes in recreational preferences.

g) Pollution - Anticipated quantities and character of waste; Technological advances.

4.4 Project Formulation:

Once the basic data and the projection of future conditions are assembled, actual formulation of project can be taken up. This is the most important phase of planning where imagination is required. The important first consideration is the compilation of a comprehensive list of alternatives. The planning process should be evaluation of all possible alternatives with respect to project feature and water use. Planning process should also consider alternative and competive uses for water as well as the various possibilities for control and delivery. The Planning process may involve considerable feedback. As project formulation proceeds, it may be evident that new data or projections are required or that some revision of back ground data is needed.

Project formulation comprises of four steps namely,

i) Fixing boundary condition with respect to project;

ii) Definition of alternative land use plan;

iii) Listing of possible projects and

iv) Preliminary cost estimate, on the basis of which many of the alternatives may prove to be undesirable and ultimately be discorded.

5.0 Case Studies:

Several case studies can be cited from a number of countries, including India wherein the concept of integrated approach to river basin management has been attempted. The earliest example is probably the setting up of Tennesse Valley Authority (TVA) in 1934 in USA for the integrated development of the Tennessee Valley. The objective was not only to perform a specified number of engineering tasks but also to undertake a wide range of regional planning activities which included afforestation, extension programmes, for better land use and soil management and a variety of community development programme.

In U.K., under the Water Act (1973) based on the principle that a single authority should plan and control all uses of water in each river basin, nine English regional autholrities and a Wash Water Authority were set-up. In particular, the Thames Water Authority managing the Thames river basin, supporting as much as 3500 abstractions - 1200 for agriculture, 500 for domestic water supplies and 1800 for industrial and other uses provides a classical example of integrated river basin management.

In the Asian region, the case of Lower Mekong Basin, provides an example of a planned and integrated development of an international river , whereby full consideration was given for the development possibilities of the delta in connection with development of upstream. The Mekong river with a catchment area of 7,95,000 Sq.km. passes through six countries namely, China, Burma, Laos, Thailand, Kampuchea and Vietnam. A committee for the co-ordination of investigation of the Lower Mekong Basin was set up by ECAFE (now ESCAP) in 1957 for formulating a proposal for the development of water and related resources of the basin. The Committee came out with an "Indicative Basin Plan" in 1970 , proposing a number of possible dams and storage reservoirs that could be implemented on the main river and tributaries.

The concept of integrated river basin planning is not

new to India. Several basin organisations have been established from time to time to accomplish the above objective either fully or partly. The Damodar Valley Corporation (DVC) was set up in 1948 as an autonomous corporation to exploit the resources Damodar Valley as a whole and to develop the region as an integral unit, on the line of TVA. The main objective of DVC was the promotion and operation of scheme for irrigation, water supply and drainage, generation transmission and distribution of electrical energy, flood control, afforestation, control of soil erosion, the promotion of public health and the agricultural, industrial, economic and general well-being in the valley and its area of operation.

In accordance with the directives of the National Water Policy for considering basin as the basic unit for Planning and Management of Water Resources, Central Water Commission has been pursuing the matter regarding setting up of multi-disciplinary planning unit in the states for preparation of river basin master plans. Comprehensive planning studies were carried out by CWC for Mahanadi basin, with a catchment area of over 1,41,000 Sq.km. and 90 potential large reservoir sites. Screening these reservoirs for arriving at the optimal combination of reservoirs for different levels of development necessitated the use of System Analysis. Studies in the field of integrated operation of System of reservoirs, using simulation models, in Mahanadi and Godavari basins revealed the advantages of integrated operation of reservoirs in place of Conventional method of operating reservoirs individually.

Sone river Commission was set up to prepare comprehensive co-ordinated and integrated river basin/regional plan for short and long range conservation, utilisation, planning, development and management of basin's water and related land resources. The total catchment area of Sone basin is 70,074 Sq.km.,48,835 Sq.km. in M.P., 4913 Sq.km. in U.P. and 16,326 Sq.km. in Bihar. Realising that such planning will need the services of specialists in multi-disciplinary fields such as economics, agriculture, hydrology, forestry, watershed, management etc., two focal technical groups namely (i) Focal Technical group on System studies and (ii) Focal Technical Group on Land Resources and Agro and Socio-economic Surveys and studies in right earnest in the fields of System Planing; Land Resources and Agro and Socio-economic aspects. The studies have once again demonstrated the usefulness of integrated river basin planning and development.

6.0 Conclusion:

As it has been increasingly becoming apparent that the available water resources will no longer be able to match the ever rising demands, the need for integrated river basin planning & development for the optimum and efficient utilisation of water resources is being fully recognised. There appears to be no other alternative to the establishment of river basin organisation for the formulation of comprehensive, co-ordinated and integrated basin plan for short and long range conservation, utilisation, planning and development and management of basins' water and related land resources in optimum and efficient ways.

References

i) Central Water Commission (1987), Systems Engineering for integrated Development of Water Resources in India, Final Report Vol.I to V.

ii) Govt. of India (1987), National Water Policy.

iii) Mars, Arthur Etal, 1982 "Design of Water Resources Systems", Harvard University Prars.

iv) Ray K Linsley Etal "Water Resources Engineering" MC Graw Hill.

v) Design of Small Dams, USBR Publication.

vi) Sinha, S.K. etal, 'Formulation of Linear optimization Model for Sone Basin', published in National Symposium on design of Water Resources Systems - 10-12 Feb.,93.

CONJUNCTIVE IRRIGATION IN THE HUMID ALLUVIAL PLAINS - IMPERATIVENESS AND ISSUES -

T. PRASAD
Professor of Civil Engineering & Director
Centre for Water Resources Studies
(Patna University)
Bihar College of Engineering
Patna-800 005, INDIA

ABSTRACT : The high agropotential of humid alluvial plains, existing largely in the eastern India, cannot be realized without appropriate irrigation. Irrigation from surface water sources, mostly by means of diversion of river flows, has been found to be grossly deficient due to its supplies being inadequate, irregular and unreliable due to hydrological and hydraulic factors. It also proves counterproductive due to its adverse effects, like raised watertable and waterlogging. Use of groundwater for irrigation is not only feasible but also has distinct benefits. However, exclusive reliance on groundwater for meeting total irrigation needs year after year is precluded due to several limitations. Thus, conjunctive use and management of the two sources of water is imperative, which will result in higher cropping intensity, better yield and more remunerative cropping patterns. It will also ensure sustainability of productive irrigated agriculture and socio-economic equity in accrual of benefits. Adoption and implementation of conjunctive irrigation have been inhibited by several issues related to planning, design and operation of the irrigation system as well as to economic, management and equity aspects involved. Unless these issues are resolved, conjunctive irrigation will not be fully practised in the region, and its agriculture and hence its economy will continue to perform poorly.

1. Introduction

Mankind can obtain water for his diverse needs primarily from two terrestrial sources, hydrosphere or surface sources such as rivers, and lithosphere or various water yielding geological formations. As the natural regime of the occurrence of water in these sources is not in spatial and temporal conformity with the pattern of requirements of water for meeting various needs, engineering measures have to be resorted to in order to bring about conformity or correspondence between the patterns of natural availability of water, on the one hand, and its multipurpose requirements, on the other. Since these measures essentially constitute an intervention in the natural regime of water resources, they are liable to disturb the ecological equilibrium leading to adverse side or long term effects. The nature and magnitude of these adverse effects depend upon the source exploited, the need met and the measures adopted. As agricultural requirements of water for mankind are most vital, next in criticality only to his drinking water needs, and are by far the most dominant in terms of magnitude, the adverse effects consequent to adoption of measures aimed at satisfying these requirements are also correspondingly high. Thus, the strategy and technology of irrigation, particularly with regard to the source of water exploited for the purpose, assumes special significance. The efficiency of irrigation as well as productivity, sustainability, remunerativeness and equity of irrigated agriculture vitally depend on it.

Conjunctive irrigation, which implies exploitation and utilization of both the surface as well as ground water sources to meet the agricultural requirements, is now widely accepted as an eminently suitable strategy of irrigation for the alluvial plains, in general,

V. P. Singh and B. Kumar (eds.), Water Resources Planning and Management, 17–27.
© 1996 *Kluwer Academic Publishers. Printed in the Netherlands.*

and humid alluvial plains, in particular. The imperativeness of adopting this strategy as well as various issues which must be resolved in order to generally implement it are discussed below. Although the discussion has been done with reference to humid alluvial plains in general, the region comprising eastern U.P. and North Bihar in India and the 'tarai' area in Nepal has been kept in view, in particular, to propound the hypotheses contained herein.

2. Imperativeness of Irrigation

A humid alluvial region, such as the one indicated above, is characterised by high agropotentiality which is predicated upon primarily the following features; fertile tracts of land, congenial agroclimate, ample water and required human resources. The physical, chemical and biological properties of the soil constituted by successive deposits of recent alluvium provide a constant base of high fertility. Although a careful choice and use of chemical fertilisers would enhance the crop productivity, no basic amendment of the soil, except in certain localised patches, is needed to retain its fertility. Similarly, its agroclimatic features such as mean monthly temperature, sunshine hours and relative humidity are suitable for year round agriculture in principally three crop seasons and for nurturing a variety of crops. As far as water is concerned, the annual rainfall of 100 cm or more in this region as well as availability of both surface and groundwaters ensure adequate supply in the aggregate to meet all crop water requirements. These favourable natural factors attracted and sustained over the centuries an increasing population with agriculture as their mainstay. It has given rise to an abundant work force steeped in the culture of agriculture in this region. While exposure to, as well as education and training in, new concepts and techniques of modern agriculture will always be necessary, a prosperous agriculture is not likely to be constrained by lack of basic manpower resources.

However, the high agropotentiality of this region cannot be, and in fact has not been, fully realized due to the dynamic factor of water in the natural resource vector for agriculture. The annual rainfall is sufficient to meet the agricultural water requirements for year round cropping only in the aggregate sense. Due to monsoon climatic features prevailing in the region, the temporal distribution of rainfall occurrences has an extremely high seasonal skewness. About 80 to 90% of the annual rainfall takes place in 4 months during June-October. The remaining 10 to 20% of annual rainfall is distributed rather unevenly over the other 8 months, with November and December usually being the driest months. Such temporal distribution of rainfall occurrences is not in conformity with seasonal requirements of water in various cropping seasons. Moreover, the variability of rainfall occurrences in smaller time intervals even in the rainfall abundant monsoon season is considerably high. If these time intervals are short enough to be agriculturally significant, the rainfall incidences may even be termed as erratic. Thus, due to either inadequacy, or variability, or both, the rainfall occurrences cannot be relied upon to provide productive and year round agriculture which is otherwise feasible in this region. Irrigation is imperative to realize the high agropotential of this region. In any case, modern agriculture with newer high yielding varieties of crops and fertiliser use, requiring more time and magnitude specific applications of water, cannot be sustained without irrigation.

3. Inadequacy and Imperfectness of Surface Water Irrigation

Irrigation which implies application of water to the crops to complement or supplement natural rainwater application requires conveyance of water from its source to the field.

This source can either be a surface water body like a river, a lake or a tank, or various water yielding geological formations called aquifers. Conventionally, surface water irrigation has been adopted in this region by diversion of river flows by means of a weir or a barrage and by conveyance of the diverted flow through a network of canals, such as through Sharda Sahayak, Gandak and Kosi Projects. Experience of several decades of irrigation through these projects has brought to light the inadequacies and imperfectness of these projects, as their limiting and counterproductive features emerged over time.

One common limiting feature of these surface water irrigation projects is predicated by the hydrology of the rivers, the source of irrigation water, as well as lack of any storage of the river flows. In conformity with the rainfall occurrences, the river flows also display a correspondingly high degree of seasonality. About 75 to 85% of the annual river discharge takes place in the four monsoon months of the year. Also, there are no suitable sites for creating storages on these rivers in India. So, unless and until dams are constructed in the upstream country Nepal in order to create storages and bring about better seasonal distribution of outflows, river flows diverted for irrigation cannot meet adequately crop water requirements in all the three feasible cropping seasons in the region. Consequently, the planned cropping intensity for irrigated agriculture seldom exceeds 140% in these projects. This is a gross under utilization of the agropotential of this region where other components of the natural resource vector, like agroclimate and land, can sustain three crops in a year, giving a feasible cropping intensity of 250% on the average.

Another distinct, and rather serious, imperfectness of surface water irrigation in a humid alluvial region is lack of sustainability of productive irrigated agriculture. As application of irrigation water imported from a river, which is essentially a drainage channel for the area, season after season, and year after year, would tend to systematically raise the water table which is already at shallow depths in humid alluvial regions, the threat of waterlogging, i.e., the encroachment of the rising water table, or of the capillary fringe above the water table, in the root zones is very real. Such encroachment interferes with the root zone aeration, an essential condition for healthy plant growth, and thus adversely affects productivity of the waterlogged land. If there is salinity in the groundwater, rising water table also brings it to the root zone, which also affects the fertility of the land adversely.

A third disadvantage of surface water irrigation is the limitation imposed by the requirement of gravity flow in distributing water to the areas to be irrigated. On account of this and depending upon the prevailing topography, certain areas in the command cannot be served by surface water irrigation, even though water is available for the purpose. In a command having an undulating nature of topography like the Kosi command, this considerably reduces the extent of benefits and hence of participation of farmers in the operation and improvement of the irrigation system.

A fourth deficiency of surface water irrigation system emanates from the inflexibility of the system and its consequent inability to respond to the farmers' sophisticated and specific needs of modern agriculture. The irrigation benefit to any farmer in the system is subject to the irrevocable laws of hydrology and hydraulics governing the availability, conveyance and distribution of water, as well as to the rules of operation of the system which has to serve the requirements, sometimes conflicting and competing, of a large number of beneficiaries. Thus, any enterprising farmer who wants to put the high agropotentiality of the region to his maximum benefits by diversifying his cropping to make it more remunerative and by modernising his agriculture by using newer high yielding verieties of seeds and other inputs cannot afford to fully depend on surface water irrigation system to satisfy his irrigation needs. This deficiency gets worse and more pronounced for farmers located more and more downstream in the system. In some systems like the Gandak irrigation project, the farmers even below the middle reach find the surface water irrigation system inadequate, unreliable and counterproductive even for their conventional agriculture, leading to their antagonism towards, or rejection of, the system (1).

4. Feasibility, Benefits and Limitations of Groundwater Irrigation

Feasibility of the use of groundwater for irrigation in this region is established principally by three factors. One is the quantity of groundwater available annually for irrigation. Groundwater exists in this region in its very thick alluvial deposits - thickness extending upto 1000m or more in several layers of unconfined, semi-confined and confined aquifers. These aquifer layers are recharged annually by both vertical percolation and horizontal flow through porous media. It is estimated that the magnitude of vertical recharge is of the order of 50 to 60 cm annually. While no detailed investigations and estimate of the rate or annual magnitude of horizontal recharge have been made, it is known that a large quantity of water enters mostly the confined and semiconfined aquifers of this region through their recharge zones lying in a highly pervious belt, 10 to 15 km wide, known as 'bhabar' all along the foothills of the Siwaliks.

Secondly, the quality of groundwater available in exploitable aquifer layers at various depths is excellent to good in most parts of the region, as indicated by the pH, total solids, alkalinity, chloride content as well as sulphate content values. Only in certain localised pockets, some caution has to be exercised in the selection of aquifers to be tapped for large scale irrigation from the point of view of quality. In general, it can safely be said that quality will hardly be a constraint in any strategy of groundwater utilization for irrigation in this region. The third favourable factor which will help promote extensive practice of groundwater irrigation in a region inhabited preponderantly by small and marginal farmers is the availability of groundwater at relatively low depths, both in its unconfined as well as confined layers. This will ensure low cost of extraction of groundwater. In many parts of this humid region, water table depths vary between 1 to 3 m in post monsoon and 2 to 4 m in pre-monsoon seasons (2). This will also ensure low cost of installation of shallow tubewells in this region. Improvised and cheap tubewell devices, like 'bamboo tubewells', have been found to be practical in considerable parts of the region, facilitating their adoption by even marginal farmers, who will hire diesel pumping sets on rent, thus getting by with affordable investment.

Apart from its feasibility, groundwater irrigation affords distinct benefits over surface water irrigation. It is much more user friendly, i.e., it can respond to the varying needs of irrigation water requirements at different stages of crop growth much more reliably than surface water irrigation, which being a larger and hydraulically more complex system is more inflexible in operation. This is particularly significant for high yielding varieties and cash crops as well as other crops which are sensitive to moisture stresses. Another distinct benefit of tubewell irrigation is its ability to serve high elevation areas which cannot be commanded by gravity-flow canal irrigation system. In a command area with undulating land topography, it is only with recourse to tubewells that all irrigable areas can be brought under actual irrigation. A third benefit of groundwater irrigation is its high duty, i.e., larger area actually irrigated by a unit discharge of irrigation water on account of its low conveyance losses. Still another benefit of tubewell irrigation is its maneuverability as an instrument of socio-economic equity. It affords an opportunity to make positive discrimination in favour of certain target socio-economic groups by providing suitable subsidies and incentives in installation and operation of tubewell. In surface water irrigation system, where hydraulics is an overwhelming factor in conveyance and distribution of irrigation water, it is difficult and impractical to do so.

Extensive utilization of groundwater for irrigation, however, is subject to certain limitations. One is the quantum of groundwater that can be annually extracted, and hence the area that can be assured full irrigation, is limited by safe yields of the aquifers tapped. As is well known, it is not the quantity of water held by an aquifer, but rather its annual recharge, which determines how much water can be extracted on an average annually without creating undesirable consequences. Secondly, groundwater irrigation is essentially subject to the externality of energy availability, preferably electricity. This becomes a severely limiting factor in an electricity starved or electricity deficient region.

The alternative of diesel as a source of energy for tubewell operation is less attractive and more inhibiting on account of cost and maintenance problems. In any case, dependence on diesel for larger scale expansion of groundwater irrigation brings in the question of competing uses of diesel, which is mostly to be imported, for other sectors, such as transport, in an economy where balance of payments is a dominant consideration.

5. Imperativeness of Conjunctive Irrigation

In view of its inadequacy and imperfectness, exclusive surface water irrigation cannot, and hence must not, be relied upon to realize the high agropotentiality of the region. More than two decades of experience of Kosi and Gandak projects, which were planned, designed and operated to provide irrigation for this region exclusively from diverted flows of the respective rivers, amply corroborate this. They have not only proved deficient to cater to the growing needs of irrigation in their commands, but their counterproductive features also began to emerge rather early in their lives, for which technical committees had to be constituted (3, 4) to review and evaluate their performances and to suggest remedial measures. The feasibility and benefits of utilization of groundwater for irrigation in this region naturally underline the necessity of its adoption to alleviate the problems of exclusive surface water irrigation. The other extreme option of exclusive groundwater irrigation is also not practicable in view of its limitations indicated earlier. The obvious option under the circumstances is to make use of both the sources of water, surface water as well as groundwater, to satisfy adequately and meticulously total requirements of irrigation for various crops in different cropping seasons of the year, aimed at realizing the high agropotential of the region predicated by its congenial agroclimate and fertile lands. As irrigations from the two sources have their advantages as well as disadvantages, prospects as well as limitations, and as the two sources are hydrologically interacting, particularly in humid alluvial regions where such interaction is quite vigorous, it is imperative to use them conjunctively in order to optimise the benefits in a sustainable manner.

Imperativeness of conjunctive irrigation in humid alluvial regions having agroclimate congenial for year round cropping but with highly skewed rainfall incidences, and consequently river flows, in various seasons is underscored by several compelling considerations. First of all, in a conjunctive irrigation system, a cropping intensity of 250% on an average can be planned to be achieved against the usual 140% which a surface water irrigation project in this region targets to achieve. Secondly, a conjunctive irrigation system will be much more reliable in delivery and supply of water and hence can be much more responsive to the irrigation requirements of various crops, in different cropping seasons and at all stages of plant growth, enabling the farmers to cultivate high yielding as well as more remunerative varieties of crops. Thus, both productiveness as well as remunerativeness of irrigated agriculture will be substantially enhanced by conjunctive irrigation. Thirdly, conjunctive irrigation provides sufficient scope and instrumentality to bring about some measure of socio-economic equity in the accrual of irrigation benefits. In a region having preponderance of small and marginal farmers and where glaring socio-economic inequity has been existing for long, this factor is of paramount significance. Lastly but most importantly for humid alluvial regions, it is conjunctive irrigation which can ensure sustainability of productive irrigated agriculture. Uses of surface water and ground water for the common purpose of satisfying the water requirements of irrigated agriculture also constitute antidotes for the adverse effects of each other. While groundwater use causes depression of water table and is thus a measure to ensure subsurface drainage and to prevent waterlogging which results from continued surface water irrigation, the latter is a measure for recharge of groundwater which gets mined and depleted on account of large scale and intensive use of groundwater for irrigation year after year. Thus, by suitably co- ordinating the two uses, a hydrological equilibrium of water table can be maintained, which not only ensures sustained agricultural productivity but is also desirable ecologically. A good example of such sustainability can be seen in the age old Ganga Canal System where introduction of tubewell for irrigation in its command in western U.P. many decades ago has stabilised productive agriculture in the area. Recourse to tubewells for irrigation in the commands of Kosi and Gandak surface waer irrigation projects in North Bihar, prompted primarily by failures of the projects to meet the irrigation needs of farmers who were exposed to the benefits of irrigated agriculture, has resurrected these projects from their more dismal performance than what it is now.

6. Unresolved Issues

In spite of such attractiveness and even imperativeness of conjunctive irrigation, it has not yet been adopted, systematically and in a planned manner, as a strategy for irrigation development in the alluvial or humid alluvial plains. Although the benefits of conjunctive irrigation have been widely recognised and although features of conjunctive irrigation have emerged over the years in an ad hoc manner in several irrigation projects, with obvious advantages, there is hardly any irrigation system which has been planned, designed and operated as a conjunctive irrigation outfit. In fact, there is yawning gap, with respect to conjunctive irrigation, between what Vincent and Dempsey (5) say 'good theory, poor practice'. Several inhibiting and limiting factors are at work for prevalence of this state of affairs. The salient factors and unresolved issues emanating therefrom are described below.

6.1 PLANNING ISSUE

The basic issue which has to be resolved at the stage of planning a conjunctive irrigation system is to consider the pattern of availability of total water - annually, seasonally and sub-seasonally - vis-a-vis the pattern of demands of water for various crops in different cropping seasons and at various stages of their growth. For this, the hydrology of both surface water source, such as the river the flow of which is to be diverted for irrigation, as well as groundwater, as it occurs in various exploitable layers of aquifers, is to be taken into account. For the latter, it is necessary to consider not only the water yielding characteristics of the aquifers, both in volume as well as rate, but also their recharge characteristics so that their safe yields are not unduly exceeded. As the various exploitable aquifers may be unconfined, semi-confined or confined, the sources and mechanisms of their recharge may be different. These must be ascertained and the rates as well as volumes of annual and seasonal recharge have to be determined to define the groundwater hydrology of the area. Another parameter of groundwater hydrology relevant for irrigation use is the salinity status of water from various aquifers.

Planning of a conjunctive irrigation system would require matching the crop water requirements with the hydrologies of the two sources of water in a co-ordinated manner. In doing this matching, it must also be kept in mind that while the hydrology of surface water relates to only the diversion point of the river, that of groundwater will have to ascertained both horizontally as well as vertically, i.e., at various locations in the total command area and for various aquifer layers.

6.2 DESIGN ISSUE

The purpose of designing any irrigation system is to bring water from its sources to the farmer's field and ultimately to the root zones of the plants therein. In case of surface water irrigation system, it will require designing the facilities for diversion, conveyance, distribution and application of irrigation water. The technology used for diversion is typically a headworks comprising a weir or a barrage, head regulator and other ancillary devices. The technology of conveyance consists of a main canal, branch canals and ancillary structures such as cross drainage works, falls, etc., primarily to convey the diverted flow from the offtake to the command area. Distribution of water in the command area is done through a network of channels called distributaries, minors, water courses and field channels and related structures such as cross regulators, outlets, etc. Once the water is brought to the farmer's field, he may employ various methods and techniques to apply the water to the crops, depending upon his convenience, irrigation efficiency objective, the crop to the irrigated and on-farm features.

Similarly, in groundwater irrigation, the facilities to be designed are for extraction, conveyance, distribution and application of groundwater. While the technology for groundwater extraction is typically a tubewell consisting of groundwater extraction structure and pumping device, technologies for other facilities, to the extent that they will be required, are the same as for a surface water irrigation system. As tubewells are installed mostly in the command area itself, conveyance is not required. In some cases, however, tubewells may have to be installed away from the command area due to quality constraints or other considerations, in which case conveyance facility has to be provided for. As the command of a tubewell is very small compared to that of even a minor in a surface water distribution system, the distribution network is much smaller in size as well as in complexity. In case of small capacity private tubewells which are installed right in the farmer's field, the necessity for distribution is altogether obviated. As far as application of water is concerned, the technology will be the same, and subject to the same factors, as in a surface water irrigation system.

While the technique, methodology and procedure for designing either a surface water irrigation system or a groundwater irrigation system are quite standard, even though there may be sufficient scope for improvement, there are quite a few unresolved issues when the system is to be designed as a conjunctive irrigation system. First of all, the issue is whether the facilities for conveyance, distribution and application of surface water and of groundwater should be designed as common facilities or separately. While there may be economic advantage in providing for these functions through common facilities, one has to contend with the complexity in their design and management. The criteria of design for an open channel to carry silt laden surface water may be different from those for an open channel to carry silt free groundwater. Also, since the surface water irrigation system and the groundwater irrigation system may be under different management set ups, as is generally the case presently, management of common facilities in a conjunctive irrigation system will give rise to problems which have to be resolved at the design stage.

An important design issue to be resolved arises from the duality of functions which the surface water and groundwater irrigation facilities perform, or must perform, in a conjunctive irrigation system. In this system, canals are a means of not only conveyance and distribution of irrigation water but also of recharge of groundwater through seepage. Similarly, tubewells, and particularly shallow tubewells, not only extract groundwater for irrigation but also tend to depress the water table and thus serve as a vertical drainage and anti-waterlogging device. It is this duality of functions which render conjunctive irrigation system sustainable in terms of hydrological equilibrium and productivity of irrigated agriculture, and hence has to be provided for and facilitated through proper design.

6.3 OPERATION ISSUE

A conjunctive irrigation system is not only to be planned and designed but also must be operated as such in order to derive desired and optimum benefits from it. In drawing up an operation schedule of a conjunctive irrigation system, three factors have to be taken into account, (i) irrigation water requirements, (ii) hydrological availability of both surface and ground waters, and (iii) water table status. Three types of models will be helpful for the purpose, one is the allocation model, the second groundwater-surface water interaction model and finally, the optimisation model. These models will be interactive or hierarchical, with the first two models acting as input to the third, which will finally yield the operation schedule. The allocation model will essentially give seasonal allocations of water to be provided by the surface water and ground water sources to meet the specified or desirable irrigation water requirements. The groundwater-surface water interaction model is principally aimed at determining the water table status resulting from a specified allocation and a given, assumed or trial operation schedule. Finally, the optimisation model will provide on optimum schedule of operating the conjunctive irrigation facilities, i.e., the canals, tubewells and their various regulating devices such that the net benefits of irrigated agriculture are maximised subject to the prevailing hydrology and the water table equilibrium criteria. Interactively, the models can also be used to arrive at most remunerative cropping patterns in the command.

6.4 ECONOMIC ISSUE

The economics of a conjunctive irrigation system is more complex than that of its constituents, i.e., surface water irrigation system and groundwater irrigation system. The complexity arises because a conjunctive irrigation system is not just the summation of its two constituent systems but is rather a system arising out of the integration of the two systems at the stages of planning, design and operation, as indicated above. In order to analyse this complexity, one has to differentiate between the economics from the viewpoint of the state and the economics from the view point of the farmer.

From the viewpoint of the state where it does not matter who pays, the cost of installation of a conjunctive irrigation system, if it is planned, designed and operated as such, will be less than the sum of the costs of installation of its constituent systems due to commonality of facilities and other advantages of integration of their technologies. From the farmer's viewpoint, the cost of installation of groundwater irrigation in the command of a heavily subsidized surface water irrigation system is considered an additionality. Similarly, from the point of view of the state, the operating cost of a conjunctive irrigation system will be less that that of its two constituent systems operated independently of each other, because of the savings on wasteful operation. As far as the farmer is concerned, it will depend upon the water rates and the energy tariff or cost, both of which are administered prices and hence do not reflect the real cost.

Another component of the economics of a conjunctive irrigation system is that of benefits. The benefits of a conjunctively irrigated system comprise the value of crops produced. Due to higher cropping intensity and more remunerative cropping patterns feasible through conjunctive irrigation, the benefits of a conjunctive irrigation system will be substantially higher. Rational computation of these benefits and hence of the benefit-cost ratio or other criteria of economic choice is still an unresolved issue of techno-economic analysis of conjunctive irrigation system.

6.5 MANAGEMENT ISSUE

The management of a conjunctive irrigation system is undoubtedly a complex issue on account of the fact that the management set ups of its two constituent systems of surface water and of groundwater irrigation are presently quite different and essentially independent. A surface irrigation system is typically managed by the state bureaucracy with its hierarchical assignment of functions and responsibilities at different levels.

Farmers' participation in the management is indirect and often ineffective. Although the rationality and need of participatary role of farmers in management has been widely reconginsed and attempts are being made to turn over the management of surface water irrigation system to water users' associations below distributary levels, a lot of legislative and administrative ground is still to be covered in this direction. As far as groundwater irrigation is concerned, its management scenario extends from private tubewells which are completely owned and managed by individual farmers, to group tubewells where a group of small number of farmers with coherent or common interests join to install and/or manage a tubewell, to co-operative tubewells where a registered co-operative society manages a cluster of tubewells for the benefits of its members, and finally to state tubewells which are owned and managed by the state. In most cases, however, the concerned department or organisation of the state government is different from, or independent of, the department managing the surface water irrigation system. Even here, the rationality and need of turning over the operation and management of state tubewells to beneficiary farmers are being increasingly recognised and attempts have been made in that direction.

The challenge in management of a conjunctive irrigation system is how to integrate the management set ups of the two constituent irrigation systems and evolve a unified management structure for the conjunctive irrigation system so that it can be operated, maintained and managed in a co-ordinated manner. When a conjunctive irrigation system is planned, designed and operated as such by resolving the various issues indicated earlier as well as once its economic issues are suitably taken care of, its full and ultimate success will hinge on whether it is under a correspondingly suitable management setup.

6.6 EQUITY ISSUE

It has earlier been pointed out that the groundwater irrigation system is more amenable to provision for and attaining the socio-economic objective of equity than a surface water irrigation system whose planning, design and operation are governed by more inflexible considerations of topography, hydrology and hydraulics. In a conjunctive irrigation system where the two consitutent irrigation systems are integrated, the opportunity for equity provision is enhanced. Although surface water irrigation system by itself does not allow equity consideration much role to play, it becomes quite significant and even potent in doing so in conjunction with a groundwater irrigation system. In a conjunctive irrigation system, the instruments of equity are no longer only the subsidies in installation and energy charges of tubewells, which being deviations from the principles of market economy do give rise to certain aberrations and misues. Now, conjunctive irrigation system can also provide equity through priority and preferences in distribution of surface water. This will create less social discontent and will have more social acceptability than in an exclusive surface water irrigation system.

7. Outlook

In the background of rising population in India, near stagnation of its total agricultural production has been a cause of concern. It is in this context that the poor agricultural productivity of eastern India has been engaging serious attention of the planner and concerned organisations, as it is keenly felt that while the potential of agricultural production in the wake of growth of irrigation and of introduction of modern agriculture had been largely realized in other regions of India, this region has lagged behind. Committees have been consitutted and studies have been carried out to analyse the paradox of high agro potential of the region and its poor agricultural performance (6).

One particularly disturbing feature of this analysis is that the agricultural productivity of this region has not only been far below its potential but has also been falling.

As far as the gap between the potential and performance is concerned, the alluvial plains of this region are most problematic as well as most prospective. It is well recognised that irrigation is 'sine qua non' for agricultural development of this region, as other favourable factors, such as fertile lands, congenial agroclimate and ample human resource cannot be fully harnessed without irrigation. Experience of more than two decades of irrigation development in this region has conclusively shown that exclusive surface water irrigation has, at best, delivered only mixed benefits and, at worst, proved counterproductive. Feasibility of groundwater irrigation and its several advantages have been discussed. Exclusive reliance on groundwater to meet all irrigation requirements in the region, however, is neither practicable nor desirable because of several limitations indicated earlier. Thus, utilization of both the sources of water is the logical option for irrigation in the region. Again, utilization of the two sources of water independently of each other and in an ad hoc manner can provide only suboptimal techo- economic benefit. In order to derive optimal benefits in terms of agriccutural productivity and production, sustainability and equity, these two sources have to be utilized for irrigation conjunctively. This, however, gives rise to problems in planning, design and operation of the irrigation system. Also, the economic, management and equity issues of a conjunctive irrigation system are quite distinct and different from those prevailing in either a surface water or a groundwater irrigation system. Unless these problems and issues are resolved suitably, it will not be possible either to practise conjunctive irrigation or to derive its full benefits.

These problems and issues are undoubtedly tractable and resolvable. The first step in the direction is to recognise them as they are and not to gloss them over. Then, concerted action through research and systematic as well as comprehensive studies is required to be undertaken. These studies will relate to the technical, technological, economic, management and social aspects involved.

Implementation of conjunctive irrigation strategy, right from the planning to operation and management stages, in the humid alluvial region will bring about revolutionary socio-economic transformation of the region. The performance of agriculture, the mainstay of its economy, will be enhanced several fold due to enhanced cropping intensity, higher productivity and more remunerative cropping pattern. Moreover, this will be accompanied by hydrological sustainability of productive irrigated agriculture and better socio-economic equity. As this will impart a quantum jump to the stagnating agricultural growth of India, Indian economy will experience positive thrust on account of a deficit region transforming itself into a surplus one. The secondary and tertiary production activities that will be engendered due to higher and surplus agricultural production will further benefit both regional as well as national economy. Thus, it will make a positive, substantial and even decisive contribution to correcting regional imbalance of growth that has been the bane of the process of planned development in India for the last four decades.

8. References

(1) "Conjunctive Management of Surface and Groundwater for Irrigation in the Lower Reaches of Gandak Command", Report of research carried out jointly by Centre for Water Resources Studies (Patna University) and Water and Land Management Institute (Govt. of Bihar) in collaboration with International Irrigation Management Institute (Sri Lanka), February, 1993.

(2) Prasad, T. "Problems and Prospects of Groundwater Utilization for Agricultural Development in North Bihar", Paper presented at a Workshop held at Institute of Rural Management, Anand, January, 1989.

(3) Ghosh Committee Report for Examining Extension of Canal Beyond R. D.704.00 of Tirhut Main Canal (Gandak Project, Bihar), Vols. I, II & III, Govt. of Bihar, August, 1985.

(4) Report of Kosi Irrigation Committee (Mandal Committee), Ministry of Irrigation and Agriculture, Govt. of Bihar, 1975.

(5) Vincent, Linden and Dempsey, Peter, "Conjunctive Use for Irrigation : Good Theory, Poor Practice", International Journal of Water Resources Development, Vol. 9, No. 3, 1993.

(6) Sen, S. R. "Agricultural Productivity in Eastern India" Vols. I & II, Report of a Committee set up by Reserve Bank of India, Bombay, 1984.

PLANNING CONSIDERATIONS FOR WATER RESOURCES DEVELOPMENT IN HAOR AREAS OF NORTH-EASTERN BANGLADESH

by

G M Akram Hossain[1] and Ainun Nishat[2]

ABSTRACT

A haor is a bowl shaped depression which are flooded every year during monsoon. It remains under water for several months of the year and as it drains out in post-monsoon, some water bodies are left in the deeper parts which are perennial in nature, locally called *beels* (natural shallow lake). This paper gives an outline of the planning considerations for the development of water resources in haor areas located in the north-eastern region of Bangladesh. The haor area is generally divided into three major areas which have similar characteristics in terms of morphology and hydrology: the piedmont area, the flood plain and the deeply flooded area. These divisions have been extended into six landform units by the Geological Survey of Bangladesh. The hydrology, morphology, fishery, agronomy and socio-economic issues of each of these units have been discussed and the impact of flood control projects have been investigated. Based on these findings, future development strategies have been prepared. The technical approaches for flood control, drainage and irrigation have been considered and at the same time social and environmental issues have been assessed. Institutional aspects have also been given attention. Boro (dry season rice) is the main crop grown in the area which is vulnerable to damage due to early monsoon flash floods just before the harvest. Present day strategy of water resources development aim at either full flood protection for the year round rice cultivation or construction of submersible embankments for protection against early monsoon flood. It has been established that projects with submersible embankment offer a higher rate of economic return compared to full flood protection projects. Submersible embankments with adequate water regulating structures coupled with the development of aquatic environment and navigation is the recommended mode of development for a region which meets the technical, economic, social and environmental considerations.

1 Graduate Student, Department of Water Resources Engineering, Bangladesh University of Engineering and Technology (BUET), Dhaka and Design Coordinator, Northwest Hydraulic Consultants Ltd., Dhaka.

2 Professor, Department of Water Resources Engineering, Bangladesh University of Engineering and Technology (BUET), Dhaka

V. P. Singh and B. Kumar (eds.), Water Resources Planning and Management, 29–54.
© 1996 *Kluwer Academic Publishers. Printed in the Netherlands.*

1.0 INTRODUCTION

A haor is a river backswamp, or bowl shaped depression, between the natural levees of a river (FPCO, 1993d). These haors are flooded every year during the monsoon. They remain under water for several months of the year and when they drain out, during the post-monsoon, some permanent water bodies, or "*beels*", remain in the deep part of the haors. Most of the haors lie in the Sylhet and Mymensingh Districts of north-eastern Bangladesh (figure 1).

Development strategy for haor areas need special considerations due to a different type of water regime and land use pattern prevailing there. Efforts were made by local people in earlier days to construct low embankments around the haors and close the drainage channels during the dry season in order to prevent pre-monsoon floods from entering the haor areas and destroying the standing dry season rice (Boro). Some full flood control embankment schemes e.g., the Manu river scheme have been completed recently.

For many years there was little interest in taking up any flood protection works in this area, mainly because it was considered costly and complicated to protect these low lying lands from the onrush of heavy floods. During the past few years, however, both Government and local authorities have considerably increased their activities in flood protection works. Therefore, an integrated and comprehensive planning approach for water development in these areas are considered essential before many isolated projects are taken up.

This paper evaluates the ways in which the development activities may proceed and identifies where problems may be expected in the future and how these may be avoided. The technical options for flood control, drainage and irrigation (FCDI) schemes have been discussed, and impact of these schemes on the development of agriculture, fishery and navigation with special attention to social and environmental problems have been analyzed.

2.0 PHYSICAL CHARACTERISTICS OF THE STUDY AREA

Figure 2 shows the main physiographic sub-divisions of the haor area and Table 1 summarizes the extent of each unit (FPCO, 1993d). Land unit boundaries have been synthesized from the classifications published by Geological Survey of Bangladesh (GSB, 1990) and (Rashid, 1991). This figure shows the six landform units that are found in the region. These units are: Uplands, Terraces, Alluvial fans, Piedmont Floodplain, Lowland Floodplain, and Flood Basins. A short description of this units are given below. The alluvial fans, piedmont floodplain, lowland floodplain and the floodbasins mostly constitute the haor area.

2.1 Uplands

Uplands total about 1,970 km^2 of land or about 8 per cent of the total area. These *tilas* and hills consist of Dihing and Dupi Tila formation rock of Pleistocene and Pliocene age and are composed of weathered, poorly consolidated sandstone, siltstone and conglomerates (GSB, 1990).

Figure 1

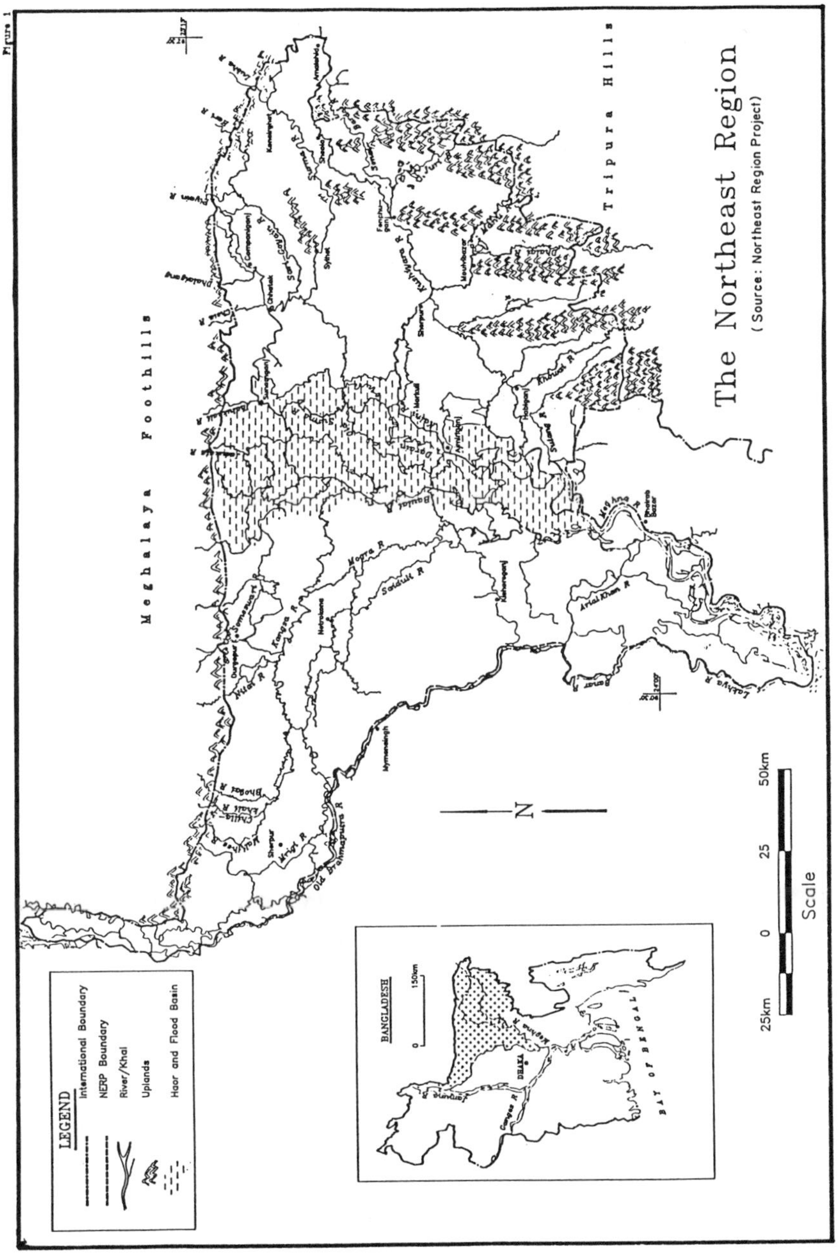

The Northeast Region
(Source: Northeast Region Project)
Meghalaya Foothills
Tripura Hills
LEGEND
International Boundary
NERP Boundary
River/Khal
Uplands
Haor and Flood Basin
BANGLADESH
BAY OF BENGAL
N
25km
0
25
50km
Scale

Figure 2

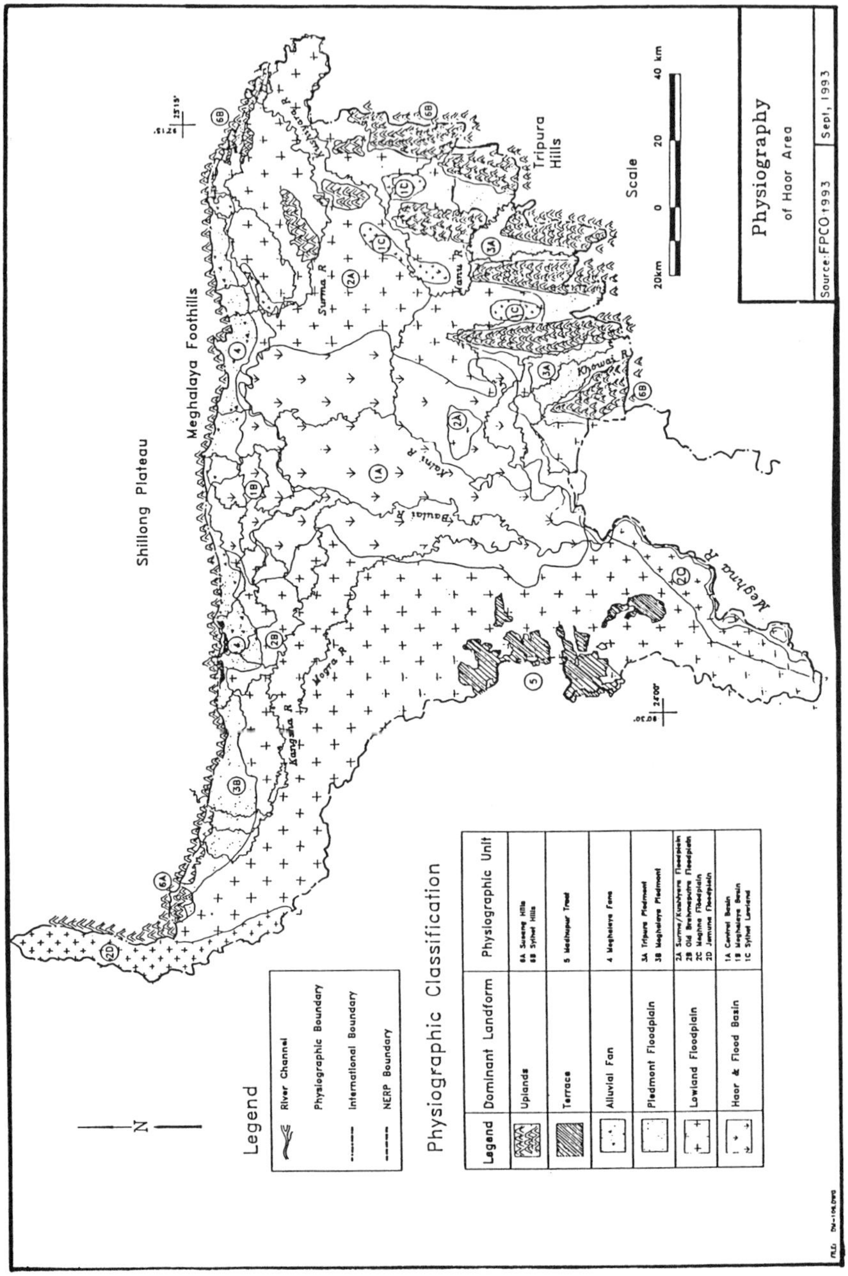
Shillong Plateau
Meghalaya Foothills
Tripura Hills
Surma R
Kangsha R
Baulai R
Khowai R
Meghna R
Scale
20km 0 20 40 km
Physiography
of Haor Area
Source: FPCO 1993
Sept, 1993
Legend
River Channel
Physiographic Boundary
International Boundary
NERP Boundary
Physiographic Classification
Legend
Dominant Landform
Physiographic Unit
Uplands
6B Sylhet Hills
Terrace
5 Madhupur Tract
Alluvial Fan
4 Meghalaya Fans
Piedmont Floodplain
3A Tripura Piedmont
3B Meghalaya Piedmont
Lowland Floodplain
2A Surma/Kusiyara Floodplain
2B Old Brahmaputra Floodplain
2C Meghna Floodplain
2D Jamuna Floodplain
Haor & Flood Basin
1A Central Basin
1B Meghalaya Basin
1C Sylhet Lowland
N

Table 1: Summary of Land Unit Areas (Source: FPCO, 1993d)

Landform Unit	Physiographic Unit	Area (km^2)	% of Region	% by Landform Unit
Uplands	Susang Hills Sylhet Hills	120 1750	0.5 7.3	7.8
Terrace	Madhupur Tract	494	2.1	2.1
Alluvial Fan	Meghalaya Fans	1486	6.2	6.2
Piedmont Floodplain	Tripura Piedmont Susang Piedmont	960	4.0	4.0
Lowland Floodplain	Surma/Kushiyara Old Bramaputra Meghna Jamuna	4636 7220 727 678	19.3 30.1 3.0 2.8	55.2
Flood Basin/Haor	Central Basin Meghalaya Basin Sylhet Lowland	5605 335	23.3 1.4	24.7

2.2 Terraces

Terraces (Madhupur Tract) occur along the western edge of the study area and part of the Old Bramaputra River. This terrace, which occupies an area of about 494 km^2, has been raised by uplifting and faulting and it is no longer subject to inundation by normal flooding.

2.3 Alluvial Fans

Alluvial are found along the northern border of the Meghalaya foothills. The fans are created when steep mountainous streams exit from their canyons and spread over the flat, unconfined land of the lowland floodplain and Sylhet depression. The decrease in channel gradient and reduction in velocity as the streams below their canyons causes deposition of sand gravel sediments in the form of a "fan-shaped' conical delta. Alluvial fans are characterized by sudden, irregular channel shifts (avulsion) which results in the periodic abandonment of some channel and development of new channels across the fan surface.

Principal streams which have developed alluvial fans include the Someswari River, Jadukata River, Jhalukhali River and Dauki/Piyain River. A review of the existing water control projects in the region indicated that most of the sediment problems occurred at sites located in the alluvial fans.

2.4 Piedmont Floodplain

Piedmont flood plains have formed along the southern and northern margins of the haor regions by relatively small, steep tributary streams. Streams in Tripura piedmont include the Khowai, Manu, Sutang, Dhalai and Juri Rivers which flow northwards from the Tripura Hills in india and join into Kushiyara River. Piedmont floodplain have also developed at the base of the Garo Hills. Piedmont streams are typically meandering sand-bed channels with prominent natural levees of sand and silt.

2.5 Floodplain

Floodplain are landform that have been created as a result of fluvial deposition and erosion. Floodplain from the Surma/Kushiyara River, Meghna River, Old Bramaputra River and Jamuna River account for about 55% of the land area in haor, totalling 13,260 km^2. Floodplain include channel deposits such as point bars and fills, over bank deposits such as natural levees and crevasse splays and fine grained flood basin and back deposits. Levees are wedge shaped ridges of sediment bordering stream channels and are mostly developed on the convex (outer) side of meanders. Levees often are the highest points of land on the floodplain and project about 3 - 5 m above the adjacent low lying back basins. Natural levees are formed by deposition of sediment when flood water of a stream overtops it's bank. During overtopping, the velocities are reduced causing deposition of sediment with the rate of deposition decreasing sharply away from the bank. On most streams, flood waters spill through distinct channels that cut across the levees. These cut channels, called "crevasse" develop their own pattern and drainage system and extend across the levee systems into the floodbasins. Occasionally, the main river may shift through the crevasse causing a river avulsion to occur. Figure 3 illustrates these landform on a typical lowland river flood plain.

2.6 Flood Basins/Haor

Flood basins and haors are the dominant landform throughout the Central Basin, Meghalaya Basin and Sylhet Lowland. These lands form a low lying, bowl-shaped depression in the middle of the north-eastern region of Bangladesh and occupy an area of about 6,000 km^2. Virtually all of the land lies below El. 8.0 m PWD and is deeply flooded during the monsoon season. This sub-region is characterized by large saucer-shaped depressions, termed *haors* and permanently inundated water bodies termed *beels*.

The Meghalaya Basin, is bordered by the Meghalaya Foothills on the north and by the Surma/Baulai River on the south and by the Old Bramaputra River floodplain deposits on the west. The main haors in this unit include Kelner Haor, Karchar Haor, Shanir Haor, Matian haor, Gurmar haor and Tangua Haor. The haors in the Meghalaya Basin have existed for a very long time inspite of their proximity to very large rivers carrying high sediment loads. The persistence of this features may be attributed to the high rates of subsidence in the region and suggests that in the long-term, sedimentation in the area have been offset by subsidence.

The Central basin is bounded by the Surma River on the north, the Surma Kushiyara floodplain on the east and the old Bramaputra River on the west. Land elevations typically range between El. 3 - 7 m PWD. The area is occupied primarily by backswamp and flood basins. Much of the land is traversed by distributary spill channels and other old partially infilled channels which at one time connected the Surma River System to the Kushiyara River. The main haors in the Central Basin include Deker Haor, Shanghair Haor, Pagner Haor, Baram Haor, Chaptir Haor and Naluar Haor. The basin is isolated from active sediment sources.

Figure 3

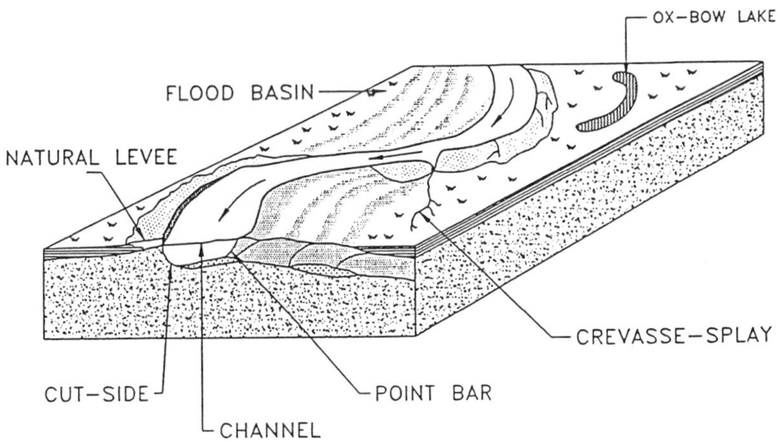

Landforms on Lowland Floodplains

Source: FPCO. 1993 | Sept, 1993

FILE: dm196.DWG

The Sylhet lowland is a discontinuous area of haor and flood basin located east of the Central Basin. This unit includes Hail Haor, south of Moulvibazar, Damrir Haor and Maijail Haor near Fenchugonj and Pakohadir Haor north of Sylhet. These features are generally surrounded by higher floodplain land and the boundary between floodplain and haor is gradational.

The hydrometric network of the area is relatively dense and records have been kept over a considerable period of time. Since 1960 the whole area has been well covered, and some stations have records which go much further back. Much information on flow is still not available. Only a few hydrological stations have flow records over a sufficient number of years, and many stations are not rated. In the flood plain, rating is sometimes difficult because a large part of the flow moves through the floodplain. In the piedmont floodplain, rating is also difficult because bed levels may change considerably from year to year. With regard to morphology little information is available. A better knowledge of the sediment transport and the sediment characteristics is essential for planning the utilization of water resources in the region. Recently a study (FPCO, 1993d) was carried out on " River Sedimentation and Morphology". An inventory of major sediment problems at major FCDI projects in the haor areas are listed in table 2 (FPCO, 1993d).

3.0 HYDROLOGICAL CHARACTERISTIC OF THE HAOR

The hydrographs of the rivers in the *haor* area show that two floods, the early monsoon flood and the monsoon flood, occur almost every year. Figure 4 shows a typical hydrograph of river in the study area. The early monsoon flood which has the characteristic of flashflood mostly occurs between the period of late April to middle of May. The gradual monsoon floods occur later on in July-August. In fact, in a typical year, the entire haor area would go under water which is considered to be normal (Nishat, 1993).

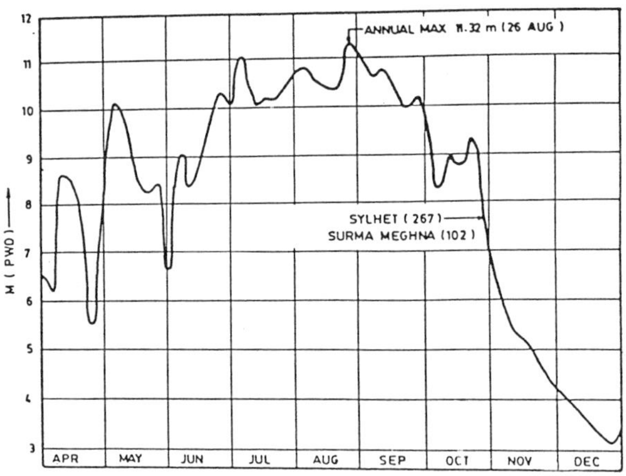

Figure 4: Typical Hydrograph of a River in Haor Area

Table 2: Sediment Problem at Major FCDI Projects in Haor Areas (Source: FPCO, 1993d)

Project	Proj. Type	Land Unit	Sediment Problem 1	2	3	4	5	6	Cause of Problem
Angurali Haor	PFC	Fan/Meghalaya Basin		X		X	X		Avulsion on Jadukata R.
Kalner Haor	PFC	Fan/Meghalaya Basin		X		X			Avulsion from Jhalukhali R.
Karchar Haor	PFC	Fan/Meghalaya Basin		X	X	X	X		Shifting on Jhalukhali R. fan
Shanir Haor	PFC	Fan/Meghalaya Basin	X	X	X	X		X	Sedimentation on Jadukata fan
Matian Haor	PFC	Fan/Meghalaya Basin	X			X	X		Shifting on Jhalukhali R. fan
Joal Bhanga Haor	PFC	Fan/Meghalaya Basin					X	X	Shifting on Jadukata R. fan
Mohalia Haor	PFC	Fan/Meghalaya Basin				X			Shifting on Jadukata R. fan
Halir Haor	PFC	Fan/Meghalaya Basin					X		Shifting on Jadukata R. fan
Chillakhali River	FC	Piedmont Floodplain	X	X			X	X	Embankments in deposition zone
Khowai River	FC	Piedmont Floodplain	X			X			Deposition in Kalni/Meghna backwater
Pagner Haor	PFC	Central Basin		X		X			Spills from Baulai R.
Baram Haor	PFC	Central Basin	X		X		X		Regime changes on Kalni R.
Bhanda Beel	PFC	Central Basin	X		X		X		Regime changes on Kalni R.
Chandra Sonathal	PFC	Central Basin			X			X	Avulsion on Someswari R.
Chaptir Haor	PFC	Central Basin	X				X		Regime changes on Kalni R.
Damrir Haor	D	Central Basin			X			X	Spill from Kushiyara R.
Dhankunia khal	PFC	Central Basin			X			X	Avulsion on Someswari R.
Gurmar haor	PFC	Central Basin			X				Avulsion on Someswari R.
Haijda Embankment	PFC	Central Basin			X				Avulsion on Someswari R.

PFC: Partial Flood Control
FC : Full Flood Control
D : Drainage Improvement

Sediment Problem Legend:
1: Aggradation in embanked channels causing increased flood levels
2: Overbank sand deposition on agricultural land, fish habitat & wetlands
3: Local siltation causing drainage congestion
4: Channel avulsion and shifting; erosion and embankment breaching
5: Region-wide impacts to water levels and flow patterns
6. Siltation causing navigation obstruction

4.0 PRESENT AGRICULTURAL PRACTICES

The soils of the haor area are silty to clayey with a low to moderate content of organic matter and are seasonally flooded. The agricultural potential of the soils in the study area has been evaluated in terms of land capability class by the Dutch funded Early Implementation Project (EIP) cells (BWDB, 1986). The lands of this area may be classified into four groups: good agricultural land, moderate agricultural land, poor agricultural land and non-agricultural land. The percentages of land falling under the different categories are 32%, 34%, 29% and 5% respectively.

Boro is the main crop in the haor area. It is generally grown in the haor basin after the flood subsides and these lands are irrigated by water which remains at the bottom of the basin. Boro is transplanted during November and December and harvested during April and early May. The high yielding variety occupy about 50% of the area covered by Boro. Aman paddy is also one of the major food crops grown in the area. This crop can either be transplanted or broadcasted. The major cropping pattern of the haor area are the: single Boro crop; Aus/Jute, transplanted aman followed by fallow; mixed aus and broadcast aman followed by fallow; single crop of broadcast aman or aus followed by rabi crops. Groundnuts, sweet potatoes, oilseeds, chilies etc. are also grown on the levee soils.
Figure 5 shows a typical cropping pattern along with the time and duration of occurrence of floods and need for irrigation water and figure 6 shows the agro-ecological divisions of a typical *haor* area.

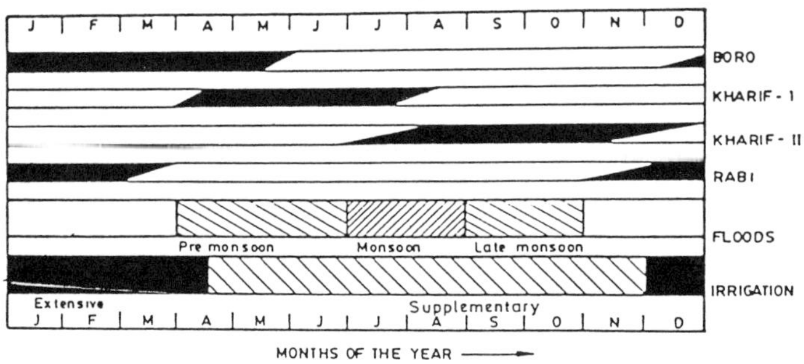

Figure 5: Typical Cropping Pattern with Floods and Irrigation

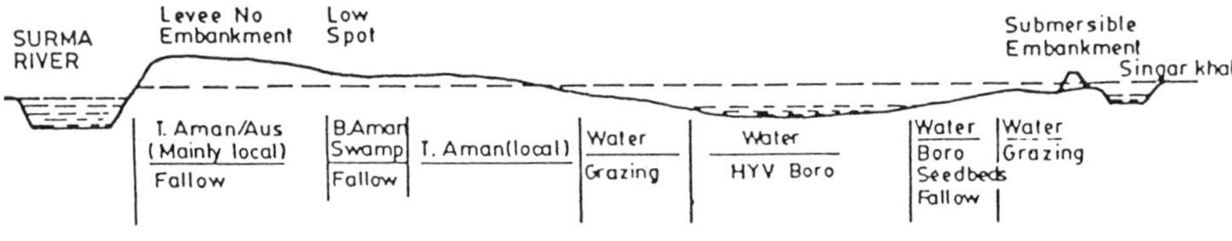

Figure 6: Agro-ecological Division of a Typical Haor (Zilkar Haor, Sylhet)

5.0 SOCIO–ECONOMIC ISSUES IN THE STUDY AREA

5.1 Land Ownership

In the haor area, as elsewhere in Bangladesh, land is the most valuable asset and conflicts over the possession of a piece of land can often be very serious and may lead to long-lasting feuds and even violent clashes. Land holdings may be taken to distinguish between different socio–economic classes. We can differentiate among at least four socio–economic classes on the size of land owned as follows (BWDB, 1986):

The first category, the landless includes poor peasants who own small amount of land approximately less than 0.4 ha. Such a small piece of land is far from sufficient for subsistence and to survive they must therefore largely rely on selling their labour to others, i.e. they are functionally landless. The means of survival is either as day labourers or as share croppers or through a combination of both activities.

The second category are small farmers who own some land which is not entirely sufficient for their survival so that they must also sell their labour to others. The upper and lower land holdings limits of this group are somewhat difficult to define, but in mainly single cropping haor areas it ranges between about 0.40 to 2.0 ha.

The third category, the medium farmers, is a group which is reasonably self sufficient. In a normal year with no crop failures they are able to survive mainly on the produce of their lands. They are in a vulnerable position, however, since they have no or only a few reserves and crop failure can therefore easily push them back into a dependent position. The landholding of this category range somewhere between 2.0 to 3.0 ha.

The fourth category are large landowners who hold more land than they actually need for their subsistence and, during normal years at least, they have a marketable surplus. An important distinction within this category is between the owner–cultivators who although they may employ people, supervise the work themselves and the absentee landlords who either let the land lie fallow or employ share croppers on various terms. This category of farmers own more than 3.0 ha of land.

A comparison of different socio–economic studies of haor areas with regard to size of landholding, shows a pattern with only slight variations. In most parts of the haor areas landholding is extremely skewed. The landless, and small farmers as defined above, constitute some 80% and medium and large farmers approximately 10 per cent each. If, on the other hand, we look at the amount of land owned by each category, a reverse picture appears and we find that medium and large landowners control some 70% of available land while small farmers own about 28 per cent and landless some 2 per cent. Of course figures vary here and there but nowhere to a degree that would invalidate this picture.

Compared with other parts of Bangladesh, the haor areas have large amount of khas land. However these lands are mostly under the control of local influential. Another important parameter from social considerations is the concentration of people living abroad (mainly U.K.) and sending back home their savings. This group of people have invested a lot in purchasing land, but are not enthusiast in further investment in agricultural production. Benefits in water resources development is obviously enjoyed by the land owners and it is extremely difficult to channelise the benefits to the poor.

With the introduction of new agricultural techniques for cultivation of HYVs, the potential income from land has increased manifold. The main benefits of these new sources of income have gone to wealthy landowners and the difference of income between a landowner and a landless day labourer has increased significantly. This is because of the fact that to make HYVs demand secure and to supply timely input of irrigation water, shallow and deep tube wells are in many cases a prerequisite. This has led to the establishment of a new wealthy people in the *haor* areas, so called water lords who have gained large fortunes by selling irrigation water from their tube wells to landowners. In other words, due to absences of equitable distribution of benefits, the HYVs have contributed to a widening gap between poor and rich.

The above are some of the many socio-economic issues which make it extremely difficult to direct benefits to the poor in development projects which aim at increasing agricultural production. If no specific efforts are undertaken, benefits will continue to go almost exclusively to the richer sections of the community, and, given existing political and social conditions they will anyway continue to reap the major share.

5.2 Activities for Women

Although women form approximately half the target group they are mostly 'invisible' in project documents of all kinds. To some extent at least, this can be explained by the fact that, although women are in charge of all post-harvest operations, they are mostly confined to the homestead and therefore never seen by male biased development agencies. Nevertheless, in many poor families women do work outside the homestead and even though this may be socially degradable, it is done out of economic necessity. But even where women remain confined to the homestead their contributions to the family economy is considerable and, in fact, women most often work harder and have longer working hours per day than men. As the contribution of women is generally labelled as household work, it is not visible in earning statistics. Therefore women's contribution to the household economy is grossly underestimated or not even considered at all in official statistics.

When planning water sector schemes as well as other development projects some special attention must therefore be given to women and projects must be designed in such a way that they consider women's needs. Bangladesh Rural Development Board (BRDB) organises rural women into women's cooperatives and several NGOs also have welfare-related programmes for women. For those women who are willing and able to do earthwork, cooperatives could be organised and, more specifically, turfing of embankments seems to be an activity well suited to women labour groups. For most women, however, various kinds of income generating activities that can be performed within the household compound would probably be preferable. One can think of activities such as mat weaving, home gardening, poultry farming, duck raising (which is specifically suitable in haor areas) pottery making, parboiling of paddy and cow fattening etc. The list can certainly be made longer but the specific activities to be included must always depend on the local situation and on the wishes of the women involved in the program.

5.2.1 **Programs Targeted to Women.** Table 3 presents gender-wise statistics on groups and their membership for five national NGOs in the study area for which data are available. For the Grameen Bank, the reporting period is the calender year 1991 and for others, it is the FY 1991-92. The Grameen Bank has the highest proportion of female enrolment within the fold of it's program at the grassroots.

Table 3: Gender Distribution of Population Covered by Selected NGOs
(Source: FPCO, 1993a)

Agency	No. of groups		No of members		Female members as % of total
	Male	Female	Male	Female	
Grameen Bank	16,011	80,053	197,275	986,373	92
BRAC	4,124	8,888	200,630	471,690	70
Caritas	3,461	3,988	73,689	79,745	52
ASA	787	3,375	22,001	73,986	77
Com. Proshika	936	638	46,525	27,805	37

5.3 Fishery in the Haor Area

When a flood protection project is undertaken, fishermen are not only unable to gain any benefit from it but most often suffer severe negative consequences. Thus, there is, a conflict between fishing and agriculture and it seems difficult to promote both simultaneously. Agriculture is the basis for survival of the people, while fish is the most important source of protein and a balance must therefore be made between the two modes of production.

In the haor areas fishing is of greater importance than perhaps anywhere else in the country, and, as already mentioned it was the abundant fish resources which first induced people to settle in this area. The first fishermen were exclusively Hindu and belonged to a low-caste (kaivarta) group and still today the majority of fishermen are Hindus. However, in the course of time, when they realized that fishing was profitable business, Muslims also adopted this profession. During the Zamindar time no tax was paid on fish caught and fishing was free for all. Sometime after the abolition of the Zamindari system in 1950, a system of leasing out of fishing rights in the haors was introduced and this system is still at work. Most of the haors are leased out for one to three years, but sometimes longer leases are available.

Since the introduction of the lease system, rich people by virtue of their financial status have been able to monopolize access to lease contracts. After the liberation of Bangladesh, the government wanted to introduce new requirements which would make it possible for genuine fishermen to enjoy the benefits of their labour. It was therefore ordained that only cooperatives formed by genuine fishermen and landless were allowed to take lease of fishing rights. However, despite such precautions, it is still the same wealthy fish merchants who are in control of the contracts. In the lease contracts lessees are required to prevent overfishing and to minimize points of conflict with the farmers. It is a common complaint from farmers that the fish merchants do not adhere to these conditions and clashes often occur. The major conflict has to do with the time when haors can be drained off for fishing. To satisfy the farmers' demand for irrigation water, the draining should not be done before the end of March, but the fish merchants want to drain off already in December or January as the price of fish are higher at that time than later in the season. As a result the paddy cultivation in areas adjacent to the haors suffer from lack of water and every year extensive areas have to be left fallow. Due to malpractice such as these, agricultural and fishing interests clash quite unnecessarily and a first requirement for further investment in a conflicted area must be that contract stipulations are adhered to. Secondly, ways must be sought to secure lease contracts for the genuine fishermen. Of course this can be easily said but extremely difficult to put into practice, since it would go against the interests of the very powerful local elite. Undoubtedly they will do anything to prevent loss of this profitable business.

5.4 Lack of Peoples Participation

A final major socio-economic issues which makes implementation of socially acceptable water sector projects very difficult is a general lack of beneficiary participation in such schemes and in particular unwillingness from the beneficiaries to promote self-help activities necessary to maintain the physical structures. There seems to be an almost total lack of communication between development agencies and people in general. When projects are decided by the elites of the society, people do not feel that this is any responsibility of theirs and when asked to maintain the structures they refer to the implementing agency as the one responsible. On the other hand when people themselves are asked to identify their problems and when they see that their experiences and advices are taken seriously, a feeling of ownership is created and in such cases people are more willing to undertake in various self-help activities. Therefore a change of present attitude of concerned agencies responsible for water resources development, will make possible an active participation of beneficiaries during project identification, planning, design as well as during implementation.
This will help in the long-term sustainability of the water resources development projects built.

6.0 WATER RESOURCES PLANNING

Attempts to develop the water resources of the haor area were made in the past and are also continuing at present. But the development approach was not systematic in the sense that projects were taken up in solving the problems of a localized area, not considering the haor area in general. Moreover, the effects of such localized water resources development on other sectors, for example fishery, navigation, ecology and also long-term effects of such development on sector parameters, like sedimentation and river morphology have not been studied in detail. The Master Plan Organization (MPO, 1984) in it's National Water Plan Project, has quantified the water resources potential of the haor area, but a strategy for total development has not been outlined. However, Flood Plan Coordination Organization (FPCO), Ministry of Irrigation, Government of Bangladesh, has started pre-feasibility study for the regional development of the haor area in the name "Northeast Region Water Management Project (NERP)". The studies so far completed by NERP are: (1) Environmental Planning: Assessment and Management (FPCO, 1992a), (ii) Draft Thematic study: Fisheries in the Haor Area (FPCO, 1992b), (iii) Draft Thematic study: Regional Water Resources Development Status (FPCO, 1992c), (iv) Participatory Development: the Role of NGO's in Haor Area (FPCO, 1993a), (v) Population Characteristics: State of Human Development (FPCO, 1993b), (vi) Wetland Resources Specialist Study (FPCO, 1993c), and (vii) River Sedimentation and Morphology (FPCO, 1993d).

A detailed plan of development of the use of water resources in the haor area cannot be done in the framework of this paper. Essential research work and data collection have to be completed before any such comprehensive planning can be carried out. In this paper, the water management projects that have been implemented in the past or are ongoing at present or will be implemented in the future, will be identified. The experience gained from these projects will form a basis for formulating a short-term future plan. Then, depending on the present data base, project planning and development for the future can be outlined.

7.0 EVALUATION OF PROJECTS IN THE AREA

7.1 On-going Projects

The list of the completed, ongoing and planned water management projects in the haor area, is presented in Annex A of this paper. The locations of these projects are shown in figure 7. The list indicates that quite a number of water management projects have been carried out or are ongoing in the area (40 projects). Most of these are small scale projects. Large scale projects include the Manu River Project near Moulvibazar and the Khowai River Project near Habiganj. The projects in the study area cover about 350,000 hectares, with an investment cost of over Tk 2,000 million (1986 prices).

Apart from Bangladesh Water Development Board (BWDB) and other specialized departments involved in water resources development projects, a number of projects have also been undertaken in the water sector by local initiatives. Sometimes these constructions, although small in size, can have considerable impact on the surface water availability (for example, closure across rivers and khals). However, very little is known about these locally initiated projects.
Another major project in the water sector (but not designed, implemented and maintained by the BWDB), is the embankment along the southern bank of Surma River, from the point of bifurcation of Surma and Kushiyara rivers at the border with India to downstream of Sylhet town. The natural levees along this part of Surma River were already quite high, thus the completed interior embankment now serves as a protection embankment against unwanted water intrusion into the haors lying south of the river. A huge flood protection, drainage and irrigation project called "Surma-Kushiyara Project" covering an area of about 100,000 hectares between the left and the right banks of Surma and Kushiyara rivers is now at the planning stage. If implemented, this project will have significant effect on the overall water regime of the haor area, which needs to be investigated in details. The Northeast Region Water Management Project (NERP) will evaluate the detailed impact of completed and proposed schemes in the region and will recommend further development needs for the entire region.

Figure 7

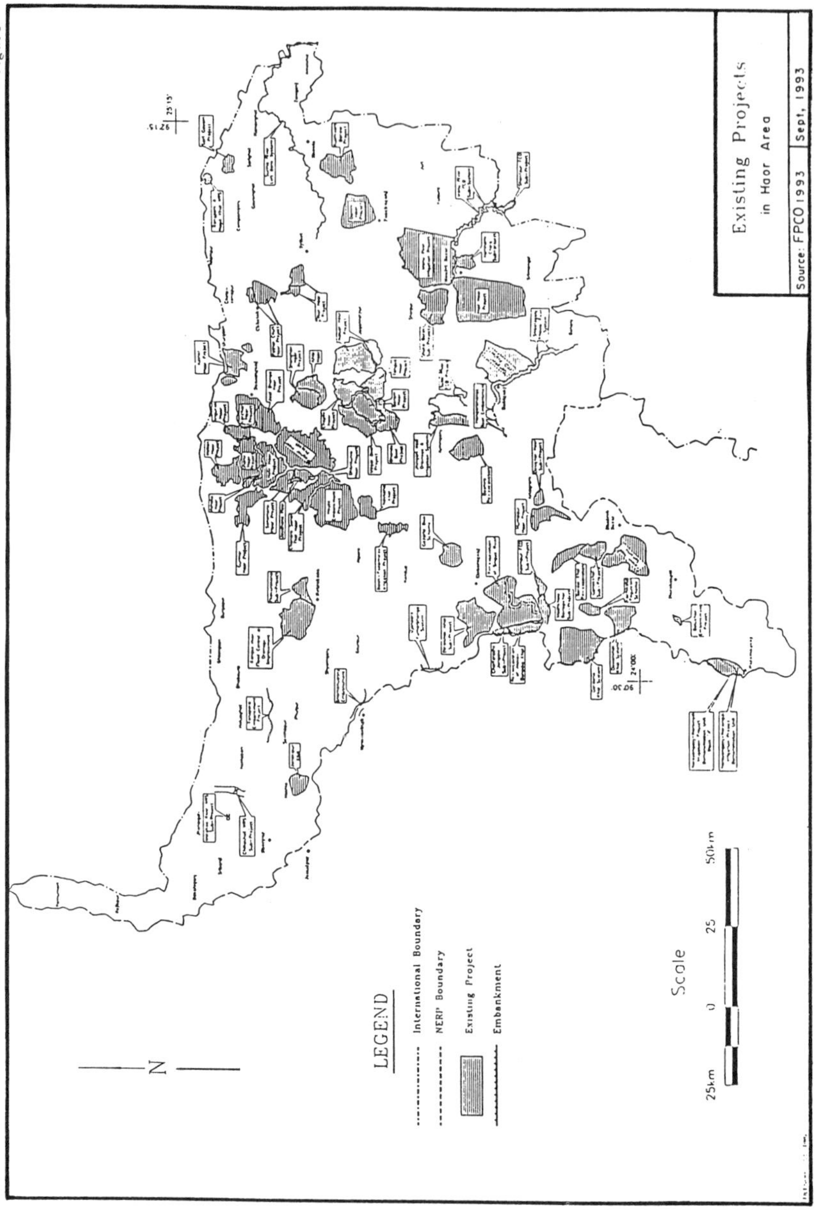
Existing Projects
in Haor Area
Source: FPCO 1993
Sept, 1993
N
LEGEND
International Boundary
NERP Boundary
Existing Project
Embankment
Scale
25km
0
25
50km

7.2 Completed Projects

Of the total 40 projects in the list, only a few of them are fully completed. This is probably the most important observation. The time required for implementation in most cases, were considerably longer than originally planned. The reasons were: i) delay in land acquisition, ii) revisions of project plans due to cost increases and iii) more importantly the short working seasons (January to April only) due to prolonged flood seasons. More careful and realistic planning and stream lining of administrative procedures can alleviate these problems.

Some of the completed projects are in reality still missing certain essential engineering works. For example, structures are missing in some already embanked haors. Cross dikes in rivers and khals and gaps in the embankments of some projects are reconstructed every year under the Food For Work (FFW) program, an assistance from World Food Program (WFP) and these often save the crop from being damaged by early floods. Although this approach does seem to work by the grace of the WFP, all efforts should be given to complete projects in the real sense within a reasonable time period.

Most of the small scale water management projects in the haor area, either completed or ongoing, are submersible embankment projects (early flood protection with post monsoon drainage). The main objective of the projects are to remove natural constraints to agricultural production. Considering this major objective, **the Shanghair Haor Project** (completed under Dutch funded EIP) was definitely a success. During the pre-project years, 20% of the crop was damaged by early spring floods every year, but after the completion of the project in 1981, the area virtually did not experience any crop damage. The crop yield has increased significantly and farmers have made a shift from local to HYV boro rice cultivation. The project has increased the rate of absorption of hired labour, thus creating higher local employment opportunities. Provisions of irrigation in the project area have increased the irrigated area by 5%. The project has created some minor problems; due to embankment construction, waterlogging has emerged in parts of the project and the navigation has been affected. It has been alleged that the project has decreased the production and reproduction of fish in the area. Finally, it can be concluded that the project has significantly increased the crop security in the area and the marginal farmers are relatively more benefited by the project. These findings should be encouraging to future haor development programme.

8.0 DEVELOPMENT OF PLANNING STRATEGIES FOR THE AREAS

Of the six areas mentioned in section 2, development of planning strategies for the three clearly defined hydrological areas have been discussed in details in this section. These are the piedmont floodplain, the lowland floodplain and the floodbasin or the deeply flooded area. The initial planning should conform with the planning area divisions proposed in the National Water Plan (MPO, 1984). But several of such planning areas are at the same time in different hydrological units, whereas the proposed planning considerations are based on hydrological boundaries. In the following paragraphs, the development strategies in the three well defined planning areas are discussed.

8.1 Development in the Alluvial fans and Piedmont Area

Based on the observations, full flood embankment projects do not seem to be a justifiable solution for the alluvial fans and piedmont region (hill foot area). Apart from the hydrological and morphological considerations, the negative effects of such full flood embankments can have on the adjacent areas, might be unacceptable. From an engineering point of view, construction and maintenance of full flood embankments in the hill foot area is hazardous. Full flood protection applying the horse shoe approach (embankment starting upstream, leaving the downstream portion open) will, in the long run, also not alleviate the problems. Negative effects at the downstream side (without embankment) will be severe because of infertile sediment deposition. Moreover siltation in the river beds will be accelerated and frequent raising of the embankment will subsequently be required.

A general feature of the piedmont regions is that the natural levees are around the average annual highest water level. To protect the winter crop, the levees are therefore high enough to withstand early flash flood. The inundation of the haors occurs through the open drainage inlets. Construction of structures which will simply prevent the early intrusion of flood water before harvesting of the winter crop, is possible. The submersible embankment will probably not have severe negative effects on the adjacent areas.

8.2 Development in the Lowland Floodplain

Several full-flood embankment projects have already been completed or are ongoing in the flood plain areas of Sylhet. The southern bank of Surma river has such an embankment. Manu and Khowai River Projects also have full-flood protection. The proposed Surma-Kushiyara project also include full-flood protection embankments along the rivers. Full-flood protection along the rivers in the flood plains is, however, not without limitations. As explained above, haors do play an important part in reducing the level of the flash flood peaks. At the same time, sedimentation processes are equally influenced by embankments along the river. At present, the possible effects of full-flood embankments on river hydrology and morphology are not fully known. But the experiences with the projects along Manu and Khowai Rivers do suggest that their impact might be considerable and mostly negative because of: (i) higher flood peaks, (ii) increased sediment deposits at unsuitable locations etc. Aggradation of river beds, is causing higher flood peaks necessitating longer and higher section of embankment. Until more is known about the effect of full-flood protection embankment, a more cautious approach can only be recommended. This does not mean, that full-flood protection embankments are to be discarded completely, but their possible negative effects have to be analyzed carefully and thoroughly. Mathematical model will provide a better insight into the limitations of full-flood embankments. Depending on the physical situation of certain haors in the flood plain (**Hakaluki Haor, Hail Haor**), development of only pre-monsoon protection schemes is a reasonable alternative (BWDB, 1963, 1976). As with the rivers in the piedmont area, the natural levees here are also high and investment cost for pre-monsoon flood protection thus might be low and mainly limited to construction of structures in the drainage outlets of the haors.

8.3 Development in the Floodbasin/Deeply Flooded Area

From a hydrologic view point, areas in the deeply flooded region can be provided with full-flood protection. Such a solution will however require very costly engineering works: high embankments with protective works against wave action and erosion, pumping stations for drainage of rain storm and seepage water and regulators etc. In the deeply flooded areas, several submersible embankments have already been implemented. Although certain technical aspects of such embankments still need further study and improvements, most of them have proved to be very effective in alleviating the main agricultural problem of the area. Few submersible embankments projects implemented under EIP program of BWDB, have performed very well and are among the most successful projects of the program. Apart from the lower project cost, the specific advantages of such projects are: (i) a secured Boro crop which can bring self sufficiency in rice in the area, (ii) a limited negative effects to navigation, fishery resources and other aquatic environment in the area. A broad economic analysis on submersible and full-flood embankments in Baram Haor project (a typical haor in the deeply flooded area under EIP program) was carried out by the EIP cell to evaluate the viability of this project. From the economic analysis, the benefit cost ratio (BCR) of a submersible embankment project was calculated as 3.8 compared to that of 1.1 for a full-flood protection scheme. The negative effect of full-flood protection schemes on fishery, ecology and navigation were not considered in the analysis.

9.0 ASSESSMENT OF ENVIRONMENTAL IMPACT

9.1 Impact on Surface Water

Any Flood Control Drainage and Irrigation (FCDI) Project has impacts, either positive or negative on the surface water environment (FPCO, 1992e). Projects may directly or indirectly affect: (i) Surface water availability (water quantity), (ii) flooded area, (iii) ground water recharge, (iv) river and channel morphology, and (v) water quality. This list is neither exhaustive nor complete, but it is indicative. Those conducting the surface water hydrologic/hydraulic analyses will need to make an assessment to identify all potential impacts. The impact analysis is based on a comparison of future "without" and "with" project conditions. As far as possible, the impacts need to be quantified for use in benefit-cost analyses. The impact analysis also must not be limited to just the project area, it must assess the impacts of the project on upstream and downstream areas as well.

9.1.1 Mitigation. Mitigation is the elimination, reduction or control of the adverse environmental impacts of a project. Mitigation strategies involve either making alterations to the project design or offering compensation. Compensation may be made in many forms, among them: restoration of damaged resources, money to affected persons, concessions on other issues, or off-site programs to enhance some other aspect of the environment or quality of life for the community. All mitigation measures cost something, and this must be quantified. A mitigation strategy would involve comparing various measures, weighing trade-offs between alternative measures, and proposing one or more actions, usually combining a number of measures. Impacts and mitigation measures for specific project studies need to be identified and evaluated.

Heavy rainfall in the monsoon season can cause flooding when drainage is impeded-drainage already is naturally impeded by the high river levels in the monsoon. Embankments tend to exacerbate this problem. Pump drainage, a mitigating measure that has been advocated in the past, is very expensive and may not be economic. A new measure, being tested under the FPCO pilot studies, is the concept of controlled flooding. In this approach, sections of land (compartments) adjoining a river are embanked on all sides, with the section of embankment facing the river being provided with structures necessary to ensure control over inflow and outflow of water, and thereby level of flooding within the compartment. Controlled flooding may well provide mitigation for other common negative impacts often associated with embankments such as waterlogging, fish movements, sedimentation, scouring, etc.

9.2 Impacts on Aquatic Habitats

Changes in aquatic habitats can be projected from the data on monthly changes in the area and depth (m) of each type of aquatic habitat and the sources of inundation water (river or rain). Based on consideration of the engineering structure and/or operation, the estimated or predicted changes are: (i) changes in land types due to changes in flooding patterns, (ii) area and depth of each type of aquatic habitat during the peak flooding period, (iii) period of flooding during the monsoon season (iv) time of water receding in the post-monsoon season, (v) water retention period and amount of the retention in beels and haors during the dry season, and (vi) possible modification of limnological characteristics.

9.2.1 Impacts on Fish and Fisheries. the impacts are; (i) loss of fish production due to possible failure of recruitment of floodplain-dependent species (adults spawn in the river and larvae or juveniles enter the floodplain with inundating river water), and (ii) decrease in the water areas, particularly beels, in the dry season to estimate possible changes (decrease) in fish production due to the decrease in recruitment of the floodplain-resident species. The changes in fish production, as the fish production in a water body is proportionate to the water surface area and depth, changes (loss or gain) in fish production of floodplain-resident species due to the limnological modification of aquatic habitats, and loss of fish production due to shortening the growth period of fish and fishing periods as the result of shortening the flooding period.

9.3 Impacts on Socio-economic Conditions

The impact of changes (loss) in fish catches of fishermen are: (i) family income, (ii) unemployment, (iii) nutritional (fish) intake, particularly for occasional fishermen, and (iv) indebtedness among religion and caste groups.

9.3.1 Mitigation. The potential mitigation strategies for reducing fishery losses can be (FPCO, 1992b): (i) establishing controlled floodplain(s) in the project area for open-water fisheries, (ii) developing fish and aquatic sanctuaries (protected from fishing) to maintain species diversity and natural fish stocks in the parts of controlled floodplain, (iii) designing fish passages in structure, particularly for fish larvae and prawn juveniles, (iv) stocking fish inside the project area to enhance production, and (v) implementing the New Fisheries Management Policy to protect fish stocks and provide a livelihood for fishermen of limited resources.

9.3.2 Recommendation. Integration of a fishery component in all future haor development projects is considered desirable, if not imperative, with a view to mitigate the adverse effect on fisheries. In planning new projects, the concerned agency should include in feasibility studies an assessment of the impact of flood control measures on fisheries and measures to minimize them. Recommendations from these studies should be implemented as an integral part of the project component and evaluated for their effects. The main measures in any project will be based on the considerations: (i) compensation for (assessed) foregone production in the open waters through pond culture or other aquaculture development in the project area, (ii) if the feasibility study shows that major spawning grounds are involved, compensation of the foregone reproduction opportunities by construction of hatcheries/nurseries for restocking purposes, and (iii) creation of jobs (e.g. in aquaculture) and job training for fishermen, so that the fishing pressure in the remaining open waters is at least not increased.

In addition to these measures, legislative measures can be taken, such as to close fishing periods in the spawning season. However these measures will only be effective if they can be enforced.

10.0 NAVIGATION DEVELOPMENT

Boat traffic is the major means of communication in the haor area during monsoon season. Farmers travel to their fields by boat and water transportation is the only reliable way of moving agricultural produce and people from village to village. Navigation during dry period is not possible in most of the rivers in haor area due to less draft available. However, proper consideration must be given in the development plan so as not to interfere with the monsoon season navigation (BWDB, 1976).

11.0 PROJECT BENEFIT

The benefits of water resources development projects are many. The primary benefit of the water resources development in haor area lies in the increase in agricultural production. The impact of the project on accelerated agricultural activity would be fur-reaching specially due to it's multiplier and linkage effect. To determine the project benefit, two alternatives with partial flood protection and full flood protection may be considered. The first alternative would be the submersible embankment for early flood protection to the period of harvesting Boro crops (to May), drainage and Low Lift Pump (LLP) irrigation and the second alternative would be with full flood protection, drainage and irrigation. The submersible embankment alternative would have not much negative impact on navigation and fishery in the haor area but there would be adverse effect on fishery and navigation with full flood protection schemes. Four types of benefit would be derived from haor development projects are: (i) direct, (ii) indirect, (iii) public, and (iv) intangible

Direct benefit is the increase in agricultural productivity. Indirect benefits would result from the increase in net income from the increased flow of agricultural products from the area. Public benefits accrue from the establishment of farming and development opportunities and from stabilizing the area's economy. Intangible benefits which are not susceptible of monetary measurements, include those arising from new and improved community facilities and services and improvement in level of living.

Evaluation of haor development project has been based on establishment of the Benefit Cost Ratio (BCR) and the Economic Rate of Return (ERR). The BCR for a submersible embankment project in the Baram Haor Project, a typical haor project in the deeply flooded area, was calculated to be about 3.8.

12.0 CONCLUSION AND RECOMMENDATION

A comprehensive development of the water resources in the haor area under the Districts of Sylhet and Mymensingh in Bangladesh is necessary before individual projects are taken up. The ongoing project in the haor area need not be stopped while waiting for the outcome of such a study as there is a pressing need for food production and to improve the standard of living. It may be recommended that water resources development schemes with submersible embankment is a good intermediate solution as they have very few adverse effect on adjacent areas and also quite high Economic Rate of Return. Furthermore, the submersible embankments may always be re-sectioned to full flood embankments at a later stage if these turn out to be feasible.

The on-going NERP will simulate hydraulic and morphologic behaviour in the haor area by mathematical modelling. The environmental impact assessment will be done in greater details. Without this, planning of water resources development in the area, would not be complete. It would also be necessary to expand the hydrometric network where information is not sufficient specially in northern Mymensingh. Fishery in haor area is very important, both as an economic activity and as an important protein source for the population. The ecological linkage and the role which the environmental considerations play on the impact of flood control schemes should be properly investigated. The desirability and feasibility of HYV crops should be properly investigated, not only technically but also socially and from the environmental point of view.

Ground water resources in the haor area has not been utilised much. It may be expected that considerable ground water resources are available in the haor area and needs proper investigation.

Acknowledgement

The authors acknowledge the support and assistance received from Dr. Hamidur Rahman Khan, Water Resources Adviser, and Mr. M.F. Rodman of Northwest Hydraulic Consultants Ltd.

REFERENCES

1. BWDB, 1976. *Hail Haor Irrigation Project, Feasibility and Definite Project Report,* prepared by Bangladesh National Consultants, Ministry of Irrigation, Government of Bangladesh, Dhaka.

2. EPWAPDA, 1963. *Feasibility Report on Hakaluki Haor Project,* prepared by Berger Engineers, Ministry of Irrigation, Government of Bangladesh, Dhaka.

3. Flood Plan Coordination Organisation (FPCO, 1992a). *Environmental Planning: Assessment and Management, Northeast Regional Water Management Project (NERP),* by Shawinigan Lavalin (1991) Inc. and Northwest Hydraulic Consultants Ltd., Ministry of Irrigation, Government of Bangladesh, Dhaka.

4. Flood Plan Coordination Organisation (FPCO, 1992b). *Draft Thematic Study: Fisheries in the Haor Area, Northeast Regional Water Management Project (NERP),* prepared by Shawinigan Lavalin (1991) Inc. and Northwest Hydraulic Consultants Ltd., Ministry of Irrigation, Government of Bangladesh, Dhaka.

5. Flood Plan Coordination Organisation (FPCO, 1992c). *Draft Thematic Study: Regional Water Resources Development Status, Northeast Region Water Management Project (NERP), prepared* by Shawinigan Lavalin (1991) Inc. and Northwest Hydraulic Consultants Ltd., Ministry of Irrigation, Government of Bangladesh, Dhaka.

6. Flood Plan Coordination Organisation (FPCO, 1993a). *Participatory Development and the Role of NGOs, Northeast Region Water Management Project (NERP),* prepared by Shawinigan Lavalin (1991) Inc. and Northwest Hydraulic Consultants Ltd., Ministry of Irrigation, Government of Bangladesh, Dhaka.

7. Flood Plan Coordination Organisation (FPCO, 1993b). *Population Characteristics: State of Human Development, Northeast Region Water Management Project,* prepared by Shawinigan Lavalin (1991) Inc. and Northwest Hydraulic Consultants Ltd., Ministry of Irrigation, Government of Bangladesh, Dhaka.

8. Flood Plan Coordination Organisation (FPCO, 1993c). *Wetland Resources Specialist Study, Northeast Region Water Management Project,* prepared by Shawinigan Lavalin (1991) Inc. and Northwest Hydraulic Consultants Ltd., Ministry of Irrigation, Government of Bangladesh, Dhaka.

9. Flood Plan Coordination Organisation (FPCO, 1993d). *River Sedimentation and Morphology, Northeast Region Water Management Project,* prepared by Shawinigan Lavalin (1991) Inc. and Northwest Hydraulic Consultants Ltd., Ministry of Irrigation, Government of Bangladesh, Dhaka.

10. Flood Plan Coordination Organisation (FPCO, 1993e). *Guidelines for Environmental Impact Assessment,* Ministry of Irrigation, Government of Bangladesh, Dhaka.

11. Geological Survey of Bangladesh, 1990. Geology of Bangladesh Map

12. Master Plan Organization (MPO), 1984. National Water Plan Project, Ministry of Irrigation, Government of Bangladesh, Dhaka.

13. Nishat Ainun, 1993. *Fresh Water Wetlands in Bangladesh: Status and Issues,* IUCN-The World Conservation Union.

14. Rashid, Haroun Er, 1991. Geography of Bangladesh, University Press Ltd., Dhaka.

ANNEX

List of Water Resources Development Projects in Sylhet and Mymensingh Haor Area

Name of Haor	Location	Area (ha)	Cost (Tk mil.)	Observation
Angurali Haor	Taherpur	2,430	11	complete
Bara Haor	Jaintapur Sylhet	4,720		complete
Baram Haor	Dirai Sunamgonj	5,260		
Chandra Sonarthal Haor	Dharampasha Sunamganj	4,050	1.9	complete;
Deker haor	Sunamganj Chhatak	30,360		
Damir Haor	Fenchuganj Gopalganj	18,220	17.3	complete
Dhankunia Haor	Dharmapasha Sunamganj	1,620	1.1	complete
Dhalai River Proj.	Habigonj	23,890		
Gangiajuri Haor	Bahubal Habiganj	8,100		
Hail Haor	Moulvibazar	24,370		ongoing
Hakaluki Haor	Sylhet			under investigation
Halir Haor	Jamalganj Sunamganj	6,480	8.5	ongoing
Haripur Khal	Habiganj	410	2.0	complete
Humaipur Haor			6.5	complete
Joydhona Haor	Dharmapasha, Sunamgonj	1,210		complete
Kalner Haor	Sunamganj	6,280	4.0	ongoing
Karchar Haor	Biswamvarpur	7,690	2.9	complete
Khowai River Proj.	Habiganj	25,510	1040.0	ongoing

Kushiyara/Bardal	Beanibazar	8,420	3.4	complete
Manu River Sluices	Kulaura	14,170	200.0	ongoing
Manu River Proj.	Moulvibazar	22,670	676.1	complete
Matian Haor	Taherpur	5,180	4.0	complete
Matikata/Islampur Khal	Azmiriganj Baniachang	1,620	3.6	ongoing
Mohalia Haor	Jamalganj	1,420	1.9	
Muria Haor	Beanibazar Sylhet	6,070		under investigation
Nainder Haor	Sunamganj	1,420		ongoing
Naluar Haor	Jagannathpur	10,730	17.5	ongoing
Nawtar Khal	Kaliajuri Detrotora	3,250	10.9	ongoing
Pagna Haor	Jamalganj Sunamganj	19,430	18.0	ongoing
Patharchuli Haor	Sylhet Companiganj	4,050	25.0	complete
Rahimpur Khal	Zakiganj	10,930	9.0	complete
Salutikar haor	Sylhet			under investigation
Sari-Goyain proj.	Jaintapur	4,210	28.9	complete
Shanir Haor	Jamalganj	6,480	8.6	complete
Shanghair Haor	Sunamganj	4,130	8.3	complete
Sukti embank. proj.	Baniachang	8,100		
Sonai River	Madhabpur	1,210	2.1	ongoing
Sunam Khal	Zakiganj	1,500	5.5	
Surma-Baulai	Itna, Nikli Kishoreganj	24,700	42.8	not started
Zilkar haor	Sylhet	5,260	53.6	complete

OPTIMAL IRRIGATION PLANNING

By, V.Ravikumar[1] and K.Venugopal[2].
Centre for Water Resources
Anna University
Madras-600 025
India

Abstract

Reservoir yield analysis needs to be performed while attempting to improve the performance of any water resource system. Reservoir yield analysis was done for a water resouce system serving Irrigation in South India using probabilistic dynamic programming. The optimizing criterion used was minimizing the expected sum of squared deviations of release from the demand. The yearly cycle was divided into four dynamic programming stages, each containing three months. The model solution gives the optimal relases for each dynamic programming stage and the reliability of those releases and the probabilities of reservoir storage states. The model results are potentially helpful in decision making.

1. Introduction

The performance of the existing irrigation systems needs to be improved in order to improve the crop production. Reservoir yield analysis needs to be performed while attempting to improve the performance of any Irrigation system. Reservoir yield depends on the reservoir size, inflow quantity and its variability and the reservoir operating policy. Dynamic programming is useful technique to carry out the reservoir yield analysis. Buras(1985) and Goulter et. al(1985) have done reservoir yield analysis for selected cases using stochastic dynamic programming. In this work a reservoir serving irrigation is taken up to study its yield under optimal operation during various periods using probabilistic dynamic programming.

V. P. Singh and B. Kumar (eds.), Water Resources Planning and Management, 55–62.
© 1996 *Kluwer Academic Publishers. Printed in the Netherlands.*

2. Study Area

The study relates to Krishnagiri Reservoir Project across river Ponnaiyar in Tamilnadu state, India. The active capacity of the reservoir is $50 Mm^3$. The areal extent of command area is 3600 ha. In that, annual crops are grown in 600 ha and the remaining 3000 ha, paddy is grown. In the 3000 ha of paddy growing area, 1800 ha is used to grow paddy in the Kharif season (July-October) using the groundwater in the command area. During the Rabi season (November-March) the reservoir is opened for irrigation and the water is supplied for 3000 ha of paddy and 600 ha of annual crops. The reservoir is closed when the rabi paddy crop comes to harvest. The command area has got very good groundwater potential. The groundwater remains at 2m below ground level during irrigation season and at 6m below the ground level during non-irrigation season. In the command area there are around 2000 wells.

In this study Reservoir yield analysis is performed while operating the reservoir optimally for the case of taking up 1800 ha of paddy in the kharif season, 3000 ha of paddy in the rabi season and 600 ha of annual crop and the supply is assumed to be made solely from the reservoir for demonstration of the methodology.

3. Model Development

The state vector Φ_i used in this study is defined as,

$$\Phi_i = (X_i, Q_i) \quad \ldots.(1)$$

where X_i is the reservoir storage at the begining of the period i, and Q_i is the reservoir inflow during period i. The crop demand is taken as deterministic.

4. State Transformation

Let the indices j and k represent the class intervals for the reservoir storage, jj and jk represent the class intervals for inflow at periods i and i+1 respectively. $X_{i,j}$; $X_{i+1,k}$; $Q_{i,jj}$; $Q_{i+1,jk}$ are the representative values of the state variables for the periods i and i+1 respectively. The inverted reservoir storage state transformation is governed by the Equation 2.

$$R_{i,j,k,jj} = X_{i,j} - X_{i+1,k} + Q_{i,jj} - E_{i,j,k} \qquad \ldots(2)$$

Where $R_{i,j,k,jj}$ is the release during the period i when the storage class interval is j, inflow class interval is jj and the storage class interval at period i+1 is k. $E_{i,j,k}$ is the evaporation loss if the storage class interval in period i is j and in period i+1 is k. Equation 2 specifies the release for a given combination of j,jj and k, for each period i.

5.Objective Function

In order to choose from among the possible values of decision variables $R_{i,j,k,jj}$ a measure of system performance has to be specified. Let the measure of system performance be G and a particular value of it namely, $G_{i,j,k,jj}$ be associated with release $R_{i,j,k,jj}$ for every period i. The system performance measure incorporated in the objective function aims at minimizing the sum of squared deviations of the release from the demand.

6.Recursive Relation

A backward moving dynamic programming algorithm was developed assuming that the reservoir operation terminates at some arbitrary year Y in the future with the total number of periods being I. Let N be the number of periods remaining till the the end of the year Y. Let $F^N_{i,j,jj}$ denote the total minimum expected value of the system performance over N periods to go in the operation of the reservoir including the current period i given that reservoir storage is $X_{i,j}$ and inflow is $Q_{i,jj}$ for period i. Then with only one period remaining,

$$F^1_{I,j,jj} = \underset{\{k\}}{\mathrm{MIN}} \left[G_{I,j,k,jj} \right] \qquad \forall \; j,jj \qquad \ldots(3)$$

where {k} denotes feasible k. In general for any period, i

$$F^N_{i,j,jj} = \underset{\{k\}}{\mathrm{MIN}} \left[G_{i,j,k,jj} + \sum_{jk} P_{i+1,jk} * F^{N-1}_{i+1,k,jk} \right] \qquad \forall \; j,jj \qquad \ldots(4)$$

where $P_{i+1,jk}$ is probability of inflow in period i+1 is jk. Equations 3 and 4 are solved until steady state policy is reached. The recursive relations are similar to the recursive relations adopted for stochastic dynamic programming by Loucks et. al(1982) except that the flows in each stage are treated as independent. This kind of formulation helps to reduce the dimensionality problem in solving the probability distribution equations as discussed in the following section.

7. Probability Distribution

The steady state optimal policy obtained from the solution of the recursive equations gives the optimal final storage volume k for each initial storage volume j, and inflow jj for each period i. Let this policy be denoted as k=fn(i,j,jj). The probability distribution of reservoir storage is obtained from the Equations 5 and 6.

$$PS_{i+1,k} = \sum_{j} PS_{i,j} * \sum_{jj} P_{i,jj} \qquad \ldots(5)$$

$$\sum_{j} PS_{i,j} = 1 \qquad \ldots(6)$$

where $PS_{i,j}$ is the probability of reservoir storage j in the period i. On solving the Equations 5 and 6 the probability distribution of the reservoir storage is obtained. For instance if the reservoir storage is discretized into NS intervals per period and inflow into NF intervals per period and number of periods in a year is NP, the number of the unknowns to be solved in this formulation is only NS*NP, whereas in the stochastic formulation the number of unknowns to be solved is NS*NP*NF. So in the probabilistic formulation as used in this study the dimension of the Equations 5 and 6 is very less and so solving them becomes easy. Goulter et. al(1985) has reported that when the number of the unknowns goes beyond 1152, solving by the available computer facilities becomes impossible. After solving the Equations 5 and 6 the reliability of releases that can be made available during each period can be computed.

8. Results and Discussion

In this study water year starts from july. Water year is considered to be four periods of 3 months each, each period being taken as a stage in dynamic programming formulation. 34 years of inflow into the reservoir was used. The serial correlation of

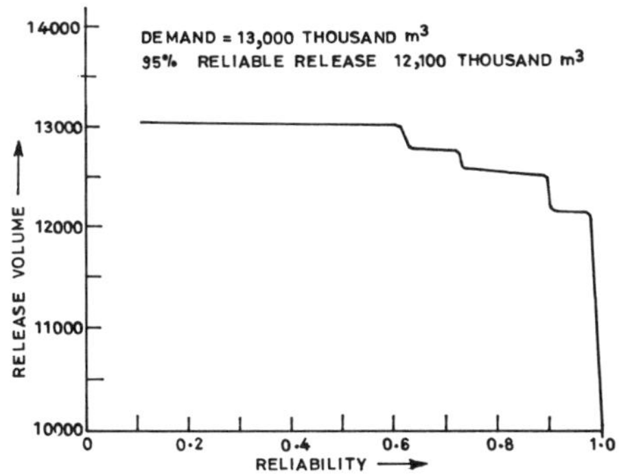

FIGURE 1 RELIABILITY OF RELEASE CURVE FOR PERIOD 1

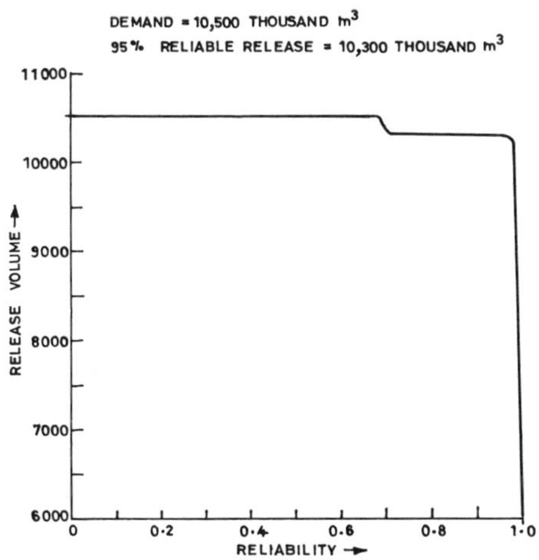

FIGURE 2 RELIABILITY OF RELEASE CURVE FOR PERIOD 2

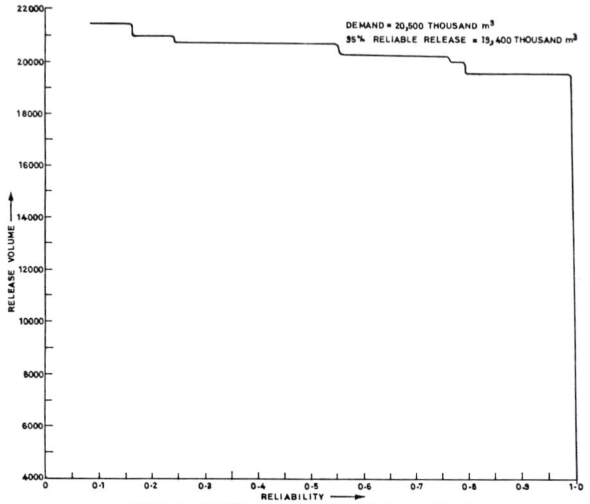

FIGURE 3 RELIABILITY OF RELEASE CURVE FOR PERIOD 3

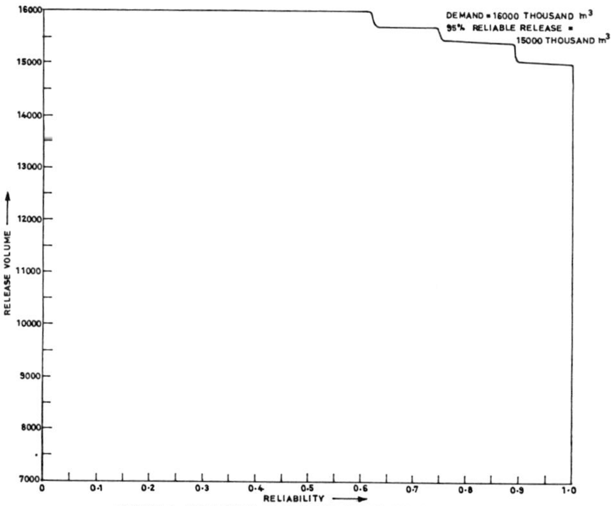

FIGURE 4 RELIABILITY OF RELEASE CURVE FOR PERIOD 4

No of flow states	PERIOD 1		PERIOD 2		PERIOD 3		PERIOD 4	
	Represent. flow state (1000 m3)	Probability	Represent. flow state (1000 m3)	Probability	Represent. flow state (1000 m3)	Probability	Represent. flow state (1000 m3)	Probability
1	1145	0.088	10501	0.324	727	0.235	3605	0.147
2	8966	0.088	34290	0.029	1427	0.088	7194	0.088
3	16787	0.206	50149	0.029	2127	0.029	10783	0.176
4	32430	0.206	81867	0.118	2828	0.088	14373	0.118
5	48072	0.118	113585	0.029	3528	0.088	17962	0.118
6	63715	0.029	145304	0.147	4228	0.147	28731	0.088
7	87179	0.029	208740	0.118	5629	0.088	25141	0.088
8	110642	0.088	272177	0.118	7029	0.088	39499	0.088
9	141927	0.059	335613	0.029	8430	0.059	53856	0.029
10	173213	0.088	399050	0.059	9830	0.088	68214	0.059

TABLE 1 : INFLOW DISCRETIZATION AND THEIR PROBABILITIES

flows between each period works out to 0.26,0.31,-0.144 and -0.122. Hence the flows in each period were considered to be independent. The cropwater requirement was computed using modified penman method. Ninety per cent of 90% dependable rainfall was taken as effective rainfall. The demand was computed by subtracting the effective rainfall from the cropwater requirement.The Table 1 shows the discretization of flows and the probability distribution of flow states. The reservoir storage is discretized into 30 uniform class intervals per stage and inflow into 10 class intervals per stage. The Figures 1 to 4 illustrate the reliability of the releases that can be made available in each period. In all the periods approximately 60% of the time releases equal the demand. The shape of the curves are not smooth because of storage and inflow discretization. The curves can be used to estimate the extent of groundwater potential to be developed for each period to meet the demand with a satifactory level of reliability. For example, for reliability of 95% the amount of groundwater to be developed is also shown in Figures 1 to 4.

The model can be used to arrive at the reliability of meeting the specified demand as explained above and also to arrive at the demand for a chosen reliability level. For example, with a reliability of 75%, the extent of area that can be irrigated without taking the groundwater contribution into account can also be arrived at.

References

1. Buras.N., 'An application of Mathematiclal programming in planning surface water storage', Water Resour. Bull., 21(6), 1013-1020, 1985.

2. Goulter, I.C., and F.K.Tai, Practical implications in the use of stochastic dynamic programming for reservoir operation. Water Resour. Bull., 21(1), 65-74, 1985.

3. Loucks, D.P., J.R.Stedinger, and D.A.Haith, 1981. Water Resources Systems Planning and Analysis, Prentice Hall Inc., Englewood Cliffs, New jersey.

LOCAL LEVEL PLANNING OF SMALL WATER RESOURCE SCHEMES IN A DEVELOPING COUNTRY

Ainun Nishat, Jahir Uddin Chowdhury,
M. Mirjahan, M. Rezaur Rahman
Bangladesh University of Engineering and Technology
Dhaka- 1000, Bangladesh

ABSTRACT: The Government of Bangladesh is pursuing a policy of decentralization to strengthen the implementation mechanism of the institutions at grass root level. The government is now giving responsibilities to local level institutions for planning and implementation of small scale water resources schemes. The objective of this work was to produce a planning guideline with the expressed intention to reach those who are legally vested with the authority to select water resource development schemes. As such, the contents, formulation and structure of the methodology reflect the objective planning environment at the local level in Bangladesh. The methodology strives to strike a balance between the desired level of comprehensiveness and available human and institutional planning resources in Bangladesh.

INTRODUCTION:

The Government of Bangladesh (GOB) is pursuing a policy of decentralization to strengthen the implementation mechanism of the institutions at grass root level. As such, the government has been contemplating to entrust planning and implementation of small scale water resources schemes at the local level. In Bangladesh, Union Parishad (UP) is the lowest level of elected representative body and represents 10 to 15 villages. In Bangladesh there are about 64,000 unions. According to Fourth Five-Year Plan (GOB, 1991) "*Very small projects covering maximum 1,000 acres for drainage and flood control will be initiated, proposed, planned and implemented by the Union Parishads.*"

In Bangladesh, earthen flood control and drainage schemes such as small scale canal excavation/re-excavation and embankment construction/re-construction are taken up at local level. Canals are excavated (re-excavated if silted up) to relieve drainage congestion in low lands or to drain out water from water bodies (locally known as beel) in-order to bring them under cultivation. Embankments are constructed (re-constructed if damaged in earlier flood) to protect agricultural lands from floods due to over-spilling of rivers.

V. P. Singh and B. Kumar (eds.), Water Resources Planning and Management, 63–71.
© 1996 *Kluwer Academic Publishers. Printed in the Netherlands.*

The objectives of these schemes are to boost agricultural production and generate local employment. Proposed schemes however, are not screened through any formal planning process. Thana Engineer (Thana is the lowest tier of administration; each thana oversees a number of unions) of the Local Government Engineering Department (LGED) usually selects and prioritize schemes himself on need-felt basis in consultation with UP chairmen and members. When Govt. allocates fund in water resources sector, schemes are implemented according to this priority. From now on, UP itself will initiate, propose, plan and implement such schemes.

The goal of this study was to develop a planning manual for selection of small scale earthen flood control and drainage schemes by the UP. The manual had to be simple in order to be acceptable and useful to the UP members most of whom have no prior knowledge, training or familiarity with a formal planning process. Attempts have therefore been made to frame the guidelines as simple as possible so that local leaders can follow them easily.

The manual has been prepared on the basis of our experiences learnt through several field visits, questionnaire survey, discussion with Thana engineers, agricultural and fisheries officers, local representatives including UP chairmen and members and others involved in planning of small scale water resources projects in Bangladesh. After the formulation of the manual, it has been tested extensively at the field level. Salient features of the manual (LGED, 1992) are presented in this paper.

OBJECTIVES OF SMALL SCHEMES:

The major considerations of the GOB behind allocating fund for very small schemes are (i) to increase agricultural production and (ii) to generate employment. A formal planning approach will require optimization of both objectives. But since that is beyond the scope of the UP and may not be worthwhile for very small schemes, a simplified approach has been chosen. Scheme with high agricultural prospect will be given preference. In case of a tie between two schemes according to this criteria, scheme with higher employment prospect will be given preference. Another desirable social objective which is that a scheme should not affect too many people with respect to number of beneficiaries has been put as a constraint. An environmental objective which is that flood control and drainage project should not harm already dwindling fisheries resources has also been put as a constraint.

CONSTRAINT:

In Bangladesh, human and institutional constraint are the major constraints in formulating an effective planning guideline for small schemes. The practice of planning at local level is still an unknown concept in Bangladesh. Therefore UP members can not be expected to carry out sophisticated planning. UP members can not be expected to spend considerable time for planning water sector alone either as the UP has many functions covering almost the total life of the community ranging from maintenance of law and order to promotion of agricultural production (see GOB, 1976). It has already been noted that the UPs are saddled with so many responsibilities of multifarious nature that they can not devote much time on detailed planning for various projects (Hye, 1982).

A pragmatic approach was therefore chosen to develop the methodology. As such, the limitations due to prevailing socio-economic condition of the rural Bangladesh were taken into due consideration. The methodology strives to strike a balance between the desired level of comprehensiveness and available human and institutional planning resources in Bangladesh.

PROFORMA:

Since the Union Parishad itself will initiate, propose, plan and implement very small schemes, so proforma in the form of questionnaire were considered the most suitable tool to abstract and use relevant information. Formal analyses such as benefit-cost analysis is not proposed as it is beyond the capabilities of UP members.

Proforma have been designed to make the UP members aware about certain aspects of planning, such as social and environmental, and at the same time ensuring the use of their extensive knowledge about the local conditions and variations. Questions of the proforma were so designed that they are easily understandable and comprehensible to the UP members who will complete the proforma.

Since the UP members themselves will fill up the proforma, it can be expected that they will try to answer the questions of proforma such a way as to bias the planning process in favor of their favorite projects. The questions were therefore, designed such that those themselves will minimize the scope of biased response and at the same time the answers to these questions can be easily cross-checked by other UP members.

Proforma for canal excavation/re-excavation and embankment construction/reconstruction are presented in Appendix I and II respectively. The original proforma are in Bengali. The english versions are presented here. So many subtleties of the Bengali version which were intentionally introduced are missing here. Considerable time and effort were spent on the wordings and language of the proforma so that they are easily understandable and they can abstract as much information as possible with as little wording as possible. In this regard, considerable cooperation and discussion were sought and received from the UP members. The logic and reasoning behind each question of the proforma is explained below.

Answer to question no. 1 will show how much area will be benefitted by the scheme. A question like what is the potential agricultural benefit of a proposed scheme may evoke biased response which can not be easily cross-checked by other UP Members. On the other hand, answer to the question like what will be the net benefitted area can easily be cross-checked by other UP members. It is true that the answer to the former question rather than that of to the later question would have been more valuable for planning purposes. But to strike a balance between value of information and minimization of biasedness, the later question in favor of the former question was considered to be more acceptable. Naming the mouzas (land units) will further reduce the scope of biasedness.

Question no. 2 wants to know how much employment will be generated from this scheme as employment generation is one of the prime objective of the Government of Bangladesh for undertaking small and labor intensive schemes at local level.

Question no. 3 seeks to know how many people will be benefitted by this scheme.

Question no. 4 seeks to know whether the scheme will affect any jalmahal (jalmahal is the local term of permanent water bodies and are source of considerable fisheries resource). Fish makes up about 70% of the protein intake of the Bangladeshis. But flood control and drainage projects have impinged to various degrees on natural fish habitats and migration routes, and production from capture fisheries have gradually declined over the years (FPCO, 1992). This question has been designed to make the local decision makers aware of this environmental fact.

Question no. 5, 6 and 7 deal with the possible negative impact of the schemes. Question no. 5, 6 and 7 elicit how many people including farmer, fishermen and boatmen will be affected by the planned scheme and how much land might be affected.

PLANNING STEPS:

The Union Parishad will identify a number of schemes covering a maximum of 1000 acres. The Union Parishad members will submit the schemes to the Parishad. In submitting the schemes, they will fill up a proforma for each scheme. Using the information in the proforma, the Union Parishad will perform the following tasks.

i. *A scheme whose benefitted area falls under the benefitted area of any proposed Bangladesh Water Development Board (BWDB) project should not be selected.* BWDB is the executing agency of large scale water resources schemes. So this step will avoid any interference with BWDB schemes.

ii. *Any scheme which is not technically feasible should be discarded.* The Thana Engineer (TE) of the LGED or his technical staff will help in assessing the technical feasibility of the schemes through field inspection.

iii. *If for any scheme, the number of people adversely affected (as calculated at the bottom of the proforma) is more than the number of people benefitted (answer to question 3 of the proforma), then the scheme should not be selected.* Such selection criteria was introduced to make the members aware of the social consequences of a scheme. It was not tried to influence the decision making process by introducing any weights based on benefits accrued or harms inflicted. Such weights were avoided as they will be difficult for UP members to calculate and may introduce the scope of biasedness. So simplicity was favored over comprehensiveness.

iv. *Any scheme which may disturb an established jalmahal should not be selected.* Such selection criteria was introduced to make the members aware of the environmental damage that a scheme may cause.

v. *Arrange the selected schemes year-wise.* This is needed as in Bangladesh, plan is made on five-year basis.

vi. *Rank the schemes for each year separately as outlined here:* The rank should be determined on the basis of net benefitted area. The scheme with the largest net benefitted area as determined in the proforma will be given rank 1, the second largest will be given rank 2, and so on. If the net benefitted areas of more than one scheme are equal then the order of preference will be determined on the basis of number of man-days required during their construction as calculated at the bottom of the proforma.

A flow chart showing the above planning steps is shown in Figure 1.

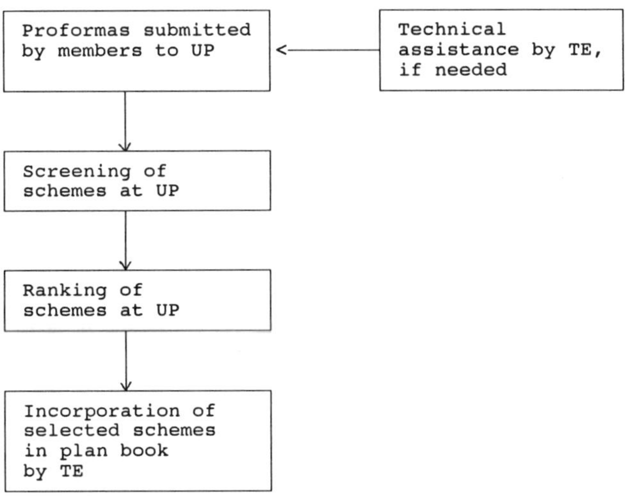

Figure 1: Flow chart for selection of small schemes

CONCLUSION:

The government is now giving responsibilities to local level institutions for planning and implementation of small scale flood control and drainage schemes. A large number of manuals are already available for water resource project planning and design in Bangladesh. However, the common denominators of all these manuals are that they are technically sophisticated, essentially design oriented and meant to be used by trained professionals.

The objective of this work was to produce a manual with the expressed intention to reach those who are legally vested with the authority to select water resource development schemes. The manual specifically addresses the local decision makers whose familiarity with formal planning process is limited and in areas where there are no local highly skilled technical staff. As such, the contents, formulation and structure of the methodology reflect the objective planning environment at the local level in Bangladesh. The main purpose was to formulate a set of verifiable criteria which the decision makers at the local levels could use while conceiving the physical problems, formulating development objectives, making priorities and finally deciding on the schemes to be selected for implementation.

In formulating the guideline, major consideration has been given to agricultural production and employment generation. However due consideration has been given to social and environmental aspects of the planning process.

Proforma for different small scale schemes have been developed by taking into consideration the existing human and institutional constraints of Bangladesh. Questions were so designed so as to use the knowledge of the local decision makers about local condition and variation as much as possible and make them aware of social and environmental aspects of planning. The main purpose was to formulate a set of verifiable criteria which the decision makers at the local levels could use while conceiving the physical problems, formulating development objectives, making priorities and finally deciding on the schemes to be selected for implementation. The methodology strived to strike a balance between simplicity given the local level planning environment of Bangladesh and the desired level of comprehensiveness.

The manual will be used in all the Union Parishads of Bangladesh. As the time passes, the members of the parishad will become familiar with the planning process and will become aware of various aspects of a formal planning process. Feedback from the members of the UP will be continuously gathered and the proforma, the questions of the proforma and the developed guideline will be continuously updated to suit the need of the time.

ACKNOWLEDGEMENT:

This work was completed as part of developing a Manual on Land and Water Use Planning for the LGED. The study was supported by Rural Employment Sector Program (RESP). The authors deeply appreciate the support provided by Mr. Jorgen Engel and Dr. Ruknuddin Ahammed of RESP and Dr. Ashraful Haq of Royal Institute of Technology, Sweden.

This is a work which was carried out mostly on the basis of field visits and discussion with numerous UP chairmen, members and Thana Engineers. The authors are indebted to each of them.

REFERENCES:

Flood Plan Coordination Organization (1992), *Bangladesh Action Plan for Flood Control: Achievements and Outlook; An Update.*

Govt. of Bangladesh (1991), Fourth Five-year Plan.

Govt. of Bangladesh (1976), The Local Government Ordinance.

Hye, Hasnat Abdul (1982), *Local Level Planning in Bangladesh*, National Institute of Local Government, Dhaka.

Local Government Engineering Department (1992), *Manual on Land and Water Use Planning*, Ministry of Local Government, Govt. of Bangladesh.

APPENDIX I

Proforma for Khal (Canal) excavation/re-excavation scheme for nontidal zone

Name of the Union:

Name of the scheme:

Year of implementation:

Name of the proposer and Ward no:

1. How much area will be benefitted by the khal? ________ acres. (name the would-be benefitted mouzas)

2a. What is the volume of earthwork? _________ cubic meter.

b. How many people will be employed during khal excavation? ________ persons ________ days.

3. How many farmers will be able to cultivate their lands when the khal will drain the beel? ________ persons.

4. Due to draining of the beel, will any established Jalmahal be disturbed? yes/no.

5a. Will there be any need to acquire land for this scheme? yes/no.

b. If yes, then how many's? ________ persons.

c. and how much? _________ acres.

6a. Due to the draining of the beel, how many farmers will be deprived of their irrigation water source? ________ persons.

b. How much land will be deprived of irrigation? _________ acres.

7. How many fishermen will lose their livelihood? _________ persons.

Net benefitted area = Answer to question no. 1 - (sum of answer to question nos. 5c & 6b) = ________ acres

Man-days required during excavation/re-excavation = multiplication of persons and days of answer to question no. 2b.= ________ .

No. of affected persons = Sum of answers to question no.5(b), 6(a) & 7.=_______.

APPENDIX II

Proforma for flood control embankment construction/re-construction scheme

Name of the Union:

Name of the scheme:

Year of implementation:

Name of the proposer and Ward no:

1. How much area will be saved from flood by this embankment? ________acres.
   (name the would-be benefitted mouza)

2a. What is the volume of earthwork? __________ cubic meter.

b. How many people will be employed? ________ persons ________ days.

3. How many farmers will be benefitted by the embankment? ________ persons.

4. Will the embankment disturb any established Jalmahal? yes/no.

5a. Will it be necessary to acquire any land? yes/no.

b. If yes, how many's? ________ persons.

c. and how much? _________ acres.

6a. Will the embankment create any drainage congestion? yes/no.

b. If yes, how many farmers will be affected? ________ persons.

c. and how much land will be affected? ________ acres.

7a. Will the embankment disrupt boat communication? yes/no.

b. If yes, how many boatmen will be affected? ________ persons.

Net benefitted area = Answer to question no. 1
- (answer to question nos. 5c & 6c) = ________ acres.

Man-days required = multiplication of persons and days of answer to question no. 2b.= ________.

No. of affected persons = Sum of answers to question no. 5(b), 6(b) & 7(b).=________.

SECTION 2:

WATER RESOURCES DEVELOPMENT

LARGE SCALE EXPLOITATION OF WATER RESOURCES: CASE OF THE SOUTHEASTERN ANATOLIA PROJECT IN TURKEY

N. B. HARMANCIOGLU
Dokuz Eylul University
Faculty of Engineering & Architecture
Civil Engineering Department
Izmir, Turkey

ABSTRACT. The development of water resources in general, and that of multi-purpose water projects in particular, is affected not only by advances in technology, but also by changes in economics, which may have unexpected effects on planning, construction, and/or operation of such projects. The case of the Southeastern Anatolia Project represents fairly well the dynamic character of multi-purpose projects with respect to technological developments and changes in economic factors. This project covers the joint development of land and water resources of two large river basins, namely the Lower Euphrates and Western Tigris basins. The first attempts at large scale development of the Euphrates basin date back to 1950's, and since then the project has undergone significant modifications through the years until it was integrated into a broader scheme, called the Southeastern Anatolia Project, together with the development of the Western Tigris basin. With its large scale input to energy production and irrigation, the project is expected to contribute not only to the economic but also to the social development of the region to eventually produce country-wide effects. From an engineering point of view, the project comprises internationally remarkable features. Accordingly, the Southeastern Anatolia Project bears today the first priority in water and land resources development in Turkey. The presented paper discusses the evolution of this large scale project within the last 40 years together with expected modifications to illustrate the dynamic nature of water resources development plans.

1. Introduction

The planning, construction, and operation of water resources systems are affected by several factors which show inherent stochasticity and/or significant uncertainty. These factors relate especially to the changes in technology and economics, that may cause unexpected changes in the development of these resources. Multi-purpose

V. P. Singh and B. Kumar (eds.), Water Resources Planning and Management, 75–90.
© 1996 *Kluwer Academic Publishers. Printed in the Netherlands.*

projects, encompassing water power systems in particular, are more sensitive to such changes (Harmancioglu and Ozis, 1983).

Large scale exploitation of water power is realized by generation of electricity, which can also be obtained from other primary energy resources. Thus, the developments in electricity generation technology, but primarily the changes in economic factors, may significantly affect water power development plans. These considerations apply first to the design phase which may undergo various modifications and revisions before technologically sound and economically the most feasible plan is selected.

During the implementation phase, the contruction of large hydraulic structures like dams, conduits, and power plants are highly influenced by uncertainties in the related geological characteristics of the environment, with respect to foundation, materials, and water tightness. Furthermore, this phase often requires a long period of time during which possible technological and economic changes may lead to further revisions.

Even the operation of water resources systems, particularly water power schemes, may be subject to considerable modifications due to unexpected changes in supply, demand and/or economic factors in the broadest sense. Thus, the development of water resources has to be considered as a dynamic phenomenon, and no development plan at any stage can be regarded as a strictly definitive one.

The presented paper demonstrates the dynamic nature of large scale water resources development plans in the case of the Southeastern Anatolia Project (SAP, or GAP in Turkish) in Turkey. This project covers the joint development of land and water resources of two large river basins, namely the Lower Euphrates and Western Tigris basins. The first attempts at large scale development of the Euphrates basin date back to 1950's, and since then the project has undergone significant modifications through the years until it was integrated into a broader scheme, called the Southeastern Anatolia Project, together with the development of the Western Tigris basin. By means of about 90 dams and 60 power plants, the project will realize the generation of 37 billion KWh/y of energy with an installed capacity of 10212 MW, which will double the current hydropower generation in the country. In addition, a total of 1.8 million hectares of land will be irrigated by completion of the project; this figure equals the total area of lands currently irrigated by state agencies in Turkey. Thus, the project will contribute not only to the economic but also to the social development of the region to eventually produce country-wide effects. From an engineering point of view, the project comprises internationally remarkable features.

With the above characteristics, the Southeastern Anatolia Project bears today the first priority in water and land resources development in Turkey. The presented paper discusses the evolution of this large scale project within the last 40 years together with expected modifications to illustrate the dynamic nature of water resources development plans.

2. Development of the Southeastern Anatolia Project

2.1. THE EARLY DEVELOPMENT PHASE OF THE LOWER EUPHRATES AND TIGRIS BASINS

The Euphrates (Firat) river basin extends over almost one-sixth of Turkey's area of 780,000 km2, comprising more than one-sixth of the water potential of 180 billion m3/y and over one fourth of the economically feasible hydroelectric potential of 120 billion kWh/y (Ozis, 1982). Considering the water power potential of the basin (approximately with head of 310 m and mean annual flow of 20 billion m3) and the presence of more than one million hectares of irrigable land, the first attempts at large scale development of the Euphrates started in 1930's, with the beginning of regular hydrometric observations by EIE, the Electrical Works Survey Authority of Turkey. The first reconnaissance report prepared in 1958 by EIE foresaw, downstream of the key project Keban, a three-dam system (Bilalusagi-Sarsap-Halfeti) for energy production only. The proposed total energy production for this system was in the order of 11,800 GWh/y (EIE, 1958; Harmancioglu and Ozis, 1981). A same report prepared by EIE (EIE, 1958) considered for the Tigris basin 3 dams on the main tributary (Selman-Diyarbakir-Delmah) and 2 dams on tributaries (Huseyinkan on Batman and Besiri on Garzan) with 5 power plants to produce 4 billion kWh/y of energy and 800 MW of installed capacity. The report also foresaw the irrigation of 20,000 ha of land.

The reconnaissance report of 1967 by DSI, the State Hydraulic Works Authority, emphasized irrigation of 870,000 ha of land in addition to energy production of 6,756 GWh/y. This report foresaw a two-dam system (High Tasustu-Hisarkoy) where the irrigation water would be supplied by the first reservoir, partially by gravity and partially by pumping. The next best scheme of this study considered two dams at Handere and Karababa, with irrigation based partially on pumping (DSI, 1967). In 1968, Garbrecht presented a modified version of this second alternative as a two-dam system (Karakaya-Karababa) with irrigation water diverted from both dams and mainly by gravity (Garbrecht, 1968). In 1968, DSI (DSI, 1968) prepared a reconnaissance report for Tigris basin which foresaw energy production in the order of 3.9 billion kWh/y by 16 power plants with a total installed capacity of 770 MW. By means of 20 dams of varying sizes, irrigation of 190,000 ha of land was planned with water supplied partially by gravity and partially by pumping.

Detailed investigations on the multi-purpose development of the Lower Euphrates basin were carried out for DSI by the joint venture of Electrowatt-Tipton and Kalmbach-S.G.I.-Gizbili until 1970. The water potential of the region had been evaluated through 22 alternative projects with different technical and economical characteristics. In each scheme, the energy production systems were analyzed under conditions with and without irrigation. Several alternatives were also considered for irrigation systems, including those which foresaw conveyance either totally by gravity, or by pumping, or partially by both. The feasibility report prepared by the joint venture in 1970 finally proposed a three-dam system (Karakaya-Golkoy-Middle Karababa), where the irrigation water would be diverted from the third reservoir and conveyed basically by pumping. The power plants were designed to have a total installed capacity of 2,700 MW, and the system would supply water to 700,000 ha of irrigable land. Furthermore, the development of the irrigation system was based on two stages, the first one comprising water supply for 300,000 ha of agricultural land in the Urfa-Harran and Mardin-Ceylanpinar regions. A different pumping scheme would serve the remaining 400,000 ha of land in the Mardin-Ceylanpinar and Nusaybin-Cizre regions at the second development phase (Electrowatt et al., 1970). Summarized versions of this proposed project are presented later by Doluca and Pircher (1971) and Gokcesu (1972).

2.2. IMPACT OF THE OIL CRISIS ON DEVELOPMENT OF THE LOWER EUPHRATES BASIN

The oil crisis of 1973/74 and especially the increase in unit prices of energy production based upon petroleum led to later modifications by Electrowatt-TAMS-SGI-DOLSAR, resulting in an increase of the installed power plant capacities from the total of 2,700 MW to 3,900 MW. This modification is actually an interesting change that the Lower Euphrates plan went through, since it reflects the first impact of the oil-crisis and thereby that of the highly increased energy prices upon projects prepared before 1973/74 (Electrowatt et al., 1974 and 1975). Another modification presented in the new report of 1975 by Electrowatt et al. was the introduction of two new hydroelectric power plants downstream of Karababa. These are the Findikli and Karkamis hydroelectric plants with installed capacities totaling 1000 MW.

The unexpectedly high increases of unit energy prices due to the oil crisis of 1973/74 led to a macroeconomic revision of the feasibility study of the Lower Euphrates project, since, until then, the economic comparisons of energy productions of hydroelectric plants have generally used to be made with those of fuel-oil fired thermal plants. In a study by Harmancioglu (1976), based upon power plant costs and energy prices of 1975, it was found that the difference between the feasibility of the proposed plan of 1970 and that of the two-dam alternative of Karakaya and High Karababa, which would produce 600 million GWh/y more energy and which would considerably reduce energy requirements for irrigation, was reduced substantially with respect to both the benefit/cost ratio and the benefit minus cost measures. The results of this study showed that such differences between the feasibility of the two alternative projects were decreased from 24% to 3% in terms of their net benefits, and from 8% to below 2% with respect to the benefit/cost ratios. Consequently, detailed analyses for the future were foreseen, considering the above figures which clearly reflected the impact of the oil crisis, with' 5 years between 1970 and 1975, on the development of the Euphrates basin (Harmancioglu, 1976; Harmancioglu and Ozis, 1978).

A detailed investigation by DSI between 1976-77 resulted in the definite superiority of the two-dam (Karakaya-High Karababa) alternative. This analysis basically involved the comparison of those parts of the two schemes which lay downstream of the Karakaya dam and power plant. Finally, the High Karababa alternative of the second scheme was found to be economically more feasible and technically more convenient for implementation and operation than the Golkoy-Middle Karababa alternative of 1970, which had proposed Bedir-Hilvan and Suruc-Baziki pumping schemes for the irrigation system. The report of 1978 by DSI (Melikoglu et al., 1978) introduced the new plan, where heightening of the Middle Karababa dam, proposed in 1970, makes it possible: (a) to eliminate completely the pumping scheme of Bedir; (b) to reduce considerably the consumption of pumping energy in the Hilvan system; (c) to irrigate by gravity 330,000 ha of land in the Urfa-Harran and Lower Mardin-Ceylanpinar regions (Harmancioglu and Ozis, 1981). In 1978, the General Director of DSI proclaimed the results of the new development plan, and the High Karababa, which will be Turkey's largest dam and power plant, was renamed as the Ataturk scheme to commemorate the centennial in 1981 of the birth of the founder of the modern Turkish Republic (Harmancioglu and Ozis, 1981 and 1983; Ozis, 1982). In the meantime, the most significant development in the Tigris basin was the preparation of the feasibility report for the Ilisu dam and power plant project.

2.3. SOUTHEASTERN ANATOLIA PROJECT BY DSI (1980)

The investigations for the development of water and land resources of the Lower Euphrates basin have now been integrated into a broader scheme, called the Southeastern Anatolia Project, together with the development of the Western Tigris basin (DSI, 1980). This new scheme was formulated by inclusion of Birecik and Karkamis projects on the main tributary of Euphrates downstream of Ataturk dam, together with two dozens of dams and half a dozen power plants on other tributaries. These changes were integrated with developments in Tigris basin projects.

DSI divides the SAP development scheme into 13 projects altogether, 7 of which belong to the Lower Euphrates basin and 6 to the Western Tigris basin. Each project includes varying numbers of dams and power plants together with various irrigation development plans. The benefit/cost ratios of these projects are in the order of 2, or above 2, for energy production schemes. The same ratio remains below 2 or even becomes marginal for irrigation projects. Net benefits assume varying values for different projects. Besides providing economic benefits, SAP will enhance the social development of eastern Turkey, which ranges below national averages with regard to various socio-economic parameters (Harmancioglu, 1985; Ozis, 1992).

Revisions and modifications of this large scale scheme continued after 1980. Between 1982-85, the final designs and feasibility reports of Ilisu and Cizre dams and power plants on Tigris river were completed. Similar studies were carried out in 1984-85 for Birecik and Karkamis dams and power plants on Euphrates to finalize the development of the main tributary near the Turkish-Syrian border (Ozis, 1992). On the tributaries, Adiyaman-Kahta projects were changed in 1985 to include 6 smaller dams and 5 power plants for generation of 195 MW of power and 0.5 billion kWh/y energy. 160,000 ha of lands to be irrigated by Adiyaman and Kahta projects were reduced to 77,000 ha, where irrigation water will be supplied partially by gravity and partially by pumping from the Ataturk dam. To compensate for this reduction, 5 new dams were planned as part of the Araban project to irrigate 72,000 ha of additional land (Ozis, 1992).

As it can be followed from the above, the Lower Euphrates and Tigris basin development plans which make up the Southeastern Anatolia Project have experienced 20 years of primary data collection and more than 30 years of project development and formulation. In essence, the development phase of the Project is still continuing with minor revisions and modifications. The last trend is to include the Eastern Tigris basin within the framework of SAP in order to cover the entire Tigris basin along with that of the Lower Euphrates. At present, SAP is changed into an "integrated regional development plan" to enforce its contribution to regional and country-wide development. Within this framework, in addition to the technical implementation of the scheme, a master plan is prepared to incorporate social, cultural, educational, agricultural and economic developments as well as those related to health, infrastructure, and transportation services in the region (Ozis, 1992).

In 1970, a development period of 30-35 years was foreseen for SAP to be realized. At the recent stage of implementation, the entire scheme is expected to be completed by the beginning of the 21st century.

3. Hydraulic Schemes of the Southeastern Anatolia Project

3.1. LOWER EUPHRATES BASIN

The Lower Euphrates is considered as the region downstream of the Keban dam, completed in 1974, representing almost half of the Euphrates basin in Turkey. Within the context of the Lower Euphrates development, four large dams on the main river and four dozens of smaller dams on the tributaries are foreseen. Figure 1 shows the dams and irrigation areas of the Lower Euphrates development plan. Major dams along the profile of the main tributary of Euphrates are shown in Fig. 2. The construction of the first dam on the main river, Karakaya (Ozis and Ozel, 1989), was completed and operation was started in 1988; the second dam, Ataturk (Ozis, Basmaci and Harmancioglu, 1990 and 1992), is almost complete; the construction of the third, Birecik dam, has been awarded in 1993; and the fourth dam, Karkamis, is completely designed. The active storage provided by Lower Euphrates reservoirs will amount to more than 18,000 million m3.

All four dams are equipped with power plants at their foot. Together with fourteen minor hydroelectric plants of mostly high-head diversion type, the total installed capacity will amount to 5400 MW and the annual average energy generation to 20 TWh/y.

The area to be irrigated in the Lower Euphrates basin, mainly from the reservoir of the Ataturk dam, supported by Birecik and the smaller dams, is in the order of 1.1 million ha. About 45% of this area will be irrigated by gravity flow and the remaining 55% require pumping. Pumping requirements for irrigation amount to 2200 MW of power and 5.2 billion kWh/y of energy (Ozis, 1992).

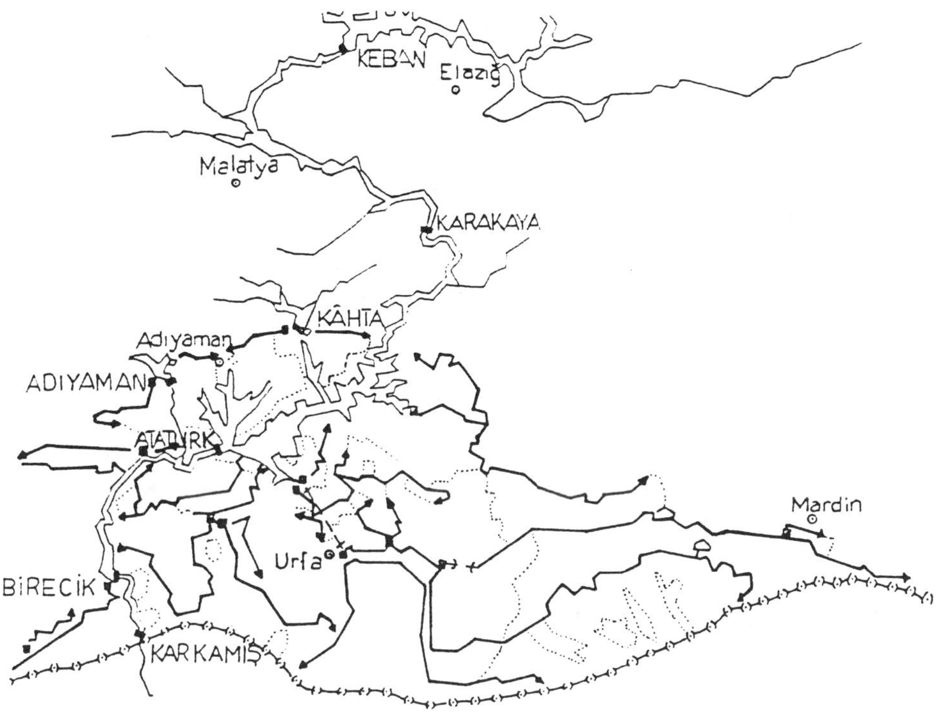

Figure 1. Dams and irrigation areas of the Lower Euphrates basin (DSI, 1980).

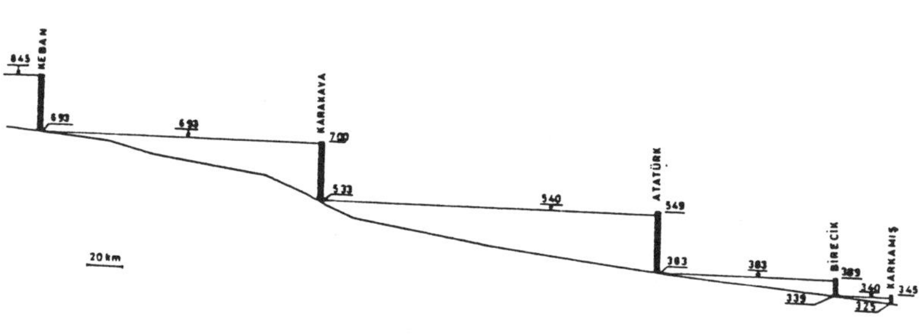

Figure 2. Dams along the profile of the Lower Euphrates river (DSI, 1980; Ozis, 1982).

3.2. TIGRIS BASIN

The Tigris basin covers almost one-twelfth of Turkey's area, provides almost one-tenth of the water potential and one-seventh of the economically feasible hydroelectric potential. Western and Lower Tigris regions of the basin, as considered in the 1980 formulation of SAP, represent about two-thirds of the Tigris basin in Turkey.

Within the framework of the Tigris development, the most important dams of the western part are Kralkizi and Dicle on the main river, Batman and Silvan on Batman tributary, and Garzan on Garzan tributary. The first three dams are in advanced construction stages; the design of the others is completed. All dams are equipped with power plants at their foot. In the lower part, two dams of prime importance, Ilisu and Cizre on the main river, have been designed. In the eastern part, ten large dams in Botan subbasin, two on Hezil and ten on Zap tributaries are planned, primarily for energy generation only. The active storage provided by Tigris reservoirs will amount to more than 7,000 million m3. Figure 3 shows the dams and the irrigation areas of the Tigris basin development plan. Major dams along the profile of the main and secondary tributaries of the river in Western and Central Tigris basins are shown in Figure 4.

The total installed capacity of Western and Lower Tigris power plants will amount to 2300 MW and the annual average energy generation to 8 TWh/y; the development of the eastern region will contribute an additional capacity of 2500 MW and energy of 9 TWh/y.

The area to be irrigated in the Western and Lower Tigris basin from the reservoirs of Dicle, Batman, Silvan, Garzan, and Cizre dams with the support of half a dozen minor dams, is in the order of 0.7 million ha. About 55 % of this area will be irrigated by gravity flow and 45 % by pumping. Pumping requirements for irrigation are approximated as 800 MW of power and 1.6 billion kWh/y of energy (Ozis, 1992).

4. Significant Structures of the Southeastern Anatolia Project

4.1. KARAKAYA DAM AND POWER PLANT

Karakaya is an arched concrete gravity dam, with power plant located under the schute of the gated overflow spillway. The main purpose is energy, and the scheme greatly benefits from the regulation effect of the upstream Keban dam and power plant. The height of the dam above river bed is 158 m, crest length is 462 m, and the concrete volume used is 2 million m3. As an arched concrete gravity dam, Karakaya ranks as the 4th highest and the 5th largest dam in the world with respect to its concrete volume. The spillway of the dam, equipped with 10 radial gates, has a capacity of 17000 m3/sec.

The installed capacity of the Karakaya power plant is 1800 MW; six units of the plant generate 7.4 billion kWh of energy per year. Karakaya power plant is the second largest hydroelectric plant in Turkey after the Ataturk power plant. It is also considered among the largest dozen power plants in the world.

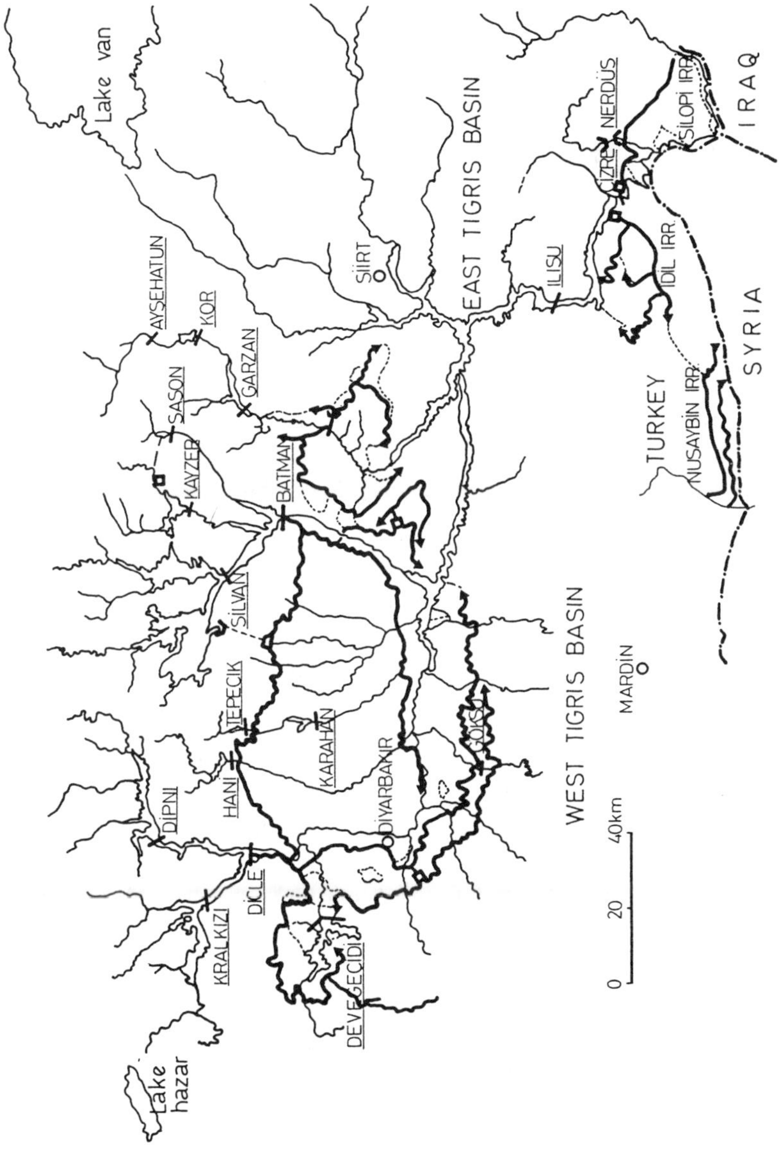

Figure 3. Location of dams, power plants and irrigation areas in the Western Tigris development plan (Ozis, 1983).

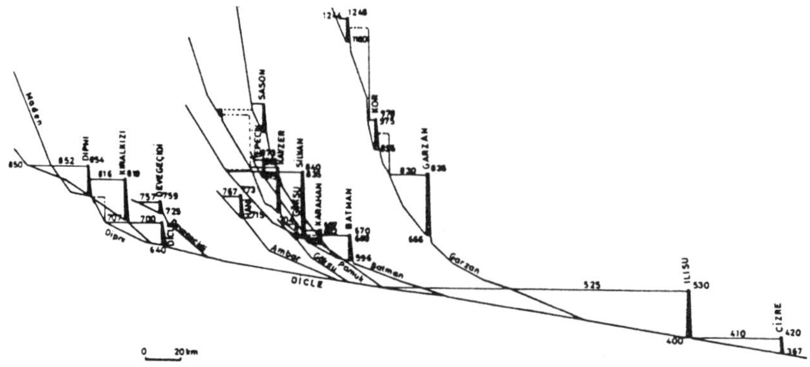

Figure 4. Dams along the profile of Western and Central Tigris (DSI, 1980; Ozis, 1983).

4.2. ATATURK DAM AND POWER PLANT

Ataturk dam constitutes the heart of the Lower Euphrates basin development plan. It is the key project that will provide water to irrigate about 900,000 ha of land in the Euphrates basin. With a total embankment volume of 84.5 million m3, this rock-fill dam ranks as the fourth largest dam in the world. It is also the 20th largest dam with respect to its active storage of 48.7 billion m3. The height of the dam is 166 m above the river bed, which makes it the 13th highest rock-fill dam in the world (third highest in Turkey). The length of the dam crest is 2 km.

Ataturk dam is equipped with 2 spillways; the larger one has a capacity of 16800 m3/sec and is operated by 6 radial gates, each with a height of 17 m and width of 16 m. The second spillway provides an additional capacity of 2000 m3/sec and is equipped with 2 radial gates of 5 m height and 8 m width.

The three derivation tunnels, each with an inner diameter of 8 m and capacity of 500 m3/sec, are operated as sluiceways. The water intake structure is a 71 m high concrete dam with a concrete volume of 530,000 m3.

The power plant at the foot of the dam, with an installed capacity of 2400 MW and energy production of 8.9 billion kWh/y, ranks as the largest plant in Turkey. It is also considered among the few dozen significantly large power plants in the world. The plant receives water via 8 penstocks with diameters varying between 6.6 to 7.25m.

4.3. SANLIURFA TWIN TUNNELS

Sanliurfa twin tunnels convey 328 m3/sec of water from Ataturk reservoir to the main channel of the Sanliurfa-Mardin-Ceylanpinar irrigation system. These are two parallel tunnels, each with an inner diameter of 7.62 m and length of 26.4 km.

Sanliurfa tunnels are the longest water tunnels in the world; with all their connection and entrance tunnels, the total length of the twin tunnels is increased to 57.8 m.

4.4. ILISU PROJECT

Ilisu is the main energy project in the Tigris basin. It encompasses a 138 m high embankment dam and a power plant at the foot of this dam, with an installed capacity of 1200 MW, capable of generating 3.7 billion kWh/y.

On completion, the dam will have a crest length of 1.7 km and an embankment volume of 32 million m3. The limitation of the dam height by the elevation of the upstream irrigation plains permits the creation of an active storage volume of only one-half of the annual natural inflow at the damsite. This volume may be satisfactory after the integral development of the basin for two reasons: upstream dams will create additional storage capacity while irrigation diversion will reduce the inflow, and the underground regulated karst spring effluents in the Western Tigris basin form almost one-third of the natural inflow at the Ilisu dam site (Ozis, 1983).

4.5. CIZRE PROJECT

The main feature of this project will be the Cizre dam and power plant, providing irrigation water for 100,000 ha of land and generating 1.3 billion kWh/y energy.

On the right bank of the Tigris, 70,000 ha land will be irrigated by pumping water up 200 m from the Cizre reservoir. According to previous planning, a large part of this subproject, called Idil-Nusaybin, had to be irrigated by the Lower Euphrates system through canals of over 400 km total length. Actually, an area of 7800 ha at Nusaybin is under irrigation by waters diverted through the Cagcag weir from the same-named tributary of Euphrates, receiving significant karst spring effluents. An additional 19,000 ha may be irrigated after the construction of the Nusaybin dam (Ozis, 1983).

At the left bank of the Tigris, 32,000 ha of land forming the Silopi subproject will be irrigated by pumping water up 100 m from the Cizre reservoir. Actually, 2400 ha of land in this area are irrigated by waters diverted through the Nerdus weir on a small tributary and conveyed by gravity (Ozis, 1983).

5. International Features of the Southeastern Anatolia Project

5.1. TECHNICAL FEATURES

By means of 85 dams (53 in Lower Euphrates and 32 in Tigris basins) and 60 power plants (19 in Lower Euphrates and 41 in Tigris basins) planned within the context of SAP, a total of 1.8 billion ha of land will be irrigated, and a total of 37 TWh/y of electricity will be generated with an installed capacity of 10200 MW (5400 MW in Lower Euphrates and 4800 MW in Tigris basins). With these features, SAP becomes one of the world's largest water resources development plans. In Turkey, it has the first priority among all other projects of its kind. This is because SAP, with the above-given figures, will double Turkey's current hydroelectric energy production and will irrigate close to 2 billion ha of land, which equals the entire amount of land currently being irrigated by state agencies.

Certain technical aspects of some schemes, like those of the rock-fill Ataturk and the arched gravity Karakaya dams, rank them among the most noteworthy examples of their kinds in the world (Ozis, 1986). In essence, the project involves 11 dams with more than 150 m of height and more than 15 million m3 of embankment volume, which classify them as a few of the world's largest dams.

SAP also includes internationally notable power plants. The installed capacities of Karakaya (1800 MW) and Ataturk (2400 MW) power plants rate among the few dozen highest capacities of the world's remarkable hydroelectric power plants.

The irrigation system in the Lower Euphrates basin involves the world's longest water conveyance tunnels. Sanliurfa twin tunnels will deliver water to 900,000 ha of land which will be irrigated by means of a single reservoir constituted by the Ataturk dam.

5.2. ECONOMIC FEATURES

Within the context of SAP, the evaporation losses from the reservoirs and the diversions for irrigation will be in the order of half of slightly over 30 km3/y inflow of Euphrates and one third of the 20 km3/y inflow of Tigris basins at Turkish borders (Ozis, 1992). The active storage volumes created by SAP dams will be almost twice the annual inflow in Euphrates and will approximately equal the annual inflow in Tigris. Thus, the resulting average dry-season outflow will significantly exceed the natural discharges in respective months. In addition to mitigation of floods, SAP reservoirs will create a beneficial storage for downstream countries (Ozis, 1992).

Water budget of Euphrates and Tigris basins worked out by Ozis (1982 and 1992) reveals that the outflow from Turkey to Syria in the Euphrates basin is 17 km3/y and that from Turkey to Syria and Iraq in the Tigris basin is 15.4 km3/y. This means that 54% of Euphrates flow originating in Turkey and 73% of Tigris flow again originating in Turkey may flow to downstream neighboring countries. Only a minor part of such outflows (18% in case of Euphrates and 7% in case of Tigris) will be irrigation return flows of lesser quality (Ozis, 1992).

On the other hand, the total active storage volume created by dams in the Euphrates basin will be in the order of 70 billion m3/y, which is approximately twice the natural inflow in Euphrates. In the Tigris basin, the active storage to be created will be 20 billion m3/y, which almost equals the natural inflow. It is impossible to realize such regulation in the downstream neighboring countries due to their topographic conditions and evaporation patterns. Accordingly, the regulation created by the Euphrates and Tigris dams in Turkey will be in favor of these countries particularly during dry seasons. Even when the irrigation projects are completed, the outflows to downstream countries are not expected to be less than the natural flows except under extremely dry conditions. For example, in the Euphrates basin, the average dry season flow under natural conditions is in the order of 200 m3/sec downstream of Keban dam. However, due to regulation created by this reservoir, this flow has increased to 500-600 m3/sec for the last ten years (Ozis, 1986 and 1992).

The active storage volumes created in Euphrates and Tigris basins in Turkey will help not only to reduce regulation needs of downstream countries but also to provide a very thorough mitigation of floods for all countries involved.

Turkey may eventually claim certain financial contributions from downstream users who will indirectly benefit from the above-mentioned discharge regulation. Furthermore, in extremely dry periods, Turkey may partly refrain from irrigation to release the respective water for exchange of additional financial contributions from downstream users (Ozis, 1992). Accordingly, the operation of energy production and irrigation systems in both the Euphrates and the Tigris basins will be of international interest and therefore will require detailed investigations.

6. Conclusions

The case of the Southeastern Anatolia Project in Turkey represents fairly well the dynamic character of multi-purpose projects with respect to technological developments and changes in economic factors. This multipurpose scheme, with special emphasis on hydropower and irrigation, underwent significant modifications until the construction of the first project, the Karakaya dam, was started in 1980. Minor revisions and changes have followed since then.

The recent scheme may be subject to further minor modifications during the implementation period, especially with respect to the construction of irrigation systems. Of special interest is the investigation of different implementation periods to be considered for the irrigation systems upon the overall economic parameters of the plan. Considering the vastness of the irrigation lands and systems, it would also be of interest to analyze the effects of staging the implementation of various irrigation projects so as to obtain the maximum benefits as earliest as possible. Furthermore, the operation of such water resources systems requires further analysis and may undergo significant modifications due to eventual further changes in economic factors.

Besides the sophisticated mathematical techniques of systems approach to water resources development, planning of projects like the Southeastern Anatolia Project must remind water resources planners and managers that in water resources development activities, there are no absolutely optimum solutions and that their dynamic character must be recognized in the design, implementation, and operation phases.

7. References

Doluca, K. and Pircher, W. (1971) 'Development of the Euphrates River Basin in Turkey', *Water Power*, vol.**23**, no.2, pp.47-55.

DSI (O. Omerbeyoglu et al.) (1967) *Firat Havzasi Istiksaf Raporu*, Ankara, DSI Etud ve Plan Dairesi, **12-46** A,B (2 volumes).

DSI (T. Inceleme et al.) (1968) *Dicle Havzasi Istiksaf Raporu*, Diyarbakir, Devlet Su Isleri X. Bolge Mudurlugu.

DSI (O. Melikoglu et al.) (1978) *Yuksek Karababa Baraji ve HES Ozet Planlama Raporu*, Ankara, DSI.

DSI (S. Uskay et al.) (1980) *Guneydogu Anadolu Projesi*, Ankara, DSI.

EIE (1958) *Data on Euphrates and Tigris Rivers Basin Development Plan*, Ankara.

Electrowatt-Tipton & Kalmbach-S.G.I.-Gizbili (1970) *Lower Firat Project Feasibility Report*, Ankara, DSI, 8 vols.

Electrowatt-T.A.M.S.-S.G.I.-Dolsar (1974) *Karakaya Hydroelectric Power Development Loan Application Report*, Ankara, DSI.

Electrowatt-T.A.M.S.-S.G.I.-Dolsar (1975) *Karababa and Golkoy Projects: Review of Installed Capacities*, Ankara, DSI.

Garbrecht, G. (1968) 'Gediz und Euphrat als Beispiele Grossraumiger Wasserwirtschaftsplanungen (Gediz and Euphrates as examples of large-scale water management schemes)', *Wasser und Boden*, vol.2, no.20, pp.42-46.

Gokcesu, S. (1972) 'Der ausbau des oberen Euphrat-Beckens (Development of the Upper Euphrates basin)', *Die Wasserwirtschaft*, vol.**62**, no.9, pp.265-270.

Harmancioglu, N. (1976) *Asagi Firat Planlamasinin Enerji Bedellerinin Degisimine Duyarligi*, Izmir, Ege University, Master's Thesis in Hydrology and Hydraulic Works, no.5, 154 pp.

Harmancioglu, N. (1985) 'Impact of Economic Factors on Development of the Lower Euphrates Basin in Turkey', Las Vegas, Bureau of Reclamation, *Water Power '85*, Section 44, paper no.2.

Harmancioglu, N. and Ozis, U. (1978) 'Enerji Bedellerinin Degisimine Asagi Firat Planlamasinin Duyarligi', Ankara, *DSI Teknik Bulteni*, no.42, pp.3-13.

Harmancioglu, N. and Ozis, U. (1981) 'Enerji Buhrani ve Ataturk Baraji', Izmir, *Journal of Civil Engineering Faculty of Ege University*, Special Issue: Centennial of Ataturk's Birth, pp.451-474.

Harmancioglu, N. and Ozis, U. (1983) 'Dynamics of Water Resources Development: Lower Euphrates Case in Turkey', Istanbul, *Workshop on Application of Systems Analysis to Water Resources Management*, May 30-June 3, 1983, 11 p.

Ozis, U. (1982) 'The Development Plan for the Lower Euphrates Basin in Turkey', *Natural Resources and Development*, vol.**16**, pp.73-82.

Ozis, U. (1983) 'Development Plan of the Western Tigris Basin', *International Journal of Water Resources Development*, vol.**1**, no.4, pp.343-352.

Ozis, U. (1986) 'Guneydogu Anadolu Projesinin Uluslararasi Boyutlari (International Dimensions of the Southeastern Anatolia project)', Izmir, W.E.C., Turkish National Committee, *Fourth Energy Congress, Proceedings*, vol.1, pp.17-29.

Ozis, U. (1992), *G.A.P. I: Guneydogu Anadolu Projesi Gelisme Plani - Karakaya Baraji ve Santrali - Ataturk Baraji ve Santrali*, Izmir, Dokuz Eylul Universitesi, Muhendislik-Mimarlik Fakultesi, n.224, pp.20-24.

Ozis, U. and Ozel, I. (1989) 'Karakaya Dam and Power Plant', London, *International Water Power and Dam Construction,* vol.**41**, n.7, (July,1989), pp.20-24.

Ozis, U., Basmaci, E. and Harmancioglu, N. (1990) 'Ataturk Nears Completion', London, *International Water Power and Dam Construction,* vol.**42**, n.9, (September,1990), pp.12-16.

Ozis, U., Basmaci, E. and Harmancioglu, N. (1992) 'Ataturk Hydro Power Scheme on Euphrates', Vienna, Technical University, *7th International Seminar - Water Energy Plants*, (ed.: A. Konigsberger, H.B. Matthias, Th. Varga), pp.15-29.

RATIONAL APPROACH TO SUSTAINABLE WATER RESOURCES DEVELOPMENT OF NEPAL

Dr. Jit Narayan Nayak
Institute of Engineering
P.O. Box 1175, Kathmandu
Nepal.

ABSTRACT. The management for the development of water resources system based on the piece-meal planning has been followed till to-date leading to failure and complexities. A rational approach to integrated planning and development with the management in conditions of DEMOCRATIC DECENTRALISM has been proposed. Despite heavy foreign aid, unusual phenomenon (concurrent floods and droughts) and abnormal phenomenon (increasing foreign aid and decreasing real GDP) have harmed national economy to a large extent. The obsolete concept of development should be dispensed with in favour of rational planning preceding foreign aid for a SUSTAINABLE level of economic growth. Development of water resources of Nepal has to be done in three phases of 10 years each starting with Terai, Mahabharat and Himalayan regions respectively providing foodgrain and water supply sufficiency, surplus electricity with balanced ecology. After 2-3 decades Nepal will develop further independently. A National Research and Design Institute for Water Resources Development (NARDIWARD) has been proposed for the solution of problems (created by unusual and abnormal phenomena) of Nepal and neighbouring countries.

1. INTRODUCTION

It is quite clear now that piece-meal planning and development has caused not only extreme environmental damage but poverty and hardships to the people and nation as a whole. Therefore, in a situation of high level of moral and social consciousness, will all the natural resources of the country be utilised rationally so that they would not last only for the present, but also for the further generation/5,6,10/.

The total population of about 19 millions and area 145,504 square kilometers varying in elevations from 60 meters in the south to 8,848 meters above MSL in the north provide Nepal highly favourable topography for the construction of a series of reservoirs to generate hydropower, regulate flow for irrigation, water supply, navigation and river control. The day to day life of millions is directly related with the intensive development of water resources of rivers alongwith

V. P. Singh and B. Kumar (eds.), Water Resources Planning and Management, 91–109.
© 1996 *Kluwer Academic Publishers. Printed in the Netherlands.*

their proper utilization/8/.

Water resources including cultivable lands and green forests are the natural base of development for prosperity and wealth of the country. The big net-work of rivers and streams is highly developed totalling the length of more than twenty thousand kilometers, i.e. 0.3 kilometer per square kilometer of the territory. The total average annual run-off has been estimated to be 200 billion cubic meters, out of which about 170 billion cubic meters come from areas within the territory of Nepal/12,14/. Almost 93 percent people depend on agriculture production. The country has limited arable land of about 4.35 million hectares, out of which the total cultivable land has been estimated to be more than two million hectares. The area under forest has now drastically decreased to two million hectares (60 percent) from 4.8 million hectares in 1975 (refer Fig.2).

Even though investment in water resources development has been significantly increasing over the years, yet the performance of agriculture, irrigation, water supply and hydropower sub-sectors alongwith environmental degradation have not been satisfactory, for example, deficiency of foodgrains as well as water supply, worsening situation of sewage disposal in Kathmandu valley and power crisis in the country.

The overall distribution of the national income is much more unequal in Nepal than that in other low income Asian countries. Nearly 70 percent of the total population of Nepal live below the absolute poverty line and the calorie intake lags behind the minimum need/2/. Therefore, considering the previous obsolete approach of development, a rational concept of "INTEGRATED" planning and development of water resources is proposed. An appropriate justification will be provided that it can definitely play a big role in energy supply, foodgrain sufficiency, water supply, river control and ecological balance not only for Nepal, but also for the neighbouring countries. A scientific solution of the development of water resources sector is urgently needed, which can be done through "Rational Approach to Sustainable Water Resources Developmet of Nepal."

2. THEORY OF RESOURCE MANAGEMENT

The complexity of planning, organization and management for the development of water resources system has been growing continuously from the past to the present due to the following factors: ad-hoc basis of previous political and social systems; obsolete scientific and technical mechanisms; ever increasing volume of the works to be done; ambiguity of technical solution to be adopted on the basis of advanced science, technology and practice; lack of specialised knowledge in Governmental staffs as well as power decentralization to all levels in the country.

The global objective of the management of any system, that is to achieve the maximum output with minimum investment, has to be followed by each specialist. For successful implementation of the modern developed science and technologies in practice, he must know his main orientation, possibility of the utilization of the technical resources and be ready for raising problems to other specialists for their further solutions.

2.1. Managerial Activities

Usually, a specialist as manager, has to face administrative, financial, engineering and social activities. It means that as the post and level of the manager become higher, the specific weight of the managerial acitivity superceeds that of specialist (refer Table I).

As the volume of works, complexity of technical solutions, development of specialization, etc. increase, the managerial works practically become too involved. The reasons may be multichannel communications, effect of their external and internal, objective and subjective factors on the process of doing work and combinations thereof. Radical increase in volume of the managerail and technical informations makes a more difficult work for managers and specialists, making it less productive and insufficiently effective. In these cases, there naturally occurs an interest to overcome this situation searching for the application of technical resources, on which some part of the work with information could be loaded.

The meaning of the management for the development of water resources system is depicted in the following. According to the fixed objective, given problem or task, the manager must get or collect data characterising a process to be organized and conditions in which

Tentative relation of activity to managerial work/15/

Table I

S.No.	Level of Manager	Component of Activity, %	
		Managerial	Engineering
1.	Overseer	10	90
2.	Engineer	20	80
3.	Executive Engineer	30	70
4.	Superintending Engineer	50	50
5.	Director General/Project Manager	90	10

it will take place. All the data to be worked out should be generalised and analysed for achieving scientifically based solutions in the concrete existing situations. On the basis of achieved solutions, the perspective/operative plan is prepared. In essential cases, optimization and correction of the plans are carefully done, considering the changing conditions.

2.2. Achievement of Managerial Solutions

It is a difficult process with no single solution. From a series of the possible alternatives, it is necessary every time to select a best optimum solution. In general form, its procedure is shown in Fig. 1.

On the basis of initial data/task, an alternative of the solutions which seems to be well, is taken into account for detailed work. It is essential to make an analysis for the evaluation of result to be expected before its solution will be achieved for implementation.

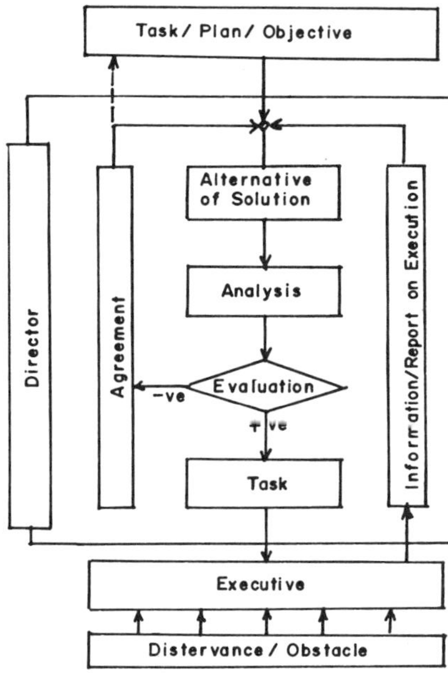

Fig. 1 Schematic process for workout and achievement of managerial solution.

The methodology of problem analysis is usually determined by the characer contents to be solved. For the solution of various technological or organizational problems, the criterion for its evaluation considered is economic indicator as well as work completion period. For positive expected result of the suggested alternative, a managerial solution is achieved with compiling the task for the Executive. In case of negative result, an agreement with concerning organization leading to make a new alternative, is practically done. All these alternatives are required to be analysed and evaluated in the same order, till a good solution is obtained. Such scientifically based procedures for achieving solutions of applied problems related to water resources systems, are essentially to be followed at all levels of the management.

2.3. Principles and Methods of Management

The principles of management/15/, in general, are the system of laws, bye-laws, rules, regulations, situation and norms of behaviour in the existing socio-economic conditions. The management for the development of water resources systems in the DEMOCRATIC conditions, in deed, is essentially to the followed on the basis of the main principles: SCIENTIFICNESS, DECENTRALISM, UNITY OF POLITICAL AND TECHNICAL RESOURCES, PLANNING OF SYSTEMS AT ALL LEVELS, MORAL AND MATERIAL STIMULIZATION, RELATIVELY CORRECT SELECTION, DISPLACEMENT AND ENCOURAGEMENT OF SPECIALISTS, DIRECTORS AND OFFICERS.

Democratic decentralism related to be above principles, obviously provides a decentralized management having broad initiative and self-dependence of the productive organizations alongwith active participation of all the workers in production management. The responsibility of all the Executive for action and result of the works as managers and their associated personnels must be withstood. Effectiveness of the management is mainly determined by the rational utilization of material and moral stimuli with respect to the Executive as well as collective as a whole. In the production management, specialists have to face a series of the socio-economic and other laws and bye-laws like Economic, Right and Socio-psychological (Moral) laws.

The basic methods of management for the development of water resources sytems including collective people are the following/15/:

Organized-order method is based on the scientific and clear division of labour, rights, duties and responsibilities of each participant, director and executive according to the existing Constitutional laws, bye-laws, directives, other rules and regulations. These are realised in practice by action to the Executives through orders, notices, instructions in written/oral forms. Material stimulization method is laid on the main principle of Democratic decentralism-"One for All and All for One." In the modern conditions, the main ways of material stimulization are: the system of salary payment with tariff-nets providing bonus for excellent work done; increment for work in special conditions; system of award for high productivity and result oriented work leading to economy of the material and technical resources. Socio-psychological method of management considers the social conditions, psychology of different executives, specifics of working groups and different workers-officers.

In the present context of Nepal, the development of water resoruces systems alongwith solution of the environmental degradation problems is still not transparent in conjunction with the scientific theory as said above and, thus, can lead to the worst conditions in the near future.

3. ANALYSIS OF AVAILABLE RESOURCES

It is necessary, now, to elaborate the available resources with their specific utilization and effect in practice in each sub-sectors: Public Health, Irrigation, Hydropower, River Control and Disproportional Ecological Balance.

3.1. Public Health

It is well understood that a healthy environment of any place/city mainly depends on the sewage disposal and cleaning process of the organizations related to their resources management. Among the municipalities of Nepal in, capital, Kathmandu the tap water supplied by the concerning organization at any time is not suitable for drinking purposes, especially during monsoon season.

Next to the environment of the Kathmandu valley has been contaminated for some years due to low flow velocity in the Bagmati river unable to transport sewage even upto the sewage-treatment plant near Chobhar/7/. As a result of it, the habitats of Kathmandu, Lalitpur and Bhaktapur, in comparision with other municipalities, have been attacked by various diseases, thereby affecting the productivity of the social and Governmental works. Millions of U.S. dollar have been spent in this valley even though the volume of water only from five rivers of Kathmandu and Lalitpur districts becomes per capita maximum to be 3,400 cubic meters and minimum 600 cubic meters annually. If the water resources of only five rivers of the two districts utilized rationally, there might not be any problem for drinking water for 70-80 years in the future, provided the ecological balance is maintained good in all sense. Out of the Kathmandu valley, some municipalities have good water supply system and some have bad too.

Water supply and sewage disposal systems constructed about 50 years ago, urgently need to be replaced by new others, but the cost of which is not affordable by the country at present. A partial work on the improvement of the water supply system carrying out by the aid of JICA/Japan, after some years, can also lead to the same situation as it is now.

3.2. Irrigation

The steep topography of Nepal has the limited availability of land for productive farming and, hence, giving pressure for efficient farming is mounting at an alarming rate on future generation. The country has 4.35 million hectares of arable land comprising of 0.25 million hectares in the Himalayan, 1.90 million hectares in the Mahabharat and Siwalik, and only 1.30 million hectares in the Terai (plain) regions/4,9/. However, according to the Department of Irrigation, these figures in terms of cultivable land account for only 2.60 million hectares, of which possibility of irrigation by the obsolete method being now in existence is only in 1.60 million hectares and the prevailing facility to the farmers is in 0.50 million hectares by the run-off river and other systems (e.g. lift irrigation and rainfed also).

It is well known that the limited cultivable land of the country remains in the natural and original stage (2.60 million hectares) that cannot be estimated for further expansion. However, its quantity is being decreased annually due to erosion, washout downstream from the river banks and the development process - highways, canals, urbanization, industrial estates and rural development. At the same time, the growth of the coutnry's population/10/ is constantly pressuring to increase the productivity of land with the help of scientific methods of cultivation, without which Nepal does/will not have any way to fulfill the foodgrain requirements (refer Fig. 2). The operation of storage reservoirs, therefore, is urgently required.

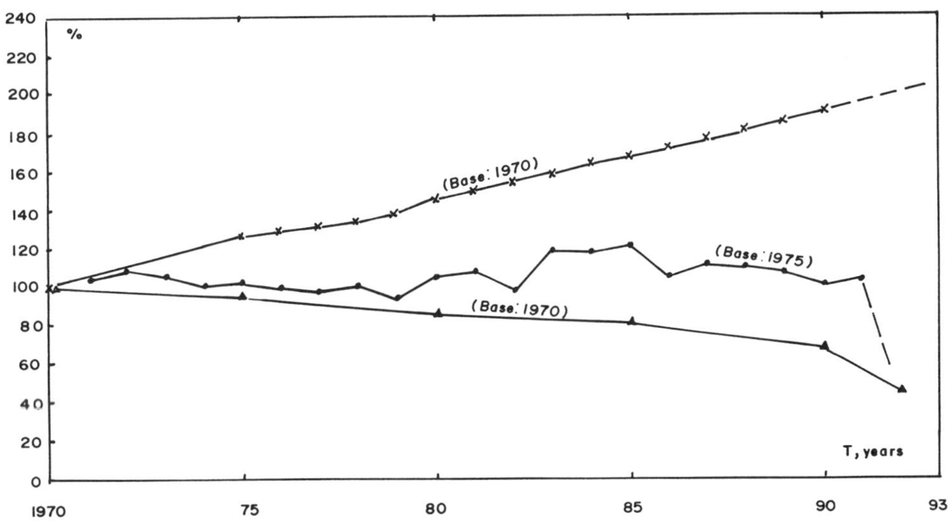

Fig. 2 Agricultural Production and Population Index /2/

Agriculture; Population; Green Forest.

In order to get the sufficiency in foodgrain, an intensive increase in the work and complexity of the irrigation and drainage hardware systems alongwith software require clear management to be inter-acted between the departments of Irrigation and Agriculture, collaborating with the academic institutions of concerning fields. For this, thus, one of the most important ways for effective expenditure on the irrigation of land is a modification of the organization and management of production at all levels of the irrigation management system.

3.3. Hydropower

Nepal, being a small hilly country, has now occupied Second position for per capita water resources (about 12 thousands cubic meter annually) next only to Brazil, whereas the country's economic growth has been poorest in the world.

From the total water resources potential of the country, mentioned above, except the storage of ground water, the total theoretical hydro-power potential has been evaluated equivalent to electrical power 83 million kilowatts (refer Table II), of which only 42 million kilowatts can possibly be generated from the economically effective point of view/11,12/. By this time, the country has hardnessed only about 0.3 percent of the gross potential, while this figure has reached 70 percent and more in the developed countries. The Government of Nepal will not have to search for alternative to hydropower potential for 80 years ahead, provided the existing potential is harnessed reasonably.

Hydropower potential of Nepal/12/

Table II

S.No.	River Basins	Hydropower potential, 10^6KW		
		Theoretical	Technical	Economic
1.	Kosi	22.35	11.400	10.860
2.	Gandaki	20.65	5.725	5.270
3.	Karnali & Mahakali	36.18	26.482	25.125
4.	Southern Rivers	4.11	0.975	0.878
	Total	83.29	44.582	42.133

The country has been facing power deficit problem for several years due to previous irrational planning and design of hydropower projects (medium/large scale), which were designed without planning by the International agencies. But this deficit was overcome by importing power from the neighbouring country - India. Actually, Nepal has started to feel the CRITICAL POWER CRISIS only from some months and if the existing trend of hydropower development continues, it will not be possible to cover the power deficit even upto 2000 A.D. as is clear from Fig. 3. At present, the contry going to establish high costly Thermal Power Plants so as to overcome this power crisis for time being, will have to bear the additional damage of its national economy.

Considering the power load forecasting of Nepal Electricity Authority (NEA) and the hydropower projects to be constructed in near future, the country might have to face a deficit of power of about 230 percent (with respect to 1977) at the end of 20th century. It is obvious that Nepal may be in dark at least upto 2000 A.D. and beyond unless the Government stress NOW on the construction of annual storage hydropower projects to be operated in the country.

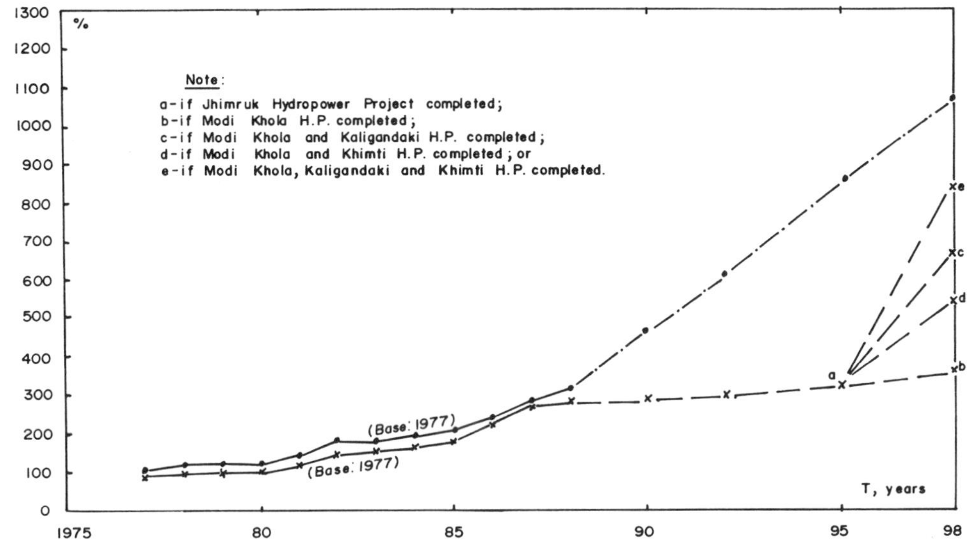

Fig. 3 Electrical Energy Consumption and Production Index/2/

Consumption — Production

Prediction of NEA — Prediction of Author

3.4. River Control

As mentioned above about the erosion and washout of limited cultivable land from the river banks, the energy contained in water flowing through steep terrain of the country, causes soil erosion with the result that about 380 million metric tonnes of surface soil (as suspended sediments) is lost annually from the territory of nepal/13/. The most sediment-laden river of the country, the Kosi river, carries suspended sediments of 10 kilogrammes per cubic meter of its water flow. The mean annual suspended semidment flow capacity of the Mahabharat region has been predicted to be 2.69 kilogrammes per cubic meter of flowing water/6/.

The total run-off, 200 billion cubic meters, from the Nepalese drainage area passes through the medium and big rivers of the country's Terai (plain) region. From the east to west these rivers are Mechi, Kankai, Kosi, Balan, Kamala, Bagmati, Lakhandei, Gandaki, Babai, Rapti, Bheri, Karnali, Mahakali and others, whose water levels rise during high flood season. As a result, every year the cultivable land and villages existing on the banks of these rivers are submerged. In addition, from these banks thousands of hectare of cultivable land are being eroded and washed out annually downstream. These affect negatively on foodgrain and cashcrop production, thereby create a serious national problem of the foodgrain deficiency for coming generation. Due to this, the damage of the national economy can apparently be evaluated in terms of millions of Nepalese rupee/8/. Hence, the flood and sediment flow (taking into account the bed sediments of different sizes due to erosion of young hill slopes) are essentially to be checked by the operation of the reservoir systems and by afforestation on the young hill slopes of the country.

3.5. Role of Ecology

The biosphere became contaminated with the intensive development of science and technology; consequently the people's surrounding environment also. In utilizing the available natural resources and in the creation of new materials in the production process at commercial level, people are searching more and more deeply for their satisfaction into the resources of nature.

The extention of communication and technological exchange taking place now between the countries, has changed the balance between society and the natural environment of the area in which it lives. All these as the result of disproportionate ecological balance of nature have created ECOLOGICAL CRISIS and scientists are now actively searching for ways of protecting nature. Success in the work of protection and of rational utilization of the natural resources can be achieved only in situation of high moral and social consciousness of people/10/.

An unusual phenomenon has arrived in the country which has started to face with the flood disaster in the one part and simultaneously drought problems in the other part/5/. Scientifically speaking, there are no rational distribution of available water resources in Nepal due to the indifference of national scientists and specialists in the process of water resources development. In any case, thus, His Majesty's Government of Nepal must have to pay a careful attention in these directions technically and educationally with the help of psychological attributes for the generation of healthy people in the country.

3.6. Assessment of Foreign Aid

Nepal is, NOW, facing a big national problem related with the development of water resources, i.e. the leading sector is running too slowly with a rapid increase in the country's population - root cause of poverty. This situation of poverty has been invited for some past years due to IRRATIONAL management of the natural resources, organizations and the manpowers engaged in the sphere of production with guidence of foreign advisors to the Government machinery of the country. Therefore, more sustained efforts for an improvement in various sectors, including specially water resoruces, on a co-ordinated basis, are called for in the coming days ahead.

For few decades the country's economic characteristics (refer Fig. 4) clear that the country's foreign aid and exchange rates have been increasing rapidly, reaching now 55 percent (plnned) of the real GDP and 500 percent in 1993 respectively (with respect to 1972), while the International reserves (except gold), declining radically in 1985, have been raised again rapidly alongwith only few (1.5) percent of increment of the real GDP in 1992. It is absolutely quite abnormal phenomenon in the development of the country's economics/5/.

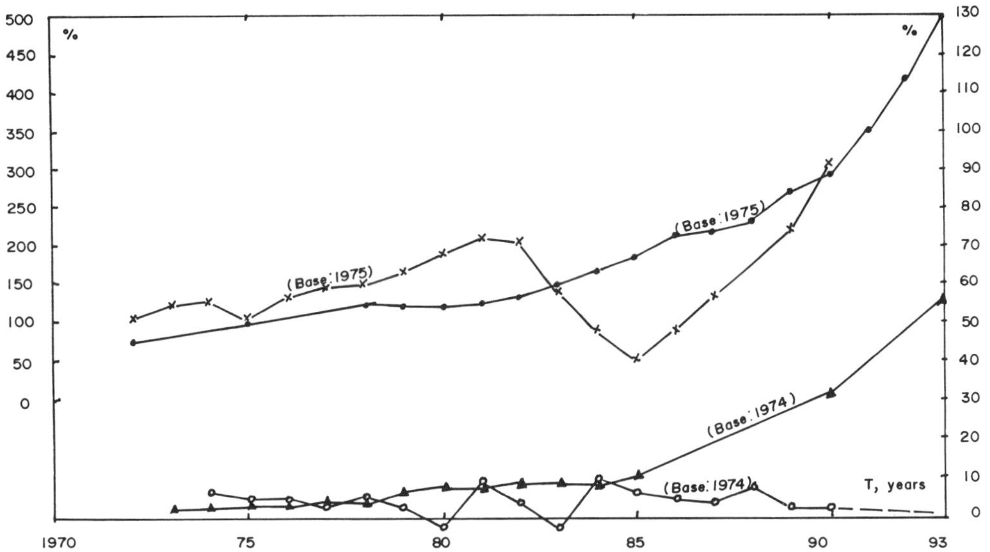

Fig. 4 Country's Economic Characteristics/2/

Foreign Exchange Rate; International Reserves (except Gold); Official Aid (% of GDP); Real GDP of Nepal.

Out of the total gross investment as 100 percent, 56.3 percent will be financed through national savings. Of the remaining 43.7 percent, 11.6 percent will be financed by the foreign grants and 32.1 percent by the external loans/1/. However, the overall actual foreign contribution to the total development expenditure under Eighth Plan (1992-97) is expected to be 65.5 percent, while it was 60 percent in Sixth Plan. Thus, though the amount of foreign aid in Nepal appears to look substantial, a major part (about 70 percent of it) flows back to the donor countries/ agencies/ other developed countries in terms of capital and running costs of imported goods and "highly" qualified specialists/5/. This typical type of aid has, however, not been able to develop hardwares and softwares within the country and all the time to dependence on the outside for design, construction, maintenance and even, in some cases, for operation.

For twenty years the effectiveness and appropriateness of foreign aid in the development (including water resources) and transfer of a suitable technology in Nepal have resulted in disappointments in all sense. Thus, dispensing with the previous traditional and unsuccessful concept of development of natural (water) resources which included the flow of foreign aid WITHOUT PLANNING OF PROJECTS, a rational approach to the sustainable development of natural resourcs should essentially be followed. This requires FOREIGN AID ONLY ON THE BASIS OF PLANNED PROJECTS to be implemented in practice for providing basic requirements of the people.

4. APPROACH TO INTEGRATED DEVELOPMENT

There are mainly the three groups of people for rational utilization of any natural (water) resources of the country, while the political sector is a major driving force.

First group including national scientists essentially searches for ways of proper utilization of the natural (water) resources. Under the guidance of scientists, second group with specialists has patriotically to check/ verify the statements/ hypothesises in the laboratory/ field for the development of appropriate technology. And the people of third group, engaged in the sphere of production, as consumer also at a time, create/produce required materials by the use of natural resources and the appropriate technologies developed. Therefore, the created/produced materials (goods) are quite essential for the existence of three groups of people including those engaged in the sphere of non-production in the country.

4.1. Selection of Development Way

Broadly speaking, for the rational development of natural resources of the country, if available, there are the two ways of development/5,7/: Extensive and Intensive.

Depending upon the condition of national economics, any one of both the ways or their combination can be applied in the country. In the country having healthy national economics, its economic growth can be increased slowly by the application of extensive way of development - evolution in science and technologies, and can be raised rapidly with the help of intensive way of development - revolution in prevailing seicne and technologies of the country.

While in the country with poor national economics, like Nepal, whose development budget more than 50 percent depends on the foreign aid, there is no way except the Intensive way of development. By its application the country's economics can only successfully be accelerated radically with revolution in science and technologies available at present in the country. This way of development is mainly based on the mobilization and proper use of national scientists, specialists and natural (water) resources with the help of only loans with MINIMUM STRINGS to be provided by the International/foreign national development agencies, but under consideration with the help of grants at this time in the world.

In context of the further developmetn of Nepal, for rapid growth in the national economics within 2-3 decades the ratio of foreign aid to the real GDP of the country should essentially be needed to be decreasing proportionally, after one decade, as the real GDP must be increased rapidly/5/. Meanwhile, there will be a period, when it will reach a SUSTAINABLE level and after which there will be no need of foreign aid. Lateron, Nepal would be able to plan its development activities independently on the basis of national economics strengthened.

4.2. Planning of Development Infra-structure

Necessity of planning of the infra-structures for the purpose of water resources development in Nepal has been realised only after the disasters, i.e. foodgrain deficiency, deficit of water supply, power crisis togetherwith imbalance of ecology (environmental degradation) in the country.

In the modern age, planning of water resources development must take into consideration the different group of water users and consumers, their objectives and criteria. In general, it is not possible to make a planning only to achieve the high income; it is necessary to know: who is going to finance the projects and who is interested to get profit? QUESTION OF RISK, DISTRIBUTION OF NATIONAL INCOME, QUALITY OF ENVIRONMENT, NATIONAL LIVING STANDARD AND ECONOMIC EFFECTIVENESS are quite essential for planning of water resources development/3/.

Based on the main principle of Democratic decentralism - "One for All and All for One", the following infra-structures must be achieved:

a) Provision of autonomy of the Institute of Engineering and Institute of Agricultural and Animal Science from the Tribhuvan University so that they might become independent universities towards 2000 A.D.;

b) Mobilization of national scientists and specialists engaged in water resources development of the country with their work responsibility and motivation;

c) Handing over the power to formulate and solve the problems of water resources development to national scientists and specialists of the water resources sector;

d) Establishment of the National Research and Design Institute for Water Resoruces Development in the country;

e) Collaboration of Institute of Engineering and of concerning H.M.G. Departments with the National Research and Design Institute proposed;

f) Collaboration of Institute of Forestry and Institute of Agricultural and Animal Science and of concerning H.M.G. Departments with the National Agricultural Research Council and the National Research and Design Institute proposed;

g) Establishment of the "Ministry for Science and Technology" in the country;

h) Let the Water and Energy Commission Secretariat (WECS) be a co-ordinating machinery of the Ministry of Water Resources, depending on the THINK TANK - Research and Design Institute proposed, which will be free for making the policies of water resoruces development; and

i) Making a policy for the International consulting firms/agencies engaged in water resources development, to work in collaboration with the above Research and Design Institute.

The proposed Research and Design Institute would have to perform not only policy making, planning, research and design, but also monitoring and evaluation of the medium and large scale projects of water resources development, while the small scale projects could be performed, after planning, by the concerning H.M.G. Departments or by national private consultancy services.

4.3. Planning of Water Resources Development

In order to plan rationally the available water resources, cultivable land, electrification and communication of the whole country, the territory of Nepal has been geographically divided into three regions: Terai (plain) with altitude from 60 meters to 300 meters above MSL, Mahabharat-from 300 to 3,000 meters, and Himalaya region exceeding altitude of 3,000 meters above MSL. Development of water resources for radical increase in Nepal's economic growth can be done in three phases according to geographic regions.

The first phase development (10 years period) from the Terai region can definitely provide foodgrain sufficiency not only for the present, but also for the future generation of the whole country. The cultivable land about 0.90 million hectares with present population of nearly 9 millions is partially facilitated by the run-off river irrigation systems and about 70 percent of the cultivable land which is rainfed but drought affected last year. Therefore, to overcome this problem, a scientific method of cultivatation on the available land must be achieved by operation of a series of reservoirs with medium head in the Siwalik mountain ranges from east to west. These storage types of the project will provide water to the farmers, downstream, throughout the year in required amount for irrigation and will partially produce electrical energy. Both will be required not only for the agricultural intensification, but also for the electrification and transportation facility of rural areas of the Terai region. The remaining energy can also be diverted to the Mahabharat retgion for its electrification that will reduce deforestation, provided the Government can supply hot-plates cheaply to the people for their cooking accessories.

Simultaneously, the research works on the geology and hydrology alongwith seismology of river basins of the Mahabharat and Himalayan regions are to be intensified for the base of further development of water resources in the country.

In this way, the country withing 10-15 years will surely get a sufficiency in foodgrain and cashcrop production alongwith small and medium scale industrialization, thereby strengthening the national economics automatically togetherwith a radical increase in the national capital. Some part of which can also be invested in the development of the Mahabharat region (second phase), reducing significant portion of foreign aid to the country. Thus, the Government with the advice of its national scientists and specialists can achieve the sustainable development process by constructing few annual storage type of the hydropower projects in the country during first phase of development.

In the second phase (10 years period) of water resoruces development, the Mahabaharat region should have impartial use of the stored water by the help of medium/large reservoir constructions, mainly for hydropower generation be partially for irrigation intensification in hills. About fifty percent of the power to be produced from this region is proposed to but utilized in the electrification and ropeway transportation of the Mahabharat and Himalayan hills alongwith electric transportation in the Terai also. It is necessary to establish some factories in the country. The remaining 50 percent of it can be exported to the neighbouring countries. Through this approach, after two decades, Nepal can sustain its economic growth with a few percent of foreign aid dependence. While within twenty years various small/medium scale industries will be developed. As a result, the country will have its own qualified technical, economic and administrative personnels who will move ahead in the direction of development process independently. Therefore, the time has essentially arrived for all Nepalese people to think "LOOK BEFORE YOU LEAP" and to stand on own legs in the development process. This equally applies to least developing countries in the world.

Finally, in the third phase development (10 years period) of the Himalayan region, by time the research base for last 20 years for water resources development at scientific level has been completed and also science and appropriate technologies developed, only then the water resources potential of the Himalayan region would be easily and cheaply harnessed, mainly for hydropower development with the help of construction of the large scale reservoirs, their water surface area not exceeding 100 square kilometers. Secondly also there will be few irrigation intensification in specific climatic conditions. Hence, the Government should not initiate, IN HURRY, any hydropower project in the Himalayan region only because the research works on Snow and Glacier of that region, specially Glacier Mechanics, have not been developed till to date for the design and construction of the medium/ large scale water resources projects.

In this way, working for two decades and strictly following above mentioned directions of water resources development, some positive results in the national economics will surely be seen and only then, after the second decade, the country may breathe easily for its further development independently. All these things, stated before, mainly depend on the consciousness of the leadership and the related machinery engaged in the sphere of production as a whole. In fact, Nepal has high potential of natural resources (water, cultivable land, forest and others), but is being made economically poorest in the world by IRRATIONALISING them. Thus, for the balance of the regional ecology and for rational utilization of the regional water resources potential for the sake of people living in this region, it is recommended to establish the regional co-operation committee - "SOUTH SCIENTIFIC AND ECONOMIC COMMITTEE FOR WATER RESOURCES DEVELOPMENT" (SOSECWARD).

5. CONCLUSIONS AND RECOMMENDATIONS

1. Considering the previous traditional and unsuccessful concept of Piece-meal planning and Piece-meal development of the country's water resources system, a rational approach to the INTEGRATED planning and development is urgently needed. This would fulfill the basic requirements of the people-foodgrain, water supply and electrical energy alongwith ecological balance not only for the present, but also for the future generations.
2. On the basic principle of Democratic decentralism - "One for All and All for One" and for successful implementation of the modern science and technologies prevailing at present in the country (but essentially to be developed), each specialist must know his main orientaiton, utilization possibility of the technical sources and be ready for raising problems to other specialists for their futher development.
3. Rapid growth of the country's population is constantly creating a pressure on the limited cultivable land, protection from the flood disaster, erosion and washout of the rivers' banks. Hence, productivity of land can only be increased with the help of scientific methods of cultivation to fulfill the foodgrain requirements for future growing population. Loss of cultivable lands (equivalent to millions of rupee) will create a serious national problem.
4. In case of ecological balance, the water resources potential of the Kathmandu and Lalitpur districts can rationally provide drinking water to the habitats of their municipalities including Bhaktapur also for 70-80 years in future and about 90 percent population of these municipalities already attacked by several virus infections and diseases would be free from them.
5. Nepal, having the gross and economic hydropower potential equivalent to, respectively, 83 million and 42 million kilowatts, will not have to search for ways of alternative to hydropower potential for 80 years in future. But the existing trend of hydropower development may lead the country to a power deficit of about 230 percent (with respect to 1977) by the end of 20th century.
6. The country has started facing unusual phenomena: concurrent flood disaster and drought problems, meaning thereby that, scientifically speaking, no rational distribution of available water resources over the territory of Nepal has been done due to the indifference of national scientists and specialists. In any case, a careful attention in these directions, technically and educationally, must be paid at the Government level.
7. For few decades, the country's foreign aid and exchange rates have been increasing rapidly, reaching 55 percent (planned) of the

real GDP and 500 percent in 1993 respectively (with respect to 1972), while the International reserves (except gold), declining radically in 1985, have been raised again rapidly alongwith only few (1.5) percent of increment of the real GDP in 1992.

8. Dispensing with the previous traditional and obsolete concept of development of the natural (water) resources which included the flow of foreign aid WITHOUT PLANNING OF PROJECTS, a rational approach to the SUSTAINABLE development should essentially be followed which requires the flow of foreign aid only on the basis of PLANNED PROJECTS, to be implemented in practice for providing basic needs of the people.

9. For rapid growth in the economics of Nepal within 2-3 decades, the ratio of foreign aid to the real GDP of the country is needed to be decreasing proportionally, after one decade, as the real GDP must be increased rapidly. A time will appear when it will reach a SUSTAINABLE LEVEL and after which no need of foreign aid will be required. Only then, Nepal would be able to plan its development activities independently.

10. The National Research and Design Institute for Water Resources Development should essentially be established for the solution of problems of water resources development of Nepal related to research and design of the medium and large scale water resources projects alongwith establishment of the Ministry of Science and Technology so as to monitor and finance the projects' performance.

11. For the balance of the regional ecology and rational utilization of the regional water resources potential, a regional co-operation committee - SOUTH SCIENTIFIC AND ECONOMIC COMMITTEE FOR WATER RESOURCES DEVELOPMENT (SOSECWARD) is quite needed to be established in this region.

12. Development of water resources of Nepal can be done in three phases (geographically divided into three regions: Terai, Mahabharat and Himalayas). First phase development for 10 years from the Terai region can definitely provide foodgrain sufficiency and electrification of rural areas and agriculture alongwith basic needs of the small and medium scale industrialization. Second phase development for next 10 years in the Mahabharat region would stress mainly on the hydropower development so as to facilitate electrification, irrigation and ropeway transportation in the hills as well as trolley transportation in the Terai region. While third phase development for the last 10 years in the Himalayan region, on the basis of research data obtained for 20 years, would comprise only for hydropower development by constructing large scale reservoirs, if necessary, alongwith agriculture intensification depending upon climatic conditions.

13. Working for two decades and strictly following the above mentioned directions of water resources development, some positive results in the national economics will definitely be seen. And only then, after the second decade, the country may breathe easily for its further development INDEPENDENTLY.

6. ACKNOWLEDGEMENT

The author acknowledges gratefully the financial and administrative assistance of the Ford Foundation Project headed by Mr. Basil S. Jaceb, advisor for Hill Irrigation Development, at Institute of Engineering (I.O.E.) in preparing, submitting and presenting this paper at the said conference. The author thanks Prof. S.R. Mathe, Dean of IOE, and Prof. R.K. Poudel, Head of the Department of Civil Engineering, without whose moral and material help this paper would not have been possible. Many thanks are due to visiting Prof. N.C. SAXENA for his editorial works.

REFERENCE

1. Eighth Plan (1992-97). National Planning Commission, H.M.G./Nepal, 1992. Kathmandu.
2. Key Indicators of Developing Asian and Pacific Countries. Vol. XXII, Economics and Development Resources Centre, ADB, July, 1991. Manila.
3. Loucks D.P. and et. all. Water Resources Systems Planning and Analysis. Prentice-Hall Ltd., 1981. Delhi.
4. Master Plan for Irrigation Development in Nepal. Canadian International Water and Energy consultants in collaboration with East Consult (P) Ltd., UNDP (NEP/85/013)/ World Bank, Cycle 1 (Main Report), 1989. Kathmandu.
5. Nayak J.N. 'Rational Approach to Flood Disaster Prevention in Nepal.' National Seminar on Information Exchange in Field of Disaster Prevention/ Preparedness, March 29-30, 1993. Kathmandu.
6. Nayak J.N. 'Planning and Design of Reservoirs with Sediment Flushing.' Regional Seminar on Methods for Preservation of Useful Reservoir Storage on Heavily Sediment-laden Rivers, October 13-16, 1992. Kuala Lumpur.
7. Nayak J.N. 'Jalsrot Sadupyog: Bikasko Purwadhar (Proper Utilization of Water Resources: Prospect of Development)'. Tech. Jr. Pravidhik, No. 11, 1992. Kathmandu.
8. Nayak J.N. 'Evaluation of Bed Forms of Alluvial Rivers'. Regional Seminar on Mathematical Modelling of Alluvial Rivers, April 14-18, 1992. Kathmandu.
9. Nayak J.N. 'Khetiyogya Jaminma Sinchaiko Avshyakta (Necessity of Irrigation in Cultivable Land)'. National Gorkha Patra, Chaitra 30, 2043. Kathmandu.
10. Nayak J.N. 'Preserving Ecological Balance'. The Rising Nepal, October 18, 1986, Kathmandu.
11. Nayak J.N. 'Jalshakti Sadupyog: Bikasko Purwadhar (Proper Utilization of Hydropower: Prospect of Development)'. National Gorkha Patra, Chaitra 06, 2042. Kathmandu.

12. Pradhan B.K. Water Resources Development and Management in Nepal. First Monthly Seminar organised by Department of Road Project Directorate/ ADB, H.M.G./ Nepal, September 13, 1990. Kathmandu.
13. Sharma K.P. and et. all. Suspended Sediment Yield Estimation Model for Nepalese Watersheds. Regional Seminar on Mathematical Modelling of Alluvial Rivers, April 14-18, 1992. Kathmandu.
14. Water Resources Development in Nepal. H.M.G., Ministry of Water Resources, 1985. Kathmandu.
15. Ysinetsky V.G. Organization, Planning and Fundamental of Management for Water Resources Construction. Coloss Publication, 1982. Moscow.

SECTION 3:

WATER-RESOURCE SYSTEM OPERATION AND MANAGEMENT

MODELLING FOR RESERVOIR OPERATION FOR IRRIGATION

S. VEDULA and P.P. MUJUMDAR

Department of Civil Engineering
Indian Institute of Science
Bangalore 560 012, India

ABSTRACT. An overview of the developments in modelling for reservoir operation for irrigation is given. The models developed in this field progressively improved from time to time to take into account more and more aspects in the complex decision making mechanism with uncertainty in such factors as water supply, rainfall and crop water demands. Stochastic considerations of the inputs and multiple crops and the need to integrate the decision making mechanisms at the reservoir and the field levels brought additional complexity into the model formulation. Recent works in this area have promised a great hope for meaningful real time applications of the developed models.

1. INTRODUCTION

Irrigated agriculture is the most important concern of the developing countries today in keeping up their food demands for the ever increasing population. Operation of reservoirs built for irrigation deserves a special consideration in this context. Reservoir release decisions for irrigation depend on such factors as the nature and timing of the crops being irrigated, crop growth stages, competition for different crops for the available water and the effect of a deficit water supply on the crop yield.

A number of publications in the literature deal with modelling the operation of a multi-purpose reservoir but studies of significance relating to the reservoir operation for irrigation wherein the crop water demands have to enter the model formulation are but few. Even fewer are the studies which take into account the stochasticity of the reservoir inflows and the crop water demands. However, steady and significant efforts in modelling were advanced in the last two decades in these respects. The most recent of the attempts in modelling have been to integrate the decision making processes at the reservoir level and the field level for reservoir irrigation of multiple crops and to develop an appropriate methodology for optimal irrigation allocations in real time. The decision should be sufficiently explicit to indicate not only how much water is to be released from the reservoir in a given period but also how much of it should be allocated to a given crop. Computational limitations sometimes set constraints on the conceptualization of the problem in terms of its practical application to a real system.

V. P. Singh and B. Kumar (eds.), Water Resources Planning and Management, 113–119.
© 1996 *Kluwer Academic Publishers. Printed in the Netherlands.*

The present paper gives an overview of the developments of the recent past in the field of modelling for reservoir operation for irrigation. A brief account of the earlier work done in the field is given followed by a rather relatively wider coverage of the recent studies in modelling.

2. EARLIER STUDIES

The uncertainty among the hydrological inputs comprising primarily of the inflows into the reservoir, rainfall in the irrigated area, evapotranspiration and soil moisture contributes significantly to the complexity in modelling the decision making process. Earlier models developed for optimal reservoir operation for irrigation dealt with various aspects of the problem in different degrees of complexity, by treating one or more of these variables as stochastic.

While variability in reservoir inflow was taken into account [Hall et al. 1968, 1969; Schweig and Cole, 1968] most of the approaches treated seasonal crop water demand for irrigation as deterministic. However, subsequently the need to take account of intraseasonal variations in crop water requirement was recognized. Burt and Stauber [1971] incorporated stochasticity in crop water demand but assumed a deterministic water supply. Stochasticity in both water demand and storage addition was considered by Dudley et al. [1971a, b, 1972], Dudley [1970, 1972], and Dudley and Burt [1973].

Dudley et al. [1971a] incorporated a plant growth-soil moisture simulation model into a two state variable stochastic dynamic programming model to determine the optimal intraseasonal allocation pattern for irrigation water in a variable environment. A single crop was considered and available water (at a given maximum value) for allocation and available soil moisture at the beginning of the time period were treated as the two state variables. Both rainfall and crop water requirement were treated as stochastic variables. The stochastic nature of the water- supply state variable arose because of the stochastic irrigation water requirements and not because of random inflows into the storage reservoir. The linking of the available water supply to a reservoir was not considered in the model. The decision variable in the model was the level to which the available water content is allowed to fall before irrigating the whole of the root zone to field capacity.

In a subsequent model the same authors [Dudley et al. 1971b] dealt with the problem of what acreage to plant with a single crop from a reservoir in which the crop water demand and additions to storage (supply) both were considered stochastic. The model does not consider different areas of crop at different intensities which virtually would have amounted to a multiple crop situation. This problem was referred to as the intermediate- run irrigation planning problem by the authors the solution to which depends on the solution to the earlier problem [Dudley et al. 1971a] which was referred to as the short-run irrigation planning problem.

The question of determining the best area to set up for irrigation (out of a given total area under a reservoir) was considered in the so called long-run irrigation planning problem [Dudley et al. 1972] using a combination of simulation and dynamic programming. Again a single crop is considered with stochastic supply and stochastic demand. An extension of the models developed by Dudley et al. [1971a, b, 1972] was presented by Dudley [1972] wherein the value of water carried over from one season to the next was recognized. A modified version of the previous approach was used to generate reservoir level transition probabilities and expected benefits. The results of an application of the models indicated that interseasonal transfer can considerably increase the present value of expected benefits from the project.

An integrated intraseasonal and interseasonal stochastic dynamic programming model was developed by Dudley and Burt [1973] for a single crop to determine an optimal decision rule with respect to intertemporal water application rates, the question of whether acreage should be 'abandoned' from further consideration for the remainder of the season, and the optimal acreage to plant for potential irrigation at the beginning of the season. An economic evaluation of the net benefits and their variability was made to determine the trade off beween them. Area available for irrigation, soil moisture level, available water in the reservoir and a measure for crop production function were treated as state variables in the model formulation, though soil moisture was deleted as a state variable in application.

3. RECENT STUDIES

Subsequently, Dudley et al. [1976] developed a hierarchy of models to aid decisions in multicrop systems for deterministic crop water requirements. These models use linear programming (LP), simulation and dynamic programming (DP). LP is used to select best crop combinations, simulation to predict changes in reservoir storage and dynamic programming to optimize interseasonal water allocation. Dudley [1988] advanced this earlier work by simulating the effects of the optimal decisions under the assumption of a single decision maker at the reservoir and field levels.

Bras and Cordova [1981] modelled the optimal temporal allocation of irrigation water, taking into consideration the intraseasonal stochastic variation of the crop water requirements and the dynamics of soil moisture depletion process. The formulated model considers only a single crop and the volume of water available for irrigation is assumed to be given at the beginning of the season. Rao et al. [1990] solved the problem of allocating a limited water supply for irrigation of several crops grown in the same season. Both seasonal and intraseasonal competition for water between crops were considered. Dynamic programming has been used as the optimization tool. What needs to be done in terms of modelling reservoir operation for irrigation is to integrate these concepts with the reservoir release decisions with inflows treated as stochastic.

Development of a single model, computationally simple enough for application to reservoir operation for multiple crops, which takes into account stochastic reservoir inflows and stochastic crop water demands with dynamic soil moisture accounting still remains a formidable task. Vedula and Mujumdar [1992], however, made a successful beginning toward modelling for reservoir operation in a multicrop environment in which the mechanisms governing the decisions at the reservoir level and the field level were integrated in a single model. Intraseasonal periods smaller than the crop growth stage durations form the decision intervals of the model to facilitate irrigation decisions in real situations. Reservoir storage, inflow to the reservoir and soil moisture in the irrigated area are treated as state variables in the dynamic programming approach used. An optimal allocation process is incorporated in the model to determine the allocations to individual crops when a competition for water exists among them. The model serves also as an irrigation scheduling model in that at any intraseasonal period it indicates whether irrigation is needed and, if it is, the amount of irrigation to be applied to each crop. The impact on crop yield of water deficit and the effect of soil moisture dynamics on crop water requirements were taken into account. The model proposed in this study was formulated conceptually in two phases. In the first phase, the model uses deterministic dynamic programming and allocates a given amount of water among all the crops to optimize the impact of the allocation within a period. This allocation is determined for all possible supplies in a given period, for all periods in a year. In the second phase, a stochastic dynamic programming (SDP) model optimizes the overall impact of the allocations among all the intraseasonal periods over a full year. The end result is a set of decisions indicating the reservoir release to be made in each interseasonal period and the distribution among the various crops of this release available at the crop level after accounting for losses. The crops and the crop calendar were assumed to be fixed. The field situation addressed is typical of a developing country like India where the land holdings are small and reservoir irrigation is designed based on a design cropping pattern in the entire irrigable area commanded by the reservoir. The objective used is to maximize the sum of the relative yields of all the crops in a year.

The model's [Vedula and Mujumdar, 1992] utility has been demonstrated through its application to Malaprabha reservoir in Karnataka State, India. Ten day decision intervals were used with a total of eight crops, three in the monsoon season, four in the nonmonsoon season and one biseasonal crop (cotton). The steady state reservoir operating policy derived specifies, for given period, the optimal end of the period storage in the reservoir for given state of the system (values of the beginning of the period storage, the known or estimated inflow during the period, the average soil moisture over all the crops at the beginning of the period in the irrigated area and estimated or given rainfall during the period) and the associated optimal irrigation allocation to each crop in that period. The discretization scheme consists of 15 class intervals for reservoir storage, four class intervals for inflow, five for soil moisture level and 12 for available water at field level for allocation to crops with 36 intraseasonal periods in a year.

The model incorporates the complex features of operation at the reservoir level and at the field level, however, with some limitations. Three limitations were identified. A major weakness of the model lies in the averaging of the soil moistures among all crops at the beginning of a period. This can be avoided in this model by defining a soil moisture state for each crop in the formulation which naturally increases the computational burden to an unmanageable degree in a practical application. Another limitation of the study, though perhaps not as serious as the averaging of soil moisture is the use of deterministic rainfall and evapotranspiration in the irrigated area. A simulation of the optimal policy practically showed insensitivity to even wide variations in rainfall. Lastly, interdependency of the crop production in a given period on water allocations in other periods is not directly reflected in the model because of the limitation caused by the use of dynamic programming as the problem solving tool.

The most recent study by Nagesh Kumar [1993] removes the limitations of the earlier study of Vedula and Mujumdar [1992] and finds a solution to the problem through the use of linear programming and stochastic dynamic programming given seasonal forecasts of reservoir inflow and rainfall in the command area. Both inflow and rainfall are treated as stochastic variables and the crop irrigation requirements computed through the soil moisture balance equation.

4. REAL TIME OPERATION

Once the modelling is done taking into account all important aspects at the reservoir and field levels, procedures for real time operation of the reservoir and crop water allocations can be developed. Real time operation requires forecasts of reservoir inflows and rainfall in the irrigated area one time period in advance. Errors due to the imperfect forecasts and discretization used in dynamic programming approach to optimization in problem solving invariably introduce difficulties in actual implementation of the optimal release and consequently the optimal crop water allocations for a specific period in real time. Corrections are to be incorporated to the optimal decisions in real time implementation in such a way that the departure from the optimal path is set at the minimum.

5. REFERENCES

Bras, R.L. and J.R. Cordova, 'Intraseasonal water allocation in deficit irrigation', *Water Resour. Res.*, **17(4)**, 866-874, 1981.

Burt, O.R., and M.S. Stauber, 'Economic analysis of irrigation in subhumid climate', *Am. J. Agric. Econo.*, **53(1)**, 33-46, 1971.

Dudley, N.J., 'A simulation and dynamic programming approach to irrigation decision making in a variable environment', *Agric. Econ. and Bus. Manag. Bull.* **9**, Fac. of Agric. Econ., Univ. of New England, Armidale, Australia, 1970.

Dudley, N.J., 'Irrigation planning, 4, Optimal intraseasonal water allocation', *Water Resour. Res.*, **8(3)**, 586-594, 1972.

Dudley, N.J., 'A single decision-maker approach to irrigation reservoir and farm management decision making' *Water Resour. Res.*, **24(5)**, 633-640, 1988.

Dudley, N.J., and O.R. Burt, 'Stochastic reservoir management and system design for irrigation', *Water Resour. Res.*, **9(3)**, 507- 522, 1973.

Dudley, N.J., D.T. Howell, and W.F. Musgrave, 'Optimal intraseasonal irrigation water allocation', *Water Resour. Res.*, **7(4)**, 770-788, 1971a.

Dudley, N.J., D.T., Howell, and W.F. Musgrave, 'Irrigation planning, 2, Choosing optimal acreages within a season', *Water Resour. Res.*, **7(5)**, 1051-1063, 1971b.

Dudley, N.J., W.F. Musgrave, and D.T. Howell, 'Irrigation Planning 3, The best size of irrigation area for a reservoir', *Water Resour. Res.*, **8(1)**, 7-17, 1972.

Dudley, N.J., D. Reklis, and O.R. Burt, 'Reliability, trade-offs and water resource development modelling with multiple crops', *Water Resour. Res.*, **12(6)**, 1101-1108, 1976.

Hall, W.A., W.S. Butcher, and A. Esogbue, 'Optimization of the operation of a multiple reservoir by dynamic programming', *Water Resour. Res.*, **4(3)**, 471-477, 1968.

Hall, W.A., G.W. Tauxe and W.W.-G. Yeh, 'An alternate procedure for the optimization of operations for planning with multiple river, multiple purpose systems', *Water Resour. Res.*, **5(6)**, 1367-1372, 1969.

Nagesh Kumar, D., 'Integrated modelling for optimal reservoir operation for irrigation', *Ph.D. thesis*, Dept. of Civil Engg., Indian Inst. of Sci. Bangalore, 1992.

Schweig, Z., and J.A. Cole, 'Optimal control of linked reservoirs' *Water Resour. Res.*, **4(3)**, 479-497, 1968.

Rao, N.H., P.B.S. Sarma, and Subhash Chander, 'Optimal multicrop allocation of seasonal and intraseasonal irrigation water', *Water Resour. Res.*, **26(4)**, 551-559, 1990.

Vedula, S., and P.P. Mujumdar, 'Optimal reservoir operation for irrigation of multiple crops', *Water Resour. Res.*, **28(1)**, 1-9, 1992.

GENERALIZED HEADWORKS SIMULATION MODELLING: THE AUSTRALIAN EXPERIENCE

George Kuczera
Department of Civil Engineering and Surveying
University of Newcastle, NSW, 2308, Australia

ABSTRACT. Recent work in Australia on the modelling of complex headworks systems is considered. This work has focussed on the development of computer software implementing the concept of a generalized objective-driven simulation model. Two models, REALM and WATHNET, are described with emphasis on their practical implementation. At the heart of both models is a network linear program (NetLP) guided by a hierarchy of objectives and constrained by physical and institutional constraints. The use of iterative convergence techniques has greatly enhanced the power of the NetLP. These NetLPs require complex input data and can produce voluminous output. To realize the potential of these models considerable effort needs to be expended on the design of the user interface. Both REALM and WATHNET are supported by user interfaces that port easily across different computer systems, yet simplify the task of complex data entry and, most importantly, communicate complex results through extensive use of graphics. Experience with a range of Australian headworks systems ranging from urban to complex irrigation-urban networks suggests that the concept of a generalized simulation model based on network linear programming is sound and provides a powerful tool for simulating complex headworks systems.

1. INTRODUCTION

A water supply system harvests, stores and transfers streamflow to demand zones serving urban, industrial, irrigation or environmental consumers. The system is typically engineered with small buffering storages which permit the supply system to viewed as two, almost independent, components:

i) Headworks system: This component deals with the infrastructure responsible for harvesting and bulk storage of water and with its bulk transfer to demand zones. During periods of drought the headworks system is usually under stress in the sense that the risk of a demand shortfall (that is, failure to meet a zonal demand) increases. The planning and operation of the system focus on the management of the economic and environmental losses arising from demand shortfalls. The time scale for modelling the headworks system depends on the interplay between streamflow and storage volume with a month usually being adequate.

ii) Distribution system: This component deals with the infrastructure responsible for transfer from distributor storages to individual consumers within the demand zone. The critical planning and operating issue is the capacity of the transfer system. The modelling time scale, dictated by consumption patterns and hydraulic constraints, may range from times of the order of an hour to days.

V. P. Singh and B. Kumar (eds.), Water Resources Planning and Management, 121–142.
© 1996 *Kluwer Academic Publishers. Printed in the Netherlands.*

In Australia the per capita investment in headworks infrastructure is large by world standards because the Australian continent is the second driest (with Antartica taking the honours) and, more importantly, has the highest variability in streamflow. Therefore, there exists a major incentive to make the most efficient use of limited water resources.

This paper describes recent Australian developments in the modelling of complex headworks systems characterized by several reservoirs and demand zones. The research literature is replete with works considering single-reservoir, single-demand systems, but is relatively barren on the topic of complex systems. Yet it is complex systems which invariably have to be dealt with in practice.

The Australian developments focus on the development of computer software to assist in the planning and management of complex headworks systems. The software implements the concept of a generalized objective-driven simulation model supported by a graphical user interface that simplifies the task of complex data entry and, more importantly, communicates complex results in a form readily understood by the engineer.

After reviewing the issues related to generalized objective-driven simulation, the development of two models, REALM and WATHNET, is described emphasizng the translation of a research idea into tools gaining wide acceptance within the Australia water industry. Some of the key ideas central to both models are described along with their computer interfaces.

2. GENERAL CONSIDERATIONS

The REALM and WATHNET models represent a significant departure from traditional headworks simulation models. Before looking at these models in some detail, it is appropriate to review some of the reasons, both philosophical and practical, that provided the impetus for the development of these models.

2.1 Simulation Models as Decision Support Tools

The best way of operating and augmenting complex headworks systems remains an unsolved problem. Two obstacles stand in the way, network complexity, and the fact that future streamflow and demand are both unpredictable. Given our limited success in overcoming these obstacles, practical reality dictates that the computer simulation model assume the role of a decision-support tool providing assistance to engineers who have to make informed judgements about what to do. The effectiveness of this approach depends very much on how well the simulation model serves the needs of the engineer, ideally relieving him of the tedium of the complex water accounting logic inherent in headworks simulation and freeing him to use his intuition and lateral thinking skills.

To be effective in the role of decision-support the simulation model must be capable of:

i) Answering quickly and reliably the ‘what if’ questions posed by the engineer; and

ii) Presenting the answer in a way intelligible to the engineer.

These two desiderata were the focal point of the developments described in this paper.

2.2 The Case for Generalized Simulation Models

The development of simulation models in the form of software has evolved in two distinctly different styles. The first involves customizing computer code to a particular headworks system by explicitly coding the network configuration and operating rules. This has been the traditional approach which has granted the programmer complete control over the model. For complex systems, however, considerable programming effort is required to develop and maintain such a model. As the system grows in complexity, program logic may become so complex that no single individual fully understands it and, more importantly, modifications can become major technical tasks. In such an environment, the model may severely constrain the engineer by limiting the range of 'what if' questions that can be answered.

The second style is motivated by the fact that the headworks system can be represented as a network of nodes interconnected by arcs. The nodes may represent reservoirs, demand centres, stream diversions, junctions and treatment plants, while arcs may represent streams and man-made conduits. High-level programming languages are particularly capable of implementing this kind of network representation in what will be called generalized simulation models.

The generalized simulation model deals explicitly with the network paradigm making no prior assumptions about network topology and storing as much information as possible about the headworks system on a data file. The pioneering generalized model was HEC-3 (HEC, 1981). Subsequently, models such as MODSIM (Labadie et al., 1986), WASP (Kuczera and Diment, 1988) (and its offspring PRISM and REALM) and IRIS (Loucks and Salewicz, 1989) have further refined the original concept.

The generalized model in its ultimate form becomes truly independent of the system under study. It should remove most, if not all, the constraints on the type of 'what if' questions the engineer may wish to have answered.

Experience suggests that it is extremely difficult to anticipate all the future questions to be asked of the simulation model. In contrast to the traditional approach, use of a generalized model largely avoids the need to anticipate by providing a robust framework which should be capable of accommodating most questions asked of it.

Apart from being an intellectually elegant solution, the generalized model confers very significant savings in development and maintenance costs. A well-designed generalized model should be capable of being applied to a wide range of headworks systems without any modification to the source code. This avoids the duplication of effort that occurs with the development of custom software.

2.3 Rule-based or Objective-driven Simulation?

Both custom and generalized simulation models require a credible, robust and computationally efficient algorithm for making assignments between the nodes in the headworks network. The algorithm must be credible meaning it describes what actually happens, not what should happen, it must be robust if it is to work in all situations, and it must be computationally efficient to provide acceptable response times to questions posed by the engineer.

Two algorithmic approaches represent extreme ends of the spectrum. The first makes assignments according to user-defined rules. An example is the reservoir release rule which tells the operator how much to release depending on present and expected future storage volume. Most systems are, in fact, operated according to such rules.

Developing software for rule-based simulation is conceptually simple but suffers from two problems that in complex systems can become serious:

i) The amount of information required to fully define the rules can be substantial making data input a cumbersome and error-prone task.
ii) Each rule adds additional degrees of freedom available to the modeller for manipulating system performance. The number of choices available can explode as the complexity of the system grows leaving the modeller wondering which set of rules are most efficient. Because it is unthinkable to enumerate all possible rule combinations, the use of heuristics becomes indispensable. Moreover, system complexity introduces interdependence and redundancy which require the use of more complex rules which, in turn, exacerbate the original problem.

At the other extreme water assignments can be totally determined by a mathematical program which is driven by user-defined objectives. We will call this objective-driven simulation. By handing over detailed control of water assignments to a mathematical program the engineer no longer has to worry about the detailed water assignment logic and, more importantly, has the number of degrees of freedom in controlling system performance reduced to hopefully a manageable number. Moreover, if the mathematical program is properly formulated, it should efficiently assign water within the network regardless of the network complexity.

The benefits of objective-driven simulation depend critically on the availability of efficient algorithms for solving mathematical programs. Yeh (1985) in his review of water resource optimization shows that almost all mathematical programming approaches are burdened by severe computational problems when applied to complex systems. To date the only mathematical program capable of making assignments in complex systems with acceptable computational speed is the network linear program (NetLP) and its closely related cousin, the network flow program. The recent literature (for example, Sabet and Coe, 1986; Chung et al., 1989; Sabet and Creel, 1991) has seen renewed interest in the use of NetLP for this reason. However, the NetLP approach is far from ideal because of the constraints it places on problem formulation. Nevertheless, its superior computational performance cannot be ignored.

The Australian developments, to be described below, have attempted to exploit the advantages of an objective-driven simulation model based on NetLP but have had to make compromises. Both the REALM and WATHNET models are hybrid models using both rule-based and objective-driven assignment algorithms.

2.4 Computer Interfaces

One characteristic common to all simulation models of complex headworks systems is that they require considerable input information and can produce huge volumes of output information, particularly if synthetic hydrology is used. Recent developments in graphics hardware and software enable the development of intuitive graphical interfaces which streamline data entry tasks and, more importantly, offer the opportunity to present complex results in a form that can be readily comprehended by the engineer.

From the perspective of headworks simulation it is important to have a computer interface that simplifies the task of asking 'what if' questions and presents the answer in a simple-to-grasp format. Without an appropriate interface the engineer's productivity may be reduced through delays in system definition, debugging input data, and, more significantly, through missed insights.

3. NETWORK LINEAR PROGRAMS AND DIRECTED GRAPHS

As background to the REALM and WATHNET models, it is necessary to briefly review the main features of a NetLP and define some notation. A NetLP is a linear program which finds the minimum cost solution for conveying a commodity (water in our case) through a network of unidirectional arcs interconnecting supply, demand and transhipment nodes. Figure 1 illustrates a simple network showing the paths water can be routed from supply nodes to demand nodes. At each node a mass balance must be satisfied; that is, the sum of commodity inflows to the node must equal the sum of the outflows. A NetLP takes the following form for a n-arc, m-node network:

$$\text{minimize } \mathbf{c}'\mathbf{x}$$
$$\text{subject to } \mathbf{A}\mathbf{x} = \mathbf{d};\ \mathbf{l} \leq \mathbf{x} \leq \mathbf{u} \tag{1}$$

where **c** is a n-vector of costs or penalties, **x** is a n-vector of commodity flow volumes along the arcs, **d** is a m-vector of known nodal inflow or outflow volumes, **l** and **u** are lower and upper bounds on **x** respectively, and **A** is a (m,n) node-incidence matrix whose $(i,j)^{th}$ element is defined as:

$$A_{ij} = \begin{cases} +1 \text{ if arc j is directed away from node i} \\ -1 \text{ if arc j is directed toward node i} \\ 0 \text{ otherwise} \end{cases} \tag{2}$$

The constraint **Ax** = **d** ensures that a mass balance is achieved at every node in the network.

An important property of a NetLP, which is exploited by computer codes, is that, if **c**, **A**, **d**, **l** and **u** are integer, then the least-cost solution $\mathbf{x}^*$ will also be integer. This means all computations can be done using integer arithmetic which is exact and faster than floating-point arithmetic. This specialized structure of the NetLP can be exploited to produce very fast computer codes over two orders of magnitude faster than revised simplex codes.

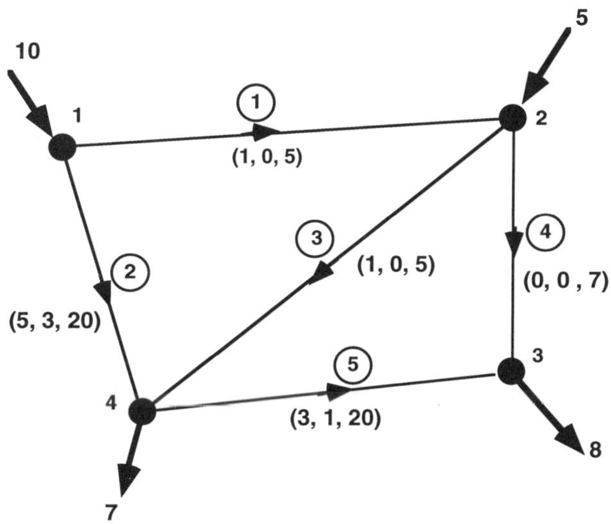

Figure 1 Directed graph for a sample network

When formulating a NetLP it is more intuitive to work with graphs depicting the network rather than the algebraic form given by eqn. (1). The node-incidence matrix **A** can be represented by the node set Y = {1, 2, ..., m} and the arc set $\Lambda = \{e_1, e_2, ..., e_n\}$ where

$$e_j = [F(j),T(j)], \quad F(j) \text{ and } T(j) \,\varepsilon\, Y, \quad F(j) \neq T(j) \tag{3}$$

with F(j) and T(j) being respectively the nodes from and to which flow is directed for arc j. Information about costs and bounds for arc j in Λ is summarized by the triplet (c_j, l_j, u_j) where c_j is the cost of conveying one unit of commodity in arc j, and l_j and u_j are respectively the minimum and maximum commodity flow volumes in arc j. The NetLP is fully specified once the vector of nodal requirements **d** is furnished.

Figure 1 illustrates the graphical convention used in this paper. The cost and bounds triplet is entered for each arc. The arrow on each arc indicates flow direction, from which the from- and to-nodes can be deduced. A known flow directed to (or from) a node represents a positive (or negative) requirement. For the network shown in Figure 1 the NetLP formulation would be:

$$
\begin{aligned}
&\text{minimize } x_1 + 5x_2 + x_3 + 3x_5 \\
&\text{subject to} \\
&x_1 + x_2 = 10 \quad \text{.... mass balance on node 1} \\
&x_3 + x_4 - x_1 = 5 \quad \text{.... mass balance on node 2} \\
&-x_4 - x_5 = -8 \quad \text{.... mass balance on node 3} \\
&x_5 - x_2 - x_3 = -7 \quad \text{.... mass balance on node 4} \\
&0 \le x_1 \le 5 \\
&3 \le x_2 \le 20 \\
&0 \le x_3 \le 5 \quad \text{..........capacity constraints} \\
&0 \le x_4 \le 7 \\
&1 \le x_5 \le 20
\end{aligned}
\tag{4}
$$

There are a number of computer codes which exploit the structure of the network linear program yielding computation times two-orders-of-magnitude faster than revised simplex codes (Kennington and Helgason, 1980). Kuczera (1993a) has compared two public domain NetLP codes, NETFLO, which implements a primal simplex algorithm (Kennington and Helgason, 1980), and RELAX, which implements a primal-dual algorithm (Bertsekas, 1991). He showed that RELAX typically outperformed NETFLO by at least a factor of two with greater gains if good initial flow solutions were used.

4. REALM: A GENERALIZED HEADWORKS SIMULATION MODEL

4.1 Genesis

The motivation for REALM had its origins in the early 1980s. The headworks system for Melbourne, a city of over 3 million people on the south coast of Australia, had nine major reservoirs delivering water through a complex transfer system to urban demand zones. In the 1970's a rule-based simulation model MODFLO was developed by Collinge (1978) to evaluate anticipated changes to the headworks system. Typical of its genre, both system configuration and operating rules were explicitly coded into the computer program. However, as the Melbourne system developed, new infrastructure had to be added and evaluated, resulting in the code becoming progressively more difficult to understand, maintain and update. MODFLO had become obsolete.

In 1983 the decision was made by the then Melbourne and Metropolitan Board of Works to abandon MODFLO. But rather than develop a clone of MODFLO, it was decided to develop a generalized simulation model which could keep pace with new and possibly unexpected developments in the Melbourne headworks system. It was realized at this early stage that this model would have the potential to be applied to other headworks systems obviating the need for other water authorities to make the considerable investment in a new headworks simulation model.

The author led a small team of engineers and programmers, which included Glen Diment who was to become the key player in REALM's development, to develop the prototype of REALM, a model called WASP, suitably named because it could sting the novice user. Two requirements had a major influence on model development:

i) The model should be general in the sense of accepting any system configuration and a wide range of operating policies.

ii) The interface between model and user should be designed to suit the user's requirements and require minimal knowledge of computer operating systems.

The need for a generalized simulation model meant that rather sophisticated rule-based algorithms for water assignments within the headworks system would be required. It was felt that NetLP would provide a much simpler basis for water assignments than rule-based algorithms. With hindsight it was this decision to proceed with NetLP that proved to be crucial in future developments. Details about WASP can be found in Diment et al. (1984) and Kuczera and Diment (1988).

4.2 NetLP Formulation

The NetLP in WASP makes water assignments according to the following hierarchy of five objectives:

1) Satisfy demand (which may be restricted according to predefined restriction rules) at all demand zones;
2) Satisfy all instream flow requirements;
3) Minimize spill from reservoirs and gravity diversions.
4) Ensure that reservoirs are balanced according to their space target curves and have volumes at or above their end-of-season target volumes;
5) Minimize operating costs.

The NetLP which implements these objectives is subtle and complicated. The reader is referred to Kuczera and Diment (1988) for a fuller discussion. Here the best insight about the formulation can be got by considering a simple example. Consider Figure 2 which depicts a two-reservoir, two-demand system. Figure 3 shows the NetLP formulation presented as a directed graph.

The dashed arcs in Figure 3 along with the balancing and bypass nodes illustrate two sub-networks which perform specific tasks:

i) **Bypass sub-network:** Nodes 3 and 6 in Figure 3 correspond to the demand zones 1 and 2 in Figure 2 which have demands equal to D_1 and D_2 respectively. Arcs [2,3] and [2,6] define the bypass sub-network. By assigning the largest cost c_d to these arcs one is guaranteed that meeting demand is given the highest priority. The NetLP solver will look for solutions which avoid using the bypass arcs. However, because of drought or transfer limitations, it may be physically impossible to meet the demands D_1 and D_2. In such a case the shortfall in demand is met by transferring water from the bypass node to the demand node via the bypass arcs. Without this feature the NetLP solver would return an infeasible solution without any explanation as to its cause. In Figure 3 two bypass arcs are connected to each demand node. Each arc has a capacity equal to half of the demand, but has a different cost. This arrangement allows sharing shortfalls between demand nodes. For example, the NetLP solver would fill the two bypass arcs with cost c_d before utilizing the bypass arcs with cost $c_d + 1$.

ii) **Storage target sub-network:** The function of the storage target sub-network is to ensure the end-of-season reservoir volumes are as close as possible to those specified by the space target curves which define how a given volume of water should be distributed spatially among the reservoirs. For each reservoir the available water AW, defined as the initial storage plus natural inflow minus target storage, is calculated. It represents the amount of water the reservoir makes available to the system if it is to finish the season on target; it may be positive indicating the reservoir should make a release, or negative indicating the reservoir should receive water from elsewhere in the system. Arcs [4,1] and [5,1] carry spills and above-target flows which occur if the system cannot accept all the available water forcing the reservoir to store water above its target storage or possibly spilling. Arcs [1,4] and [1,5] carry below-target flows. They are used when a negative AW cannot be satisfied by transfers from elsewhere in the system and represent a drawdown in storage below the target volume. Costs, smaller than the bypass costs, are assigned to these arcs to give the NetLP solver an incentive meeting the storage targets.

4.3 Enhancing the Basic Formulation

The formulation illustrated in Figure 3 simulates the primary features of a headworks network but is incapable of simulating secondary features such as reservoir evaporation based on average surface area, percent transmission losses, conditional operating logic, arc capacities dependent on system state variables such as flows, volumes and demands elsewhere in the network. Nevertheless, simulation of these secondary features is important, if not essential, in any practical implementation.

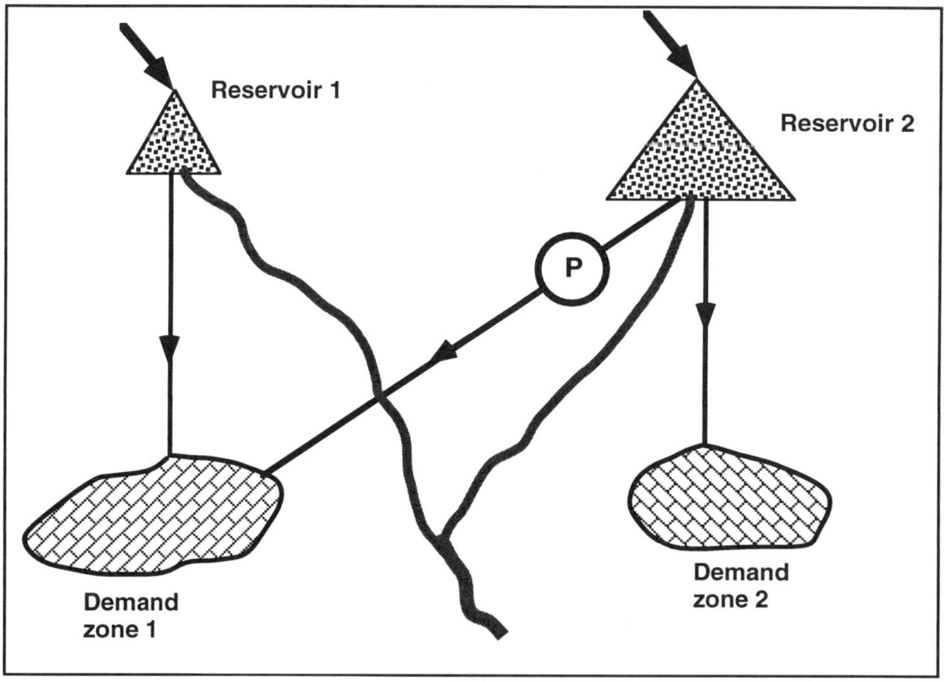

Figure 2 Sample water supply headworks system

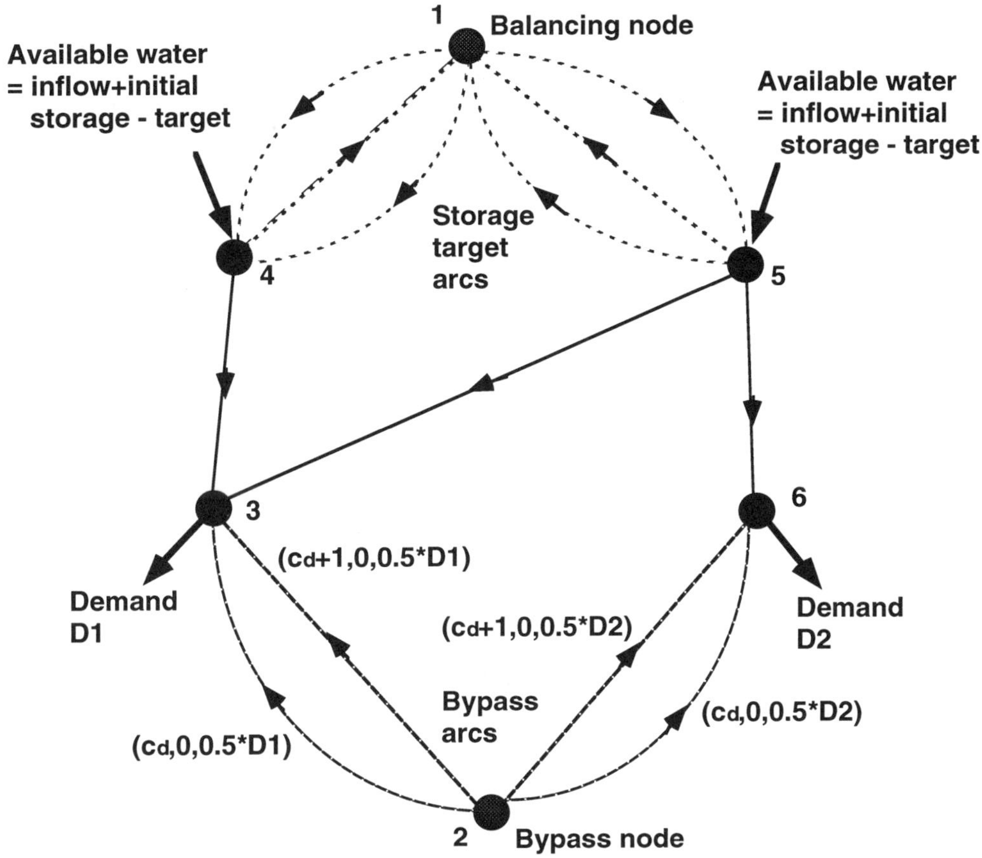

Figure 3 Directed graph of WASP NetLP for sample water supply headworks system

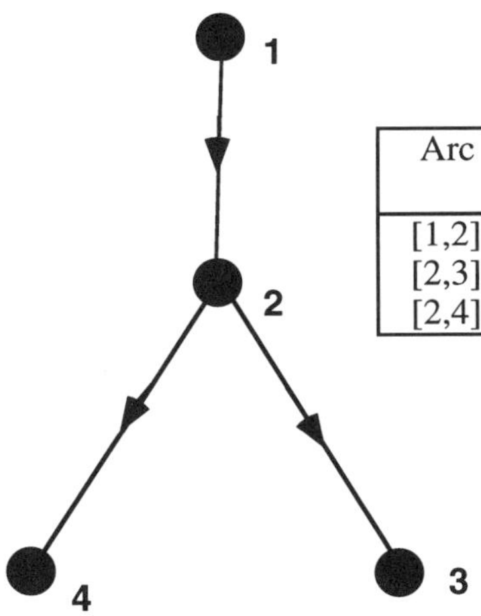

Arc	Iteration 1		Iteration 2	
	Capacity	Flow	Capacity	Flow
[1,2]	∞	100	∞	100
[2,3]	∞	23	70	70
[2,4]	∞	77	30	30

Figure 4 Illustration of flow splitting using iteration

Fortunately the very quick computation time of NetLP solvers makes it practical to use the technique of iterative convergence to find approximate solutions to problems involving constraints that cannot be directly described in the NetLP. Figure 4 illustrates the technique for a simple flow splitting problem. It is required to split the flow in arc [1,2] so that 70% flows in arc [2,3]. In the first iteration the capacities of arcs [2,3] and [2,4] are made very large. Suppose the flow in arc [1,2] is found to be 100. In the second iteration the capacities of arcs [2,3] and [2,4] are set to 70 and 30 respectively. Provided the network still wants to send 100 units to node 2 the flows will be forced to split in the desired proportions. If the flow down arc [1,2] is less than 100, the desired flow splitting may not result. In such a case further iteration would be required until satisfactory convergence occurs. Experience has shown iterative convergence greatly enhance the power of the NetLP and is essential for successful implementation of the NetLP concept. Sometimes, however, it does not work requiring careful analysis to track the cause of the problem and modifying the network.

4.4 From WASP to REALM

In 1985 the author left the Melbourne and Metropolitan Board of Works for academic pastures leaving Glen Diment to oversee continued development of WASP. Subsequently, he left to join the then Victorian Department of Water Resources which was responsible for strategic planning of Victoria's water resources. For a variety of reasons, the decision was made by the Department to rewrite WASP from scratch. Hence the birth of REALM.

Although REALM embodies the key NetLP features of WASP, its revised software architecture reflects the lessons learnt from WASP. Diment (1991) provides an overview of the REALM features, some of which are mentioned below:

i) Source code development and maintenance was simplified using a modular program architecture using ASCII files for communication. One program is responsible for editing the system file which contains a description of the system network and operating parameters. Using streamflow and demand files along with the system file, another program performs the NetLP-based simulation. ASCII output files consist of columns of data which can be readily accessed by commercial software for statistical and graphical analysis and data manipulation. In addition REALM provides a series of utilities for processing the output files, performing generic X-Y graphics and drawing a system schematic.

ii) REALM is written in standard FORTRAN 77 and uses a simple menu and graphics environment which depends on a small set of system-dependent procedures. These procedures can be readily implemented on a range of computer systems, and presently range from PCs to Unix workstations.

iii) REALM provides a tool box of nodes and arcs which the user can assemble to construct a realistic representation of the headworks system. This concept has proved successful in all applications. The ability to directly assign arc costs overriding inbuilt costs and the provision of flexible mechanisms for describing arc capacity have been central to this success.

Experience to date with REALM suggests that the generalized NetLP-based simulation concept is sound. REALM has been distributed to about 10 water supply authorities. Its applications range from urban systems to joint irrigation-urban systems such as the Goulburn irrigation scheme, to the largest of all Australian systems, the Murray river system, which is jointly operated by three states. Adams and Perera (1993), Perera et al. (1993) and James et al. (1993) document experience with REALM and modifications to the model.

5. WATHNET: YET ANOTHER GENERALIZED HEADWORKS SIMULATION MODEL

5.1 Genesis

In 1989 the author whilst on study leave at Cornell University was inspired by the work of Pete Loucks to commence development of WATHNET, another generalized headworks simulation model, primarily to be used as a teaching and research tool. Details on WATHNET can be found in Kuczera (1990, 1992). WATHNET differs from REALM in several respects:

i) Like REALM WATHNET uses a NetLP formulation to make the simulation model independent of system configuration, to free the engineer from detailed specification of operating rules, and to keep computational times within acceptable limits. However, the formulation adopted a simpler form based on the networks described by Labadie et al. (1986) and Kuczera (1989). This simpler formulation satisfactorily handles spatial balancing of reservoir volumes through the use of filling priorities and multiple target arcs and can be adapted to quasi-optimization of system operation.

ii) It is designed to fully support and perform dynamic simulation (in which the system is allowed to change over time) using synthetic streamflow and climatic data. It implements Ng and Kuczera's (1993) scheme for incorporating uncertainty in future demand as well as future streamflow. This allows future performance to be evaluated (taking into account the present state of the system), and provides a rational basis for planning future changes to the system. Dynamic simulation can only be implemented using multiple streamflow and demand replicates necessitating the use of synthetic streamflow and demand data. Static simulation in which average demand and the system network remain unchanged, is only useful for preliminary ranking of options.

iii) It makes extensive use of a mouse-based graphics interface to define the system network and operating criteria and to present results. The objective is to simplify as much as possible complex data entry and to promote better understanding of the results.

iv) It uses a simple device-dependent interface to enhance portability between computer systems, presently supporting PC and Macintosh computers, and any UNIX workstation supporting X Windows. Details about porting this interface can be found in Kuczera (1993b).

5.2 NetLP Formulation

To gain insight about the differences between the NetLPs in WASP and WATHNET, the simple example of Figure 2 is considered once again. Like WASP, WATHNET makes water assignments according to a hierarchy of five, somewhat different, objectives. The principal difference, however, arises in the way the storage target objective is implemented.

Consider Figure 5 which shows the NetLP formulation presented as a directed graph for the system shown in Figure 2. Three sub-networks are illustrated:

i) **Shortfall sub-network:** This network is functionally identical to the WASP bypass network.

ii) **Instream sub-network:** Two arcs are used to meet instream flow requirements. One arc has a negative cost c_i, second highest in absolute terms in the network, and a capacity equal to the instream requirement I. The negative cost provides an inducement to the NetLP solver to meet the instream requirement. The second arc is uncapacitated and carries any excess flow along the stream.

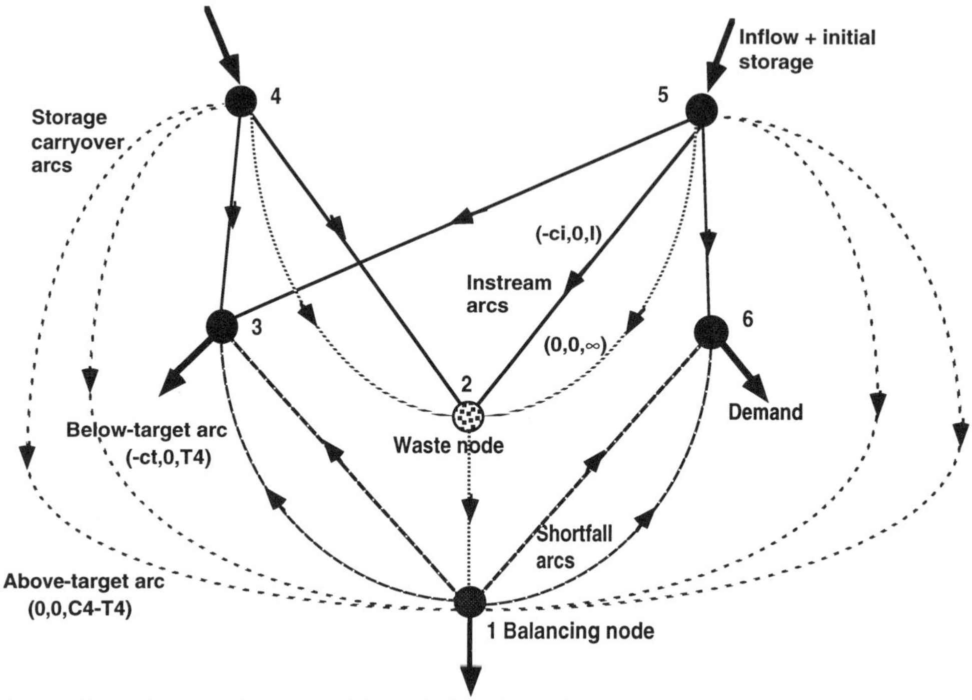

Figure 5 Directed graph of WATHNET NetLP for sample water supply headworks system

iii) **Storage carryover sub-network:** This sub-network differs markedly from that used in REALM. Carryover storage is simply defined as the flow in arcs [4,1] and [5,1]. In general, at least two carryover arcs are assigned to each reservoir. The sum of the arc capacities equals the reservoir capacity. One arc carries above-target flows and has a zero cost assigned to it. The remaining arcs have a combined capacity equal to the target volume and a negative cost, third highest in absolute terms in the network, which gives the NetLP solver an inducement to fill the arcs up to the target volume. This is illustrated in Figure 5 for the reservoir at node 4 where C_4 is its capacity and T_4 its end-of-season target volume.

This sub-network can be readily adapted to multi-period networks (Kuczera, 1989). Instead of the carryover arcs being directed to the balancing node, they are directed to the reservoir node in the next season. This provides an optimization capability similar to that offered by deterministic dynamic programming, but does not suffer from the so-called curse of dimensionality. The main limitation of this optimization formulation is that it has perfect knowledge of future streamflow and demand, and, therefore, can anticipate droughts.

5.3 Using WATHNET

Regardless of the power of the NetLP the ultimate utility of the model depends very much on the user environment which consists of the graphics interface and the tools to which it provides access. An example is used to illustrate significant aspects of the WATHNET environment.

Figure 6 presents a schematic of the WATHNET suite of programs, in which the oval shapes represent files, the rectangular shapes represent programs and the arrows denote the direction information flows. The suite consists of five programs which are briefly described below.

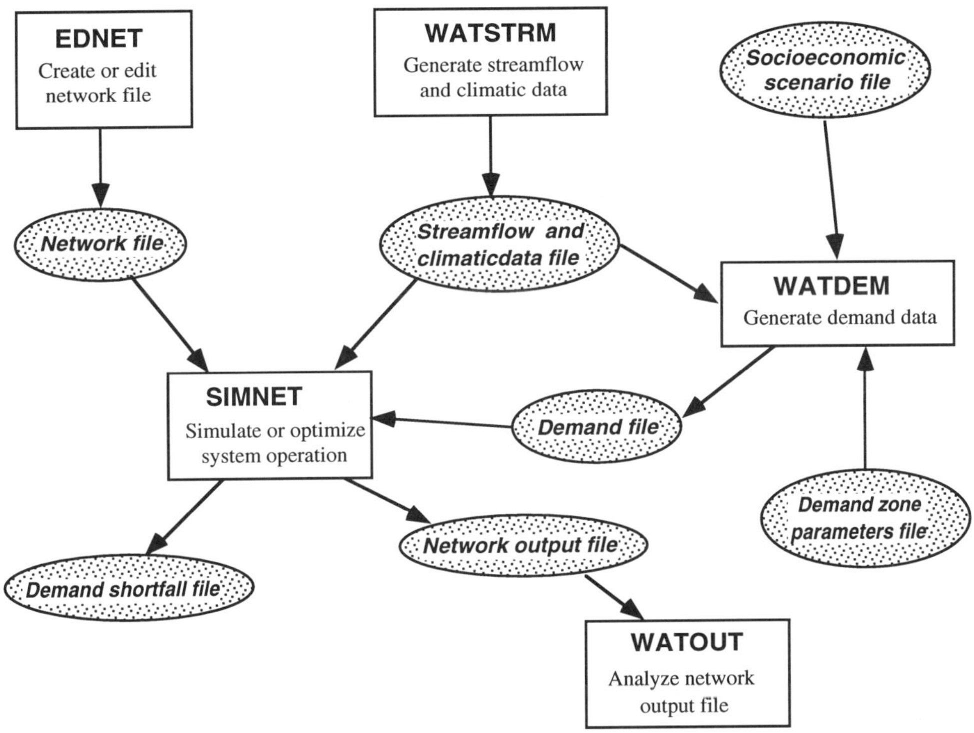

Figure 6. Schematic outlining program and file dependence in the WATHNET suite.

5.3.1 ***EDNET: Configuring the System.*** EDNET is used to interactively configure a new headworks network or edit an existing one. The operation of EDNET is best explained by an example. Consider Figure 7 which shows a schematic of a future headworks system for Newcastle located near Sydney. Figure 8 displays the WATHNET schematic for this system in which nodes are identified by the letter inside their box, solid arcs represent transfer conduits, and dashed arcs represent streams. The important point to observe is that the WATHNET schematic bears close resemblance to the system schematic with which the engineer works. This visual association is exploited when presenting the results of a simulation.

The mouse is used to draw the network schematic. New nodes or arcs are created in add mode which is selected by clicking the mouse on **Add** in the menu bar. Figure 8 shows a new arc terminating at the mouse cursor. If the mouse is clicked when its cursor is inside an existing node box, the arc is created and a window requiring data for this arc appears; otherwise the new arc is cancelled. Existing nodes can be repositioned in move mode in which the mouse is used to drag the node to a new position.

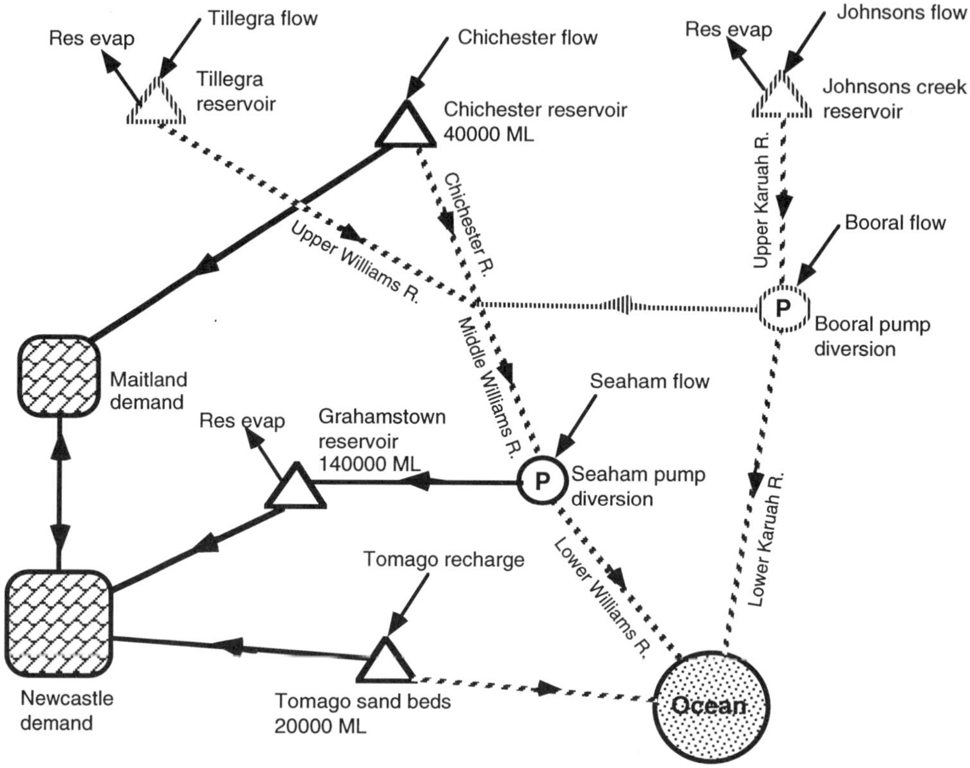

Figure 7. Schematic of Newcastle water supply headworks system

The menu bar in Figure 8 shows all the available options. Of these **Edit** requires some comment. Clicking on **Edit** puts EDNET into edit mode in which clicking on an existing arc or node brings up windows containing all information pertaining to that element. Any of this information can be edited. Of particular significance is the ability to change the network over time: Conduits, reservoirs and demand centres can be commissioned at nominated years with conduits also capable of being decommissioned.

5.3.2 WATSTRM: Managing streamflow, demand and evaporation data. Common to all headworks simulation models is the need to prepare data files containing historic streamflow, evaporation and demand data. WATSTRM was developed to help create and manipulate such files. Among its options are ones allowing files to be merged, column names to be edited, and statistical summaries generated. However, its main function is to provide a robust, virtually automatic means for generating synthetic multi-site streamflow and climatic data. After selecting appropriate data transformations, it implements Kuczera's (1987) maximum likelihood methods for the Matalas (1967) multi-site lag-one model to generate annual values which are then disaggregated into monthly values using the method of fragments first proposed by Svanidze (1960).

5.3.3 WATDEM: Generating demand data. WATDEM generates a demand data file in a format that can be read by SIMNET. It is designed to implement the simulation methodology presented by Ng and Kuczera (1993) which explicitly allows for uncertainty in future demand forecasts as well as in future streamflow and climatic inputs. WATDEM accesses three files:

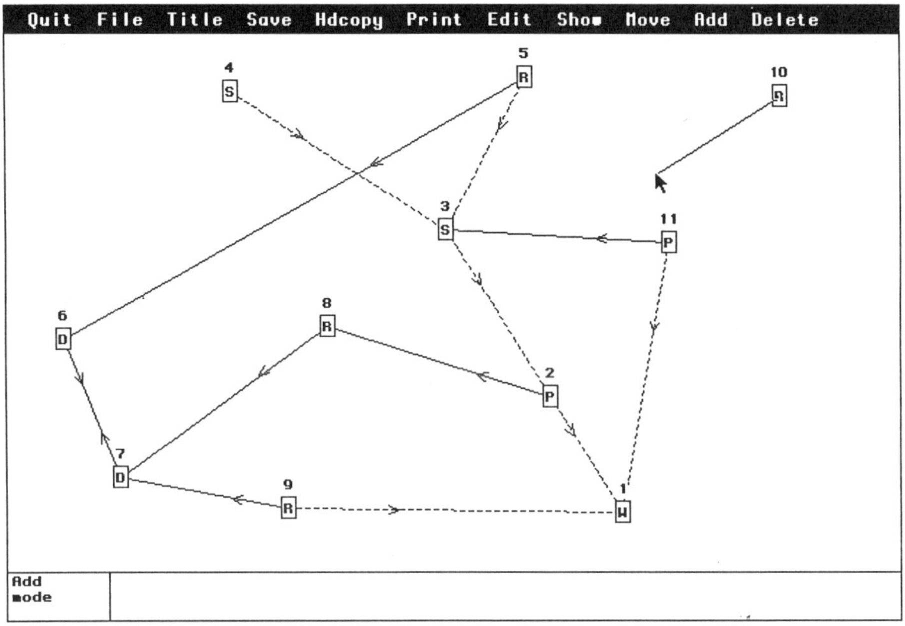

Figure 8. Creating a new arc in a headworks network

i) The streamflow and climatic data file is read to obtain climatic data that influences seasonal demand variations. This same data file is used by SIMNET to simulate system behaviour. This guarantees that the important and significant cross-correlation between streamflow and climatic variables (such as rainfall) is preserved.

ii) The socioeconomic scenario data file contains time series of the socioeconomic variables used to predict demand at each zone. The file format allows multiple scenarios for each variable. By specifying a (subjective) probability of occurrence for each scenario WATDEM will generate a demand data file that correctly allows for uncertainty in future socioeconomic scenarios. For example, future population, a major determinant of future demand, could be described by three scenarios consisting of low, medium and high growth rates. The probabilities assigned to each population scenario reflects the analyst's judgement about the likelihood of that scenario actually occurring.

iii) The demand zone parameters file contains information about the parameters to be used for each zone. This format allows uncertainty about these parameters to be explicitly expressed.

WATDEM is a shell program requiring to be linked with a subroutine called MODELS which estimates monthly and yearly zonal demand given demand zone parameters, climatic and socioeconomic time series data.. This approach has been adopted because there is no standard format for demand models.

5.3.4 SIMNET: Simulating system operation. SIMNET simulates the operation of the headworks system using a NetLP to make all the assignment decisions. It uses a menu based system to create and edit a run file which defines all the information required to formulate the NetLP. This includes, inter alia, initial conditions, run options, demand restriction rules and reservoir end-of-season target and flood control air space volumes. SIMNET produces two binary output files, one for use by WATOUT, and the other for use by custom programs needing information about demand shortfalls such as when estimating economic losses.

5.3.5 WATOUT: Analyzing simulation results. SIMNET creates an output file which stores all the simulation results. WATOUT is used to examine the contents of this file. The usefulness of WATHNET depends very much on how well WATOUT communicates the results to the engineer. Program design was based on the belief that no assumption should be made on how the engineer wants to examine the results. To this end, emphasis was put on providing the ability to browse through the results in an unstructured manner, and to present results at different levels of detail.

The use of WATOUT is best described by considering an example. For the system shown in Figure 7, WATSTRM was used to generate 500 15-year streamflow replicates with each replicate representing a sequence of flows that could occur in the next 15 years. SIMNET then simulated the operation of this system from 1990 to 2004 for each of the 500 streamflow replicates. The main purpose of the simulation was to investigate the impact of commissioning a new reservoir at Tillegra in 1992. At the start of the simulation all reservoirs were initially full. Annual demand grew from 500 ML/day in 1990 to 613 ML/day in 2004.

WATOUT provides three output option categories, two of which we shall consider. The first enables a detailed analysis for a selected replicate, while the second provides a statistical analysis of system performance:

Replicate analysis: The results for each individual replicate can be graphically studied in three ways:

i) Time series plots of selected group reservoir storage and demand shortfall which occurs when actual consumption is less than required demand, provide an overview of how the system performs. Figure 9 illustrates the window for such plots. The demand shortfalls at the end of the simulation are due to storage-triggered restrictions. At the bottom left-corner a numerical summary is provided for the time step nearest to the vertical line which tracks the mouse cursor; this feature is common to all plots produced by WATOUT.

ii) Time series and cumulative frequency plots can be viewed for any arc and reservoir or demand node. Starting with a window displaying the network schematic, plots are obtained by clicking the mouse on the desired element.

iii) The playback display, illustrated in Figure 10, provides a detailed picture of system operation for any time step. It has been found particularly useful in providing an intuitive feel for how the system is operated by the NetLP and for identifying those system characteristics adversely impacting performance. Referring to Figure 10, the leftmost box provides a global numerical summary. Demand nodes experiencing a shortfall, reservoirs and conduits below a nominated percent capacity and streams at or below their instream requirement are highlighted on the schematic. Clicking the mouse on any node or arc brings up a detailed numerical summary in the bottom box. Clicking on **Step** or **Back** in the menu bar moves respectively the playback one time step forward or backwards, while clicking on **Goto** moves it to a nominated time step.

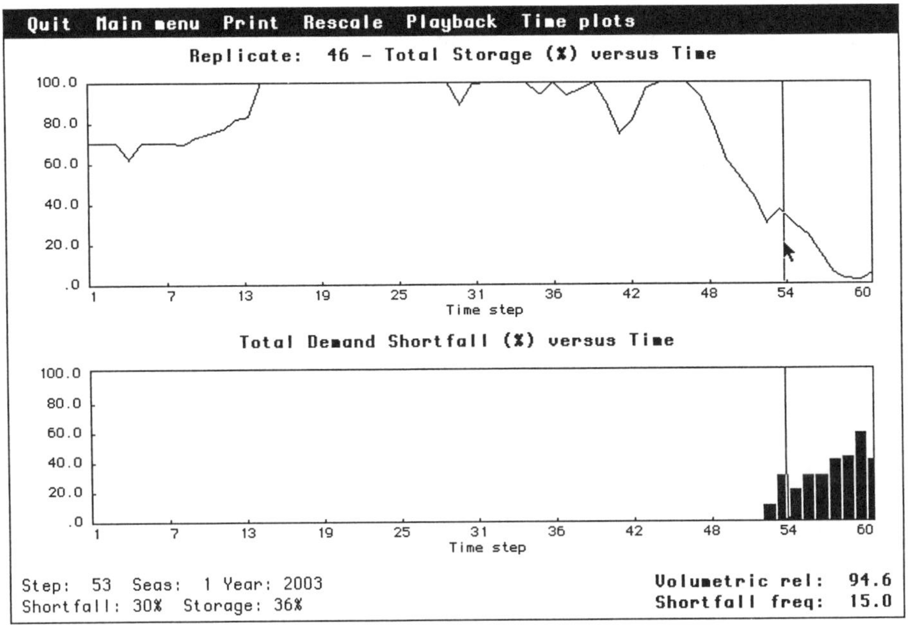

Figure 9. Time series of total system storage and demand shortfalls

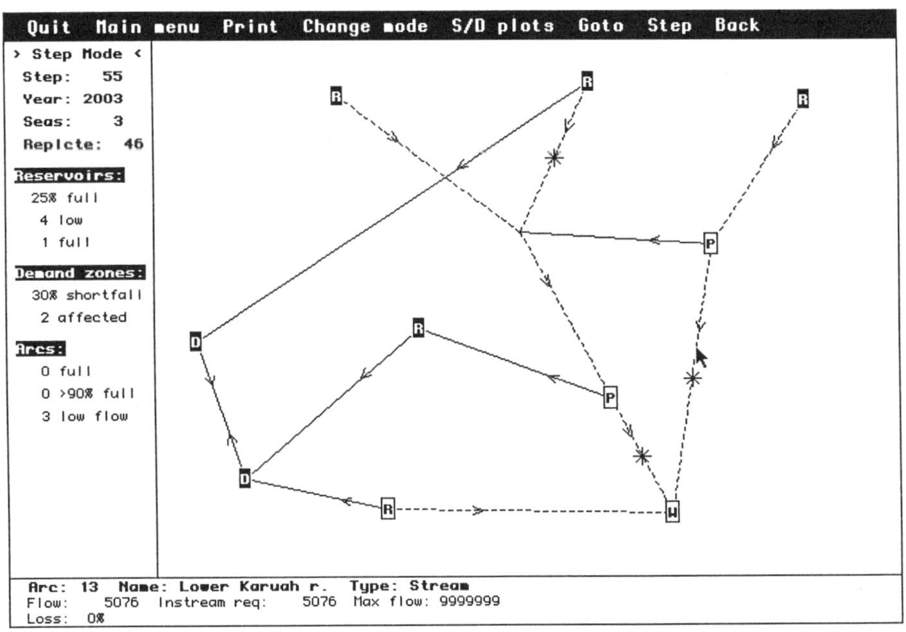

Figure 10. Playback window

Statistical performance analysis: Analysis of individual replicates, derived from either historical or generated streamflows, conveys insufficient information on how the system is expected to perform in the future. This is because future streamflows and, to a lesser extent, future demands are not known. In fact future behaviour can only be statistically described. Using the results of many simulated replicates a statistical picture of future behaviour can be constructed.

WATOUT focuses on the future behaviour of demand centres and reservoirs. For demand centres the main interest is on the temporal and spatial distribution of demand shortfalls. Such shortfalls may be due to the imposition of restrictions, capacity constraints or unavailability of water. One option in WATOUT displays a schematic of the system similar to Figure 8. Clicking on any demand node in this schematic will display one of four plots.

The first two deal with time series of annual or seasonal shortfall probabilities. Figure 11 presents a plot of expected annual shortfall probability for the Newcastle demand node. The dynamic nature of the system is evident. Because the system is initially full the shortfall probability starts at virtually zero and climbs to about 10% before the commissioning of Tillegra reservoir brings about a recovery. Observe that the numerical display associated with the line cursor includes 90% probability limits on the true shortfall probability. This information is considered essential for interpreting time series probability plots. It helps the engineer to appreciate the significance of sampling variability on any apparent trends. The only way to reduce this variability is to use more replicates.

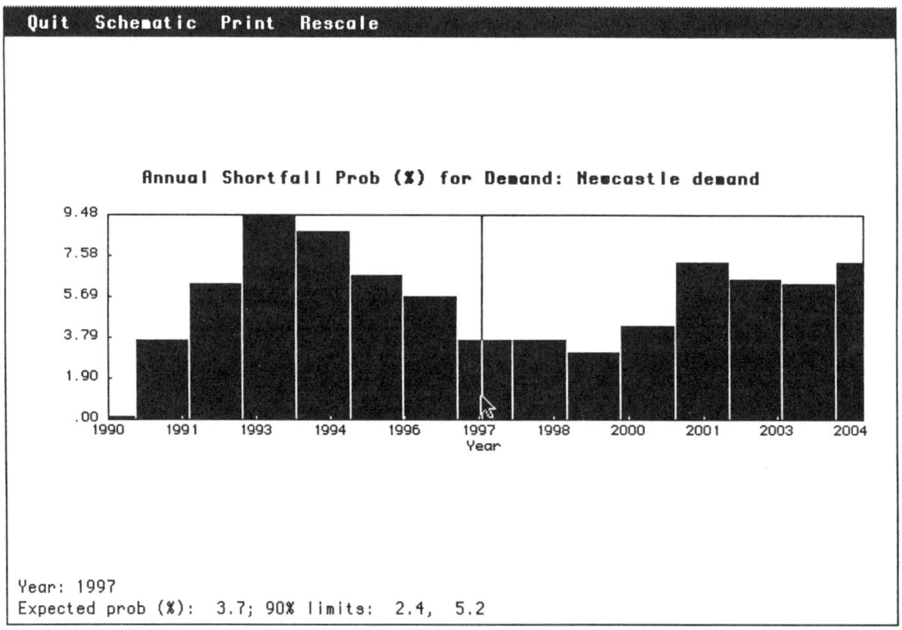

Figure 11 Shortfall probability time series for selected demand

The third plot presents information about the severity of demand shortfalls in the form of time series plots of the probability of demand shortfalls having a value less than or equal to some specified level. Hashimoto et al. (1982) have argued that knowledge of shortfall severity provides insights that reliability measures alone fail to provide. For example, consider an urban demand node for which shortfalls beyond 40% impose unacceptable social and economic costs. If the probability of shortfall is 5% for a particular season then all we know is that there is one chance in twenty of experiencing a shortfall. We know nothing about how severe such a shortfall could be. However, if, in addition, we know that there is a 3% chance that a shortfall will exceed 40%, then action would be required to reduce this risk to an acceptable value.

The fourth plot displays a time series of average and maximum shortfall for the selected demand node. For any time step the replicate experiencing the maximum shortfall is identified. This allows the engineer to study the behaviour of the system under conditions of maximum stress.

For reservoirs the main interest is on how full the reservoir is likely to be in the future. WATOUT offers an option which displays a window with the network schematic. Clicking on any reservoir node will display up to three user-selected percentile traces for reservoir volumes. Figure 12 shows such a plot for Tillegra reservoir. Observe that the traces start in 1992 when the reservoir commences to fill and that near the end of the simulation the 5-percentile trace shows a downward trend reflecting the stress due to the growing annual demand.

Another option provided by WATOUT presents scatter plots of any two selected network variables (i.e. flow, volume or demand) for all replicates over a particular range of time steps. This option is most useful when reviewing the results of a quasi-optimization run in the search of improved simulation rules.

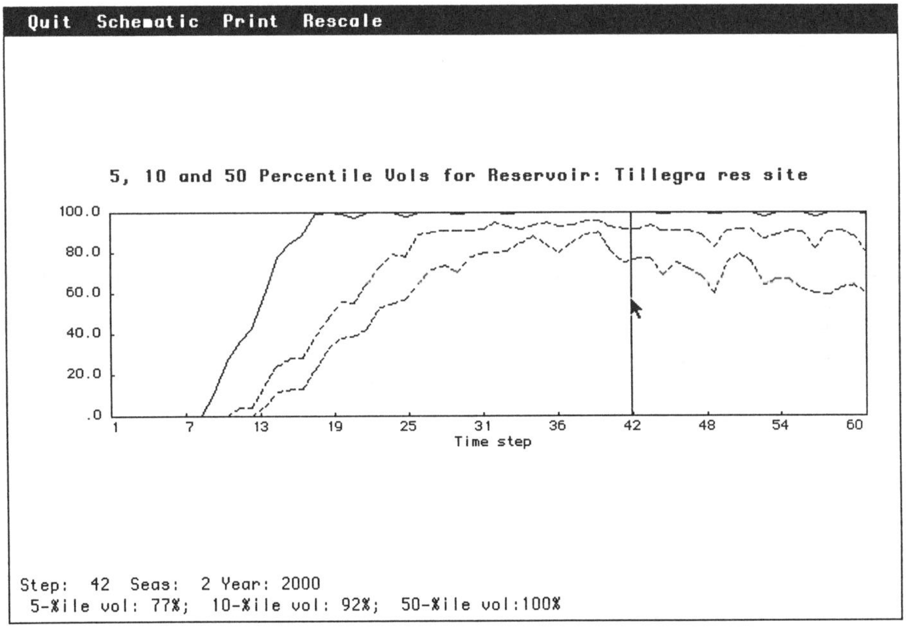

Figure 12 Storage percentiles for selected reservoir

6. CONCLUDING REMARKS

The experience with a range of Australian headworks systems ranging from urban to complex irrigation-urban networks suggests that the concept of a generalized simulation model based on network linear programming is sound and provides a powerful tool for simulating complex headworks systems. Though the REALM and WATHNET models are at an advanced stage, there is scope for further development. The development of efficient algorithms for modelling hydropower at multiple sites, more explicit consideration of water quality constraints, the potential for quasi-optimization of operating rules, and the modelling of different institutional structures present challenges for the future.

The issue of institutional structures is particularly interesting. The NetLP formulations in REALM and WATHNET simulate an institutional structure in which there is one manager. Perera et al. (1993) and Kuczera and Hill (1993) show how the tools available in REALM and WATHNET can implement reservoir and arc capacity sharing concepts which provide an institutional mechanism for divesting control and ownership of resources from a central authority to individual water users. However, the yield available to individual consumers depends partly on the behaviour of other consumers thereby requiring simulation of the full system.

There is no doubt that for complex models, well-designed computer interfaces are important. Without such interfaces the full potential of a model is not likely to be realized. REALM and WATHNET were one-man productions and, by present standards, have crude but functional interfaces. The trend in commercial technical software development is to make better use of specialist skills by employing a team of engineers and programmers. These productions can produce quite spectacular user interfaces, but it is well worth remembering the import of the aphorism 'Beauty is only skin deep'.

Most of the REALM applications have involved constructing networks which adequately mimic the existing rule-based operation of the headworks system. This typically requires imposing non-physical constraints on the NetLP to force it to mimic the existing modes of operation. Is an opportunity for more efficient operation being missed? In pure objective-driven simulation the NetLP would only be constrained by physical not institutional constraints. It would then be given maximum freedom to optimize its objectives. In principle this could lead to improved system performance but the NetLP assignments would probably be no longer capable of being translated into simple rules for operators to use. In such instances actual system operation would have to be handed over to the NetLP. The move to 'real-time' computer-controlled operation would, in most cases, require substantial institutional change. It would be fair to say that awareness of this issue is still in its infancy.

7. REFERENCES

Adams, G.A. and Perera, B.J.C., 'The development of a general water supply system simulation modelling package for multipurpose systems', *Watercomp 93,* The Institution of Engineers, Australia, Nat. Conf. Publ. 93/2, 1993.

Bertsekas, D.P., *Linear network optimization: Algorithms and codes*, M.I.IT. Press, Cambridge, Mass., 1991.

Chung, F.I., Archer,M.C. and DeVries,J., 'Network flow algorithm applied to California Aqueduct sImulation', *Journal of Water Resources Planning and Management*, American Society of Civil Engineers, 115(2), 131-147, 1989.

Collinge, R.C., 'A computer simulation model of the Melbourne water supply system', *Hydrology and Water Resources Symposium*, The Institution of Engineers, Australia, Canberra, 1978.

Diment, G.A., 'Wide use of a generalized headworks and resources model: REALM', International Hydrology and Water Resources Symposium, The Institution of Engineers, Aust., Nat. Conf. Publ. 91/22, 579-583, 1991.

Diment, G.A., Guttman, P.M., Kesari, N. and Kuczera, G., 'Case study in development of a headworks and transfer simulation model', in*Short Course in Planning of Water Resource Systems*, Department of Civil Engineering, University of Adelaide, 11-14 December, 1984.

Hashimoto, T., Loucks, D.P. and Stedinger, J.R., 'Reliability, resiliency and vulnerability criteria for water resource system performance evaluation,' *Water Resources Research*, 1982, 18(1), 14-20.

Hydrologic Engineering Center, *HEC-3 Reservoir System Analysis for Conservation, Programmer's Manual*, U.S. Army Corps of Engineers, Davis, California, 1981.

James, B., Erlanger, P.D. and Hansen, W.R., 'Modelling a complex irrigation restriction policy using a general simulation package', *Hydrology and Water Resources Symposium Newcastle,* The Institution of Engineers, Australia, Nat. Conf. Publ. 93/14, 1993.

Kennington, J.L. and Helgason, R.V., *Algorithms for Network Programming*, Wiley-Interscience, New York, 1980.

Kuczera, G., 'On maximum likelihood estimators for the multi-site lag-one streamflow model: Complete- and incomplete-data cases', *Water Resources Research*, 23(4), 641-645, 1987.

Kuczera, G., 'Fast multi-reservoir multi-period linear programming models', *Water Resources Research*, 25(2), 169-176, 1989.

Kuczera, G., WATHNET, *Generalized headworks water supply simulation using network linear programming*, Department of Civil Engineering and Surveying, University of Newcastle, NSW, 1990.

Kuczera, G., Water supply headworks simulation using network linear programming, *Advances in Engineering Software*, 14, 55-60, 1992.

Kuczera, G., 'Network linear programming codes in water supply headworks modeling', *Journal of Water Resources Planning and Management, American Society of Civil Engineers*, 119(3), 412-417, 1993a.

Kuczera, G., 'A portable FORTRAN graphics interface applied to model calibration and headworks simulation software', *Watercomp 93, The Institution of Engineers, Australia*, Nat. Conf. Pub. No. 93/2, 103-107, 1993b.

Kuczera, G. and Diment, G.A., 'A general water supply simulation model: WASP, *Journal of Water Resources Planning and Management*, American Society of Civil Engineers, 114(4), 365-382, 1988.

Kuczera, G. and Hill, I., 'Implementing capacity sharing using network linear programming, *Hydrology and Water Resources Symposium Newcastle,* The Institution of Engineers, Australia, Nat. Conf. Publ. 93/14, 1993.

Labadie, J.W., Bode, D.A., and Pineda, A.M., 'Network model for decision-support in municipal water raw water supply, *Water Resources Bulletin*, **22(6)**, 927-940, 1986.

Loucks, D.P. and Salewicz, K.A., *IRIS, An interactive Rrver system simulation program*, Department of Civil and Environmental Engineering, Cornell University, Ithaca, New York, 1989.

Matalas, N.C., 'Mathematical assessment of synthetic hydrology', *Water Resources Research*, 3(4), 937-945, 1967.

Ng, W.S. and Kuczera, G., 'Incorporating demand uncertainty in water supply headworks simulation', *Water Resources Research*, 29(2), 469-478, 1993.

Perera, B.J.C., Hansen, W.R. and Turner, G.K., 'Operational planning of a capacity shared water supply system using a general water resources simulation package', *Watercomp 93,* The Institution of Engineers, Australia, Nat. Conf. Publ. 93/2, 1993.

Svanidze, G.G., *Mathematical Modelling of Hydrologic Series* (translated from Russian), Water Resources Publications, Fort Collins, Colorado, 1960.

Yeh, W. W-G., 'Reservoir management and operation models: A state-of-the-art review', *Water Resources Research*, 21(12), 1797-1818, 1985.

Sabet,M.H. and Coe,J.Q., 'Models for water and power scheduling for the California State Water Project', *Water Resources Bulletin*, 22(4), 587-596, 1986.

Sabet,M.H. and Creel,C.L., 'Network flow modeling of Oroville Complex', *Journal of Water Resources Planning and Management*, American Society of Civil Engineers, 117(3), 1991.

OPTIMUM RESERVOIR SYSTEMS OPERATIONS, REVIEW OF STOCHASTIC MODELS

Taha B. M. J. OUARDA and Fahim ASHKAR
Faculty of Sciences, University of Moncton,
Moncton, N.B. CANADA E1A 3E9.

ABSTRACT. Although an overwhelming amount of research has been accomplished for the optimization of the operation of reservoirs, only a few of the developed techniques have been adopted for implementation in real-world systems, and the gap between theory and practice remains pronounced in this field. More effort must be directed towards closing this gap and overcoming communication problems between researchers and practitioners. The purpose of this research is to provide an updated review, classification and evaluation of the state-of-the-art of optimization of large-scale multi-reservoir systems. This research is oriented toward the stochastic environment and focuses on solution methods that account for the uncertainties related to the water system. Several models and solution procedures incorporating the stochasticity factor have been followed for the operation of reservoir systems. Such formulations include stochastic simulation, explicit and implicit stochastic optimization (Monte Carlo analysis), linear decision rules, chance constrained models, reliability programming, stochastic dynamic programming, stochastic control algorithms, or different techniques using Markov processes. Some of the models mentioned offer the advantage of keeping the "risk probability" under control, or provide the capability to assess the trade-off between risk and benefits, and hence might be more attractive. These different approaches and concepts are more detailed in this work.

Key words: Chance constrained optimization, stochastic control theory, simulation, Monte Carlo analysis, linear decision rule, reliability programming, stochastic dynamic programming, Markov processes, risk benefit trade-off.

1. Introduction

Optimal planning and operation of large multi-reservoir systems is a complex problem, with difficulties compounded by the presence of diverse and conflicting beneficial uses; such as, water supply for domestic and industrial needs, irrigation, hydroelectric production, navigation, flood control, water quality control, fish and wild life maintenance, and recreation. The basic technical difficulties associated with this optimization problem are represented by the characteristics of high

V. P. Singh and B. Kumar (eds.), Water Resources Planning and Management, 143–159.
© 1996 *Kluwer Academic Publishers. Printed in the Netherlands.*

dimensionality, stochasticity, nonconvexity and nonlinearity. These characteristics pose serious difficulties to existing optimization techniques.

A great deal of approximation and simplification has been used by researchers to fit this problem in the general structure of existing algorithms. Other researchers attempted various improvements and adaptations to available procedures to allow them to effectively handle the optimization of complex multi-reservoir systems. Recent improvements in modeling techniques, synthetic hydrology, and computational ability spurred also the development of an overwhelming number of new models and solution procedures. Although a great deal of research has been accomplished, only a few of the developed techniques have been adopted for implementation in real-world systems, and the gap between theory and practice remains pronounced in the field of water resources systems analysis. This gap is enhanced by the lack of confidence by practitioners in the capabilities of developed techniques to assess real problems, as well as by previous failures of some of these models.

2. Objectives of the study

More effort must be directed towards narrowing the gap between theory and practice, and overcoming communication problems between researchers and practitioners. Recent use of expert systems is aiding communication between users and analysts to work well as a team. Yeh (1982) states that it is extremely important and timely to review and evaluate the state-of-the-art of both theories and application development. The purpose of this research is to provide an updated review, classification and evaluation of the state-of-the-art of optimization of large-scale multi-reservoir systems. This research is oriented toward the stochastic environment and focuses on solution methods that account explicitly for the uncertainties related to the water system. Several models and solution procedures incorporating the stochasticity factor have been followed for the operation of reservoir systems. Some of these techniques offer the advantage of keeping the "risk probability" under control, or provide the capability to assess the trade-off between risk and benefits, and hence might be more attractive.

3. The Stochastic Approach

Natural hydrologic processes are typical random phenomena. Meteorological variables such as precipitation, evaporation, infiltration, and temperature, whether observed in time or space, are random variables. Morel-Seytoux (Labadie, 1978) asserts that the natural processes associated with precipitation are so complex and dependent upon so many meteorological, topographic, regional, and local variables, that it is much easier to treat rainfall as a stochastic event rather than as a long chain of deterministic events conditioned by an infinite array of multiple inputs. Reservoir inflows are mainly related to precipitation and are consequently uncertain, which significantly complicates the analysis of reservoir systems. In addition to the hydrologic uncertainty, there are other types of uncertainties, such as economic, political, or modeling uncertainties, that introduce difficulties of various types. For a reservoir system, uncertainties are not only present in future inputs , but also in future demands. The latter can only be estimated based on projected future populations and water consumption rates. Because the future is unknown, future benefits cannot be determined with certainty. This limitation in our knowledge of the future makes it impossible to guarantee that a given solution represents the optimal operating strategy, even with the use of tools of systems analysis. In fact, a supposedly optimal policy (in a probabilistic sense) can become infeasible if future events shift considerably from their expected estimations.

Deterministic models have been used extensively in the field of water resources systems operations, but more recently researchers started challenging these models by stressing their limitations in representing the real systems, giving way to stochastic models. Stochastic models have their own limitations too. In general, for the same water resources system, a stochastic formulation has a considerably larger number of variables than its deterministic counterpart, and requires higher computer memory and computational time, this, in addition to the mathematical difficulties it introduces. These limitations have restricted the success of stochastic models to relatively small problems.

It is now commonly accepted that reservoir systems operations can be decomposed into a short-range deterministic problem and a long-range stochastic problem, since uncertainties in inflow forecasts become much more significant as the horizon of the forecast gets longer. Generally, the long-term model considers the real-time reservoir operation over a time period of one year or more in usually monthly time increments. Monthly operation policies generated by this model represent inputs to the short-term model which provides daily operation policies. Both models need to be updated with new information in each time increment. The short-term model is operated daily while the long-term model is executed every month. This approach allows for some kind of "information gathering" during the decision process.

The following sections provide an overview of the methods that incorporate stochasticity of future inflows in the solution procedure for the problem of optimal operation of multi-reservoir systems. It should be noted that the lines which separate these different techniques are not always rigid and the procedures presented in the literature often represent combinations of two or more of these tools. This review does not represent a complete reference of all previous work related to the subject, and is instead restricted to covering the most prominent stochastic optimization procedures.

4. Stochastic Linear Programming and Stochastic Dynamic Programming

Linear programming (LP) and Dynamic programming (DP) are probably the most extensively used solution procedures for the management of reservoir systems. One of the powerful properties of these two techniques, is their aptitude to easily incorporate the stochastic nature of inflows in the formulation of the optimization problem. However, linear programming results in an excessive increase in the number of variables and computational requirements, and requires the linearization of the nonlinear hydroelectric generation objective function. It is also possible to solve the problem by constrained nonlinear programming approaches. These methods generally ignore the multi-stage decision structure of the problem and iteratively optimize the solution over the total number of decision variables. This approach is not as efficient as dynamic methods in dealing with constraints and requires typically a much larger effort to reach the solution. Stochastic linear and nonlinear programming methods have not therefore been directly included in this review. Nevertheless, it is important to notice that some LP-based stochastic models have been widely used by the reservoir management community, while others [such as stochastic LP for Markov processes (Manne, 1962) or stochastic programming with recourse (Dupacova, 1980)] have led to the development of more promising stochastic optimization tools. Linear Decision Rules and Chance-Constrained LP are discussed in the next section.

DP is an optimization technique particularly adapted to problems with a multi-stage decision structure such as reservoir systems operation. Along with the aptitude to include stochasticity, DP has the advantage of efficiently and easily handling such characteristics as nonlinearity, nonconvexity, and nondifferentiability which cause major problems for other optimization techniques. The primary advantage of dynamic programming is its ability to decompose complex optimization problems involving large numbers of variables into an equivalent sequence of smaller problems that must be solved recursively. This characteristic causes a linear increase in the computational burden with the number of stages in DP, as opposed to an exponential increase with the number of stages for linear and nonlinear programming techniques.

It is relevant to point out that the field of water resources management stimulated the development of the theory of dynamic programming itself by the work of Massé (1946). Another interesting fact is that this early work presented a stochastic form of dynamic programming based on the functional approach for the optimal control of a single reservoir. The general theory of dynamic programming has, however, been presented by Richard Bellman (1957) who formulated the principle of optimality as the theoretical foundation of dynamic programming. The major drawback of DP is known as the "curse of dimensionality", a term coined by Bellman (1961). It is a reference to the exponential increase in high-speed memory and computational time required by the technique with the number of state variables. Various modifications to the conventional DP technique have been presented by researchers to deal with its computational limitations.

After the work of Massé (1946), other stochastic DP-based reservoir models have been presented in the literature. Little (1955) studied the optimal operating strategy for a single reservoir on the Columbia River, taking a time lapse of two weeks and an horizon of 26 time intervals. Considering the independence assumption to be non-acceptable for river flows, Little chose the Markov assumption and determined the transition matrix for $p(I_t/I_{t-1})$ from a historical of 39 years. The optimal operating strategy computed by stochastic discrete DP was compared to the rule curves used for the operation of the hydro-generation plant, and displayed substantial potential benefits; however, the computational aspects of the model appeared to be unattractive. Gessford and Karlin (1958) chose the independence assumption for river flows and derived information about the structure of the optimal reservoir operation policies. The results obtained showed the computational advantages introduced by the independence assumption. Their work was later extended by Russell (1972). Other models using the Markov process assumption have been presented after its use by Little (1955). Howard (1960) presented the Value Iteration (VI), and the Policy Iteration (PI) variants of DP, and introduced the concept of rewards corresponding to the probability transition matrix in the Markov process. The value iteration model uses stage subscripts to denote iteration number rather than decision-stage value. The policy iteration model is a recursive solution technique especially developed for systems operated over long horizons, such as reservoir systems. The PI technique locates the optimal policy for the stationary infinite horizon problem in a finite number of iterations, but has computational limitations related to the size of the problem that has to be solved at each iteration.

Arunkumar and Yeh (1973) applied stochastic DP for the maximization of firm hydropower, and presented a decomposition approach for multi-reservoir systems, consisting of successively optimizing with respect to one reservoir while keeping a stationary policy fixed for the other reservoirs. The approach has some strong limitations related to its applicability to general configurations of reservoir systems, since it does not handle reservoirs in series or correlated river inflows. Su and Deininger (1974) considered an approach similar to Little (1955). Using historical data from Lake Superior, they divided the ranges of inflows and storages into different intervals and derived the probabilities of occurrence of the flow intervals. The operating policy obtained by their stochastic DP approach showed the potential of 15% improvement over the strategies being used then. Mawer and Thorn (1974) adopted a similar methodology but divided flows and storages into equivalent intervals.

Roefs and Guitron (1975) compared the computational efforts required by stochastic DP and Howard's policy iteration method and concluded that stochastic DP is a more efficient technique. Bogle and O'Sullivan (1979) used Stochastic DP to optimize the operation of a water supply system. They described the inflows by a piece-wise linear approximation to the probability density function, and optimized over a set of release rules rather than a set of discrete releases. Each linear rule defines the reservoir release rate as a function of reservoir storage.

Lag-1 Markov process representation fails to reproduce the long-term dependence that can be encountered in some river flow time series. The adoption of higher lag Markov processes or other "long memory" hydrologic models is restricted by computational considerations. Buchanan and Bras (1981) presented a Markov chain DP model that uses real-time observations. In this adaptive procedure the transition probabilities are updated by a multi-lag stochastic model using real-time information. The stochastic DP model is run every time step to provide the optimal strategy for the immediate following period. Can and Houck (1985) and Papageorgiou (1988) discuss the effects of forecast reliability and optimization horizon length on the performance of real-time reservoir systems operations models. Even though the authors did not use DP in their optimization models, their results are of great interest to work using forecast future inflows with a deterministic DP model.

Kelman et. al. (1990) presented a technique they named Sampling Stochastic Dynamic Programming (SSDP). In this approach, the complex characteristics of streamflows are not explicitly included in the method, but rather the complex temporal and spatial structure of the process is implicitly modeled by using a large number of historical or synthetically generated streamflow sequences. The authors applied the technique to the case study of the North Fork Feather River Hydroelectric System in California, with nine reservoirs and ten powerhouses and illustrate its usefulness. The SSDP approach was first used by Araujo and Terry (1974) in a hydrosystem study, then by Dias et.al. (1985) in a power generation and flood control study.

Even though some authors claimed that Differential DP (DDP) cannot deal with stochastic problems, Gjelsvik (1982) presented a discrete time version of the DDP algorithm that takes into account the stochastic nature of inflows to solve a five-reservoir system. The solution procedure was restricted to the case of hydraulically parallel power stations, and the approximate inflow model used was described as crude. The author suggested that the algorithm offers computer time and memory savings as compared to Linear Programming (LP) but may not be as robust. Druce (1990) presented a methodology using stochastic DP to include flood control with energy production in the operations planning of a hydroelectric complex on the Peace River in Northern British Columbia.

Saad and Turgeon (1988) combined the use of principal component analysis (PCA) and stochastic DP for the long-term operation of multi-reservoir power systems. This approach has been shown to be efficient, but requires the assumption that high correlation exists between the reservoirs trajectories. Trezos and Yeh (1987) developed an optimization model for multi-reservoir systems by extending differential DP to the case where inflows are considered to be stochastic. The methodology is an iterative procedure which transforms a stochastic operation problem with a quadratic objective function and linear dynamics into a series of quadratic programming problems at each iteration. This methodology was improved by Ouarda (1991). Other DP-based techniques, involving reliability and chance-constrained formulations or implicit stochastic approaches, have been presented by various researchers. These formulations are examined in subsequent sections of this work.

5. Stochastic Optimal Control

Stochastic Optimal Control (SOC) or Stochastic Control, is the extension of Optimal Control Theory (OCT) in presence of uncertainty. It deals with the solution of the "Optimal Control problem" when the dynamic system to be optimized is subject to disturbances and parameter variations. An optimal control problem is the problem of finding a set of control variables associated to the dynamic system, to optimize a performance function over a number of stages, subject to constraints on the state and control variables. Mathematical Programming and OCT are two optimization approaches that share many similarities. The most closely associated mathematical programming (MP) technique to OCT is DP which is particularly adapted to multistage problems. In general, MP techniques describe the system with sets of linear and nonlinear algebraic equations, while OCT involves mainly the use of ordinary and partial differential equations in continuous formulations (does not discretize the state or control spaces), which reduces the computational requirements of the method and makes it applicable to large-scale problems.

Control theory (deterministic and stochastic) has wide application in many fields for the solution of practical control problems. Unfortunately, it is still not well incorporated and utilized in the field of water resources, mainly because of the lack of knowledge of the water resources systems analysis community to the theory and the potential for its application. The problem of optimal reservoir systems operation with random inflows can be fitted into the structure of SOC. OCT and SOC have developed from Pontryagin's minimum principle (McShane, 1939; Pontryagin et al., 1962) which represents the fundamental conditions of optimality for continuous-time optimal control problems. It introduces the "Hamiltonian" function (referring to William

Hamilton, 1805-1865, who derived the principle of varying action) which is formed by adjoining to the objective function the constraints of the system multiplied by undetermined variables called "Lagrange multipliers".

After the formulation of the minimum principle by Pontryagin, many researchers worked on the extension of the theory to the stochastic case (Kushner and Schweppe, 1964, for example). within the field of water resources planning and management, the few applications have been restricted to hydrothermal energy generation scheduling. All these applications had difficulties handling state variable constraints, and employed penalty functions to replace them in the formulation of the problem. The computation effort of OCT and SOC is directly governed by the efficiency of the gradient search procedure it adopts to integrate the gradient of the Hamiltonian. This optimum search direction has been usually based on first-order gradient methods such as steepest descent, or occasionally, on more complex second-order methods such as the conjugate gradient method.

Wasimi and Kitanidis (1983) presented one of the leading applications of control theory to the stochastic optimization of reservoirs in their model for the daily operation of a multi-reservoir system in the DesMoines River basin during floods. They modeled the short-term forecasting of river flows and the dynamics of the system by a set of linear state-space differential equations and used quadratic penalty functions to control the trajectory of the reservoir storages. The optimization problem is formulated as a discrete-time Linear Quadratic Gaussian (LQG) control problem and solved analytically which minimized the computational time and memory space. The method was concluded to be efficient, flexible and very adapted to solving large dimensional optimization problems. The linear quadratic gaussian optimal control problem is treated in Stengel (1986). The combined control-estimation model developed by Loaiciga and Marino (1985) also uses a state-space formulation. Inflows are modeled as a first order autoregressive model and the objective criteria is again a quadratic penalty-type function.

Georgakakos (1983) presented an iterative solution technique using an actual-conceptual reservoir system configuration. At each iteration a LQG problem is constructed to approximate the original problem. The method allowed for nonlinear dynamics and general form objective functional, and presented an overall good performance. This work has been continued by Georgakakos and Marks (1985, 1987, 1989) who presented a relatively new method for optimal operation of multi-objective reservoir systems, and named it Extended Linear Quadratic Gaussian (ELQG) controller. ELQG extends stochastic control theory results and appears to be well-suited for the optimization of large-scale constrained dynamical systems. It is a sequential algorithm which represents the system by a set of nonlinear differential equations that describe the reservoir and river dynamics in state-space form. The method is applicable to control problems of any time scale (monthly to daily, or shorter) and handles nongaussian reservoir inputs. ELQG control seems to be a powerful solution technique for reservoir operations.

McLaughlin and Velasco (1990) presented an algorithm in which the hydropower system operation problem is simplified to fit in the framework of Linear-Quadratic Stochastic Control (LQSC). Inflow and measurement uncertainties are directly included in the formulation. The real-time optimal control approach was applied to a two-reservoir system in the Caroni River basin in Venezuela, and preliminary results seem to be successful even when the assumptions of the model are not completely met. The authors hope that the adverse effects of deficiencies in the controller (mainly that storage and release inequality constraints are ignored) will always be compensated for by the choice of a good nominal trajectory and by improvements in the more realistic hydrologic model. Ouarda (1991) developed a chance-constrained optimal control (CCOC) solution procedure and applied it to a four-reservoir hydropower system. The performance of the procedure was compared to other stochastic solution techniques. It was concluded that CCOC has the highest potential for use in the optimization of multidimentional hydropower systems. The robustness of the technique, its flexibility and its low computational requirements make it a good candidate for implementation in real world problems. CCOC controls risk probability and provides the capability to assess trade-offs between risk and benefits. Singularity arcs are known to cause convergence problems to optimal control theory, but can be avoided with judicious formulation of the objective function (Ouarda, 1991). The technique is also insensitive to initial solutions, and results in smooth policies (a characteristic of gradient-based techniques).

6. Chance-Constrained Models and Reliability Programming

Chance-Constrained formulations are widely used to include uncertainty in the optimization process, and can be considered as an extension of methods employing Linear Decision Rules. The first use of Linear Decision Rules (LDR) coupled with Chance-Constrained Linear Programming (CCLP), was by Charnes et al. (1958) in a refinery operation problem. Young (1967) was the first to apply LDR in reservoir management comparing its performance to other release rules. ReVelle et al. (1969) presented a LDR for a single reservoir design and operation and distinguished two cases: the deterministic and the stochastic case. ReVelle and Kirby (1970) modified the model to consider several objective functions and to include evaporation losses from the reservoir.

Loucks (1970) discussed LDR's and recognized them as a first stochastic modeling technique that can be extended to a multi-reservoir problem. Nayak and Arora (1971) applied a modification of ReVelle's LDR to a multiple reservoir system. Since these first modeling efforts, many authors explored the use of LDR in a variety of reservoir operation problems (LeClerc and Marks, 1973; Houck et al., 1980; Stedinger, 1984; Halliburton and Sirisena, 1984). Loucks and Dorfman (1975) have shown that LDR's lead to conservative rules when applied to reservoir operation. That is, LDR models generate operation rules which exceed the reliability limitations specified in the LDR model itself. Houck (1979) tried to remedy this conservative tendency by developing an LDR that is conditioned on streamflows in the two previous time periods. The solution lay in another approach that was being followed by other teams of researchers (Eisel, 1972; Curry et al., 1973) and which consider more general operating policies and are most often associated with the use of linear programming.

In order to provide release rules with more general structures, and to incorporate nonlinear objectives and constraints, another approach has been followed. This approach considers the use of chance-constraints but in connection with other mathematical programming techniques rather than linear programming. Askew (1974a, 1974b) presented the first models in the line of Chance-Constrained Dynamic Programming (CCDP), with objective functions including both benefits and penalties. Chance-constrained formulations have the advantage of explicitly stating the risk by selecting a priori the probability of violation of stochastic constraints.

The model presented by Askew (1974a) featured a recursive equation with a penalty for system failure. A failure is defined by the case where the system is unable to supply a target release. Since one of the principles of DP is that the decision made in a certain stage is independent of the history of the operation of the system, the model had difficulties incorporating constraints that consider probabilities which extend over many stages (e.g., probability of failures in a year, average number of failures during the life of the project). The model used an iterative approach to handle those types of constraints. Sniedovich and Davis (1975) commented on Askew's (1974a, 1974b) work and suggested that true chance-constraints can be formulated if the state variable set is augmented by introducing additional system variables representing the expected number of failures of the system up to the current stage. Askew (1975) replaced the failure penalty function by the discount factor.

Rossman (1977) used Lagrangian duality theory to indirectly incorporate the chance-constraints and solved the problem with DP. Hinks (1977) improved the work of Rossman by using a general stochastic multi-stage DP formulation to include a probabilistic "reliability" constraint that extends over the entire life of the system. Sniedovich (1980) reconsidered the problem solved by Askew (1974a) using a two-stage optimization procedure, but in a different approach termed "variance constrained reservoir control". Using a Lagrangian function to modify the problem and chance-constrained DP as a solution procedure, additional set of state variables were defined representing the total number of failures up to the current stage. This approach enabled stage to stage transfer of information regarding previous failures of the system through the DP procedure.

All the foregoing chance-constrained models have been presented for the long-range planning of a single-reservoir, and have shown some limitations for the extension to real-time control of general multi-unit systems. Moreover, these formulations require prior knowledge of the reliability (or risk) associated with the constraints. In fact, the choice of this "a priori" reliability measure is strongly based on the modelers experience and attitude towards risk, and is affected by economic considerations. To cope with these limitations, "Reliability Programming" (RP) models have been presented by different authors. RP extends the use of chance-constraints in the case where reliabilities are not prespecified but are included in the optimization model as extra decision variables to consider the trade-off between benefits and risks.

Extending Sengupta's (1972) reliability programming approach, Colorni and Fronza (1976) determined the optimal monthly contracts of a reservoir operation using a reliability program that considers a single reliability constraint. The authors made additional assumptions concerning the convexity of the benefits and risk losses to develop a computationally efficient algorithm, and suggested that the extension of the results to a multi-reservoir case or to problems characterized by more constraints is possible. The concept of risk-loss functions, for the evaluation of the value of the system failure and its probability, is one of the keys to the RP approach and was later accessed by Simonovic and Marino (1981a). Simonovic and Marino (1982) applied the RP technique to a multi-purpose reservoir system formed of three reservoirs and represented random inflows and demands by conditional distribution functions. Simonovic and Marino (1981b) presented a comparison of results based on the use of Unconditioned and Conditioned Cumulative

Distribution Functions (CDF) in a reliability program for reservoir management, and concluded that Conditioned CDF's represent a superior approach.

As noted by Marino and Mohammadi (1983), existing reliability programming models have some major drawbacks. The use of risk-loss functions, a key to the RP approach, is based on pure economic data, which in most cases are scarce and unreliable. Furthermore, the use of constant reliability levels throughout the year for floods and droughts should result in substantial inconveniences such as unnecessary releases during the summer. The third drawback is the linear assumption for the hydroelectric generation function, and the assumption that the head in reservoirs is constant (Simonovic and Marino, 1982). We know that the hydro-energy generation equation is nonconvex and nonlinear with respect to reservoir storage or reservoir release. It is our opinion that RP is a strong potential solution algorithm, but at the time being, little has been accomplished in the direction of overcoming its drawbacks.

7. Implicit Stochastic techniques

The early explicit stochastic formulations, using probability distributions of the inflows (Little, 1955; Gessford and Karlin, 1958) led to several computational difficulties, especially if serial correlation is considered in the inflow sequence. To cope with these computational disadvantages, Hall and Howell (1963) proposed the use of several inflow series with deterministic discrete dynamic programming to obtain samples of numeric operating strategies. Later Young (1967) suggested the use of regression analysis to identify the optimal operation policy. This formulation was later referred to as Implicit Stochastic Dynamic Programming, in opposition to Explicit Stochastic DP techniques which utilize inflow probabilities directly in the optimization.

In general, implicit stochastic optimization consists of three steps. In the first step, a time series model of the uncontrolled inflows is determined based on the statistical characteristics of the historical data, and a large number of synthetic equally probable streamflow sequences is generated. During the second step, the system is optimized for each inflow sequence using a deterministic optimization technique, to provide a whole set of "optimal" release trajectories. The last step involves the use of regression analysis on these optimal policies to deduce optimal reservoir operating strategies. Implicit stochastic DP, or Monte Carlo DP (MCDP), is the name given to this procedure if the deterministic optimization technique used is dynamic programming.

The foregoing approach has been used by many authors for the optimization of reservoir operation (Young, 1967; Croley, 1977; Ambrosino et al., 1979). Roefs and Bodin (1970) used the technique to define operational rules for a three reservoir system and North et al. (1977) applied it on a four reservoirs hydroelectric system in the Saguenay region of Quebec. Gal (1979) uses a similar iterative procedure by determining optimal policies for a set of synthetic streamflow sequences, which are used to iteratively compute the value of the objective function approximated by a polynomial. Bhaskar and Whitlatch (1980) applied the implicit stochastic DP on a single reservoir to derive linear and nonlinear release policies. The approach proposed by Croley (1974) and named Alternate Stochastic Optimization (ASO) is a small modification of the previous one. The procedure is based on the repeated optimization, at each time period, for a set of deterministic sequences, allowing then the use of any new information. Since results are generated at each time step for the immediate following period, only the time horizon having an effect on that period is considered.

The main idea behind the implicit stochastic approach is to reduce the computational difficulty by replacing a stochastic problem with a series of deterministic problems. In fact, this approach creates a new difficulty represented by the regression analysis which can be a laborious task, especially in the case of multi-reservoir systems, unless the linear assumption is made. Another difficulty is represented by the excessive memory and computational time requirements of the approach. However, it must be mentioned that the operational rules derived with the implicit stochastic approach are more useful for practical reservoir control purposes. These rules have also the advantage of representing closed-loop policies. Butcher (1968) mentioned that implicit stochastic techniques tend to generate unreasonable solutions in extreme hydrologic situations. Labadie et al. (1981), in a study on sewer systems control, observed that simple rules reflecting the stochasticity of the future may produce more efficient policies than the repeated optimization with deterministic forecasts.

Another implicitly stochastic approach is the Monte Carlo simulation (Hufschmidt and Fiering, 1966) which consists of generating a set of equally likely input realizations and using them in the deterministic simulation model. The statistical characteristics of the results are then analyzed. Stochastic simulation, on the other hand, considers the analysis of the input data priorly to the use of the simulation model. Neither of these techniques include any internal optimization.

The last two techniques investigated under this section are the Response Surface Method (RSM) and the Stochastic-Approximation Methodology (SAM). RSM is a heuristic approach, that is, it does not guarantee the global optimum. Response surface methodology intensively utilizes intuition and experience of the user in the solution process. It is a sequential investigation based on predicting values of the objective function for different release strategies by Monte Carlo Simulation. This generates a "response surface" which is explored by sampling or search techniques and regression analysis to arrive at the optimum or its neighbourhood following the direction of improvement (Myers, 1971; Kleijnen, 1987; Khuri and Cornell, 1987; Gates et al., 1989). RSM provides very useful information on the behaviour of the objective function, not only in the vicinity of the optimum, but also in other areas of the feasible domain. This method tends though to be computationally very intensive and might end in a local optimum instead of the global optimum. Stochastic-approximation methodology is a similar approach that reaches the vicinity of the optimum by considering a series of successive iterations. The method considers a very limited number of the random variables at each iteration to find out about the direction of the next step, and the errors are smoothed until final convergence to the vicinity of the optimum (Ermoliev, 1983; Rubinstein, 1986; Gates et al., 1989). At this stage Monte Carlo simulation can be used to provide a better estimation of the optimum solution. The Stochastic Quasigradient technique (SQG) falls under this approach.

Gates et al. (1989) applies RSM and SQG to the problem of optimal irrigation and drainage planning on a shallow saline water table. On the average, the stochastic quasigradient approximation technique required about 15% of the CPU time needed by RSM, to determine optimal solutions within 5% to 10% of the values determined by RSM. Although SQG is more efficient than RSM, it is still considered computationally very intensive, and it provides much less information about nonoptimal solutions. Gates et al. (1989) asserts that an alternative, more efficient approach consists of the use of stochastic optimization via scenario analysis and policy aggregation, but this approach is still under development (Rockafeller and Wets, 1987).

8. Conclusions

The large amount of literature available for the stochastic optimization of reservoir systems operations has been reviewed and classified into various groups. Each group corresponding to an approach that incorporates stochasticity in optimization algorithms. This is an important step towards understanding the potential of using stochastic algorithms in control of large-scale reservoir systems. A range of aggregation/disaggregation and heuristic procedures have been excluded since these techniques often rely on the special configuration of the problem in hand. This work was not intended to present a complete review of all previous work on the subject of reservoir management models and only relevant stochastic techniques have been investigated. For a more complete perspective the reader is referred to Yakowitz (1982), and Yeh (1985) whose works are considered as excellent state-of-the-art reviews, despite the need for some updating since new techniques have been presented in the last few years.

None of the models that have been investigated in this study have brought the final answer to the problem of the stochastic optimal operation of multi-reservoir systems, and at the present state-of-the-art, no universally recognized technique is available for the solution of this problem. However, the investigation leads to the selection of SOC-based techniques (such as CCOC or ELQG) as efficient procedures for the solution of multidimensional system operations. These techniques overcome some of the technical difficulties of this problem and are characterised by their robustness and relatively low computational effort. In reservoir operations, closed loop policies are usually preferred since they provide the possibility to adjust the control variables as other system variables change. This is even more crucial in short-term scheduling or real-time control. Among the techniques reviewed, implicit procedures are the only ones that can provide closed-loop control policies. These operating policies reflect considerations of inflow stochasticity and should yield satisfactory performance over a long horizon. Monte Carlo procedures are useful in providing reservoir operating rules but are limited to relatively small systems because of their large computational effort.

A great deal of additional research is still required in this area which, is still at an early stage of development. It is proposed that further research should attempt to incorporate stochastic elements in successive approximation procedures. Such techniques have been reported to yield satisfactory results in the optimization of large-scale multidimensional systems, and their extension to the stochastic framework is a natural expansion. Finally, it is necessary to point out the need for a standard hypothetical multi-reservoir testing problem. This would provide a common ground for the testing and evaluation of various solution techniques.

The water resources planning and operation community is certain today that it is possible to achieve significant benefits from an improved and more efficient operation of reservoir systems through the use of modern systems analysis techniques. It is hoped that this work will help strengthen the partnership between water resources management and systems analysis, and contribute towards closing the gap between theory and practice by narrowing the large literature available, and by providing a preliminary evaluation of solution techniques falling under completely different theories.

9. References

Ambrosino, G., G. Fronza and G. Guariso, "Real-Time Predictor Versus Synthetic Hydrology for Sequential Reservoir Management", Water Resources Research, Vol. 15, No. 4, Pages 885-890, 1979

Araujo, A.R., and L.A. Terry, "Operation of a Hydrothermal System", Braz. J. Electr. Energy, 1974

Arunkumar, S. and W.W-G. Yeh, Probabilistic Models in the Design and Operation of a Multipurpose Reservoir System, Contrib. 144, Calif. Water Resour. Cent., University of California, Davis, December,1973

Askew, A.J., "Optimum Reservoir Operating Policies and the Imposition of a Reliability Constraint", Water Resources Research, Vol. 10, No. 1, Pages 51-56, February, 1974a

Askew, A.J., "Chance-Constrained Dynamic Programming and the Optimization of Water Resources Systems", Water Resources Research, Vol. 10, No. 6, Pages 1099-1106, December, 1974b

Askew, A.J., "Use of Risk Premiums in Chance-Constrained Dynamic Programming", Water Resources Research, Vol. 11, No. 6, Pages 862-866, December, 1975

Bellman, R.E., Dynamic Programming, Princeton University Press, Princeton, N.J., 1957

Bellman, R.E., Adaptive Control Processes: A Guided Tour, Princeton University Press, Princeton, N.J., 1961

Bhaskar, N.R. and E.E. Whitlatch, Jr., "Derivation of Monthly Reservoir Release Policies", Water Resources Research, Vol. 16, No. 6, Pages 987-983, 1980

Bogle, M. and M. O'Sullivan, "Stochastic Optimization of a Water Supply System", Water Resources Research, Vol. 15, No. 4,Pages 778-786, 1979

Buchanan, R. and R. Bras,"Study of Real-Time Acaptive Closed-Loop Control for Reservoir Operation", Ralph M. Parsons Lab. for Water Res. and Hydrology., Dep. of Civil Eng., MIT., Cambridge, Mass., 1981

Butcher, W.S., "Discussions of Finding Reservoir Operating Rules by G.K. Young, Jr", Journal of Hydraulic Div., ASCE, Vol. 94, Hy6, Pages 1570-1573, 1968

Can, E.K., and M.H. Houck, "Problems with Modeling Real-Time Reservoir Operations", Journal of Water Resources Plan. and Mgmt, ASCE, Vol. 111, No. 4, Pages 367-381, October, 1985

Charnes, A., W.W. Cooper and G.H. Symonds, "Cost Horizon and Certainty Equivalents: An Approach to Stochastic Programming of Heating Oil", Management Science, Vol. 4, No. 3, Pages 235-263, April, 1958

Colorni, A. and G. Fronza, "Reservoir Management Via Reliability Programming", Water Resources Research, Vol. 12, No. 1, Pages 85-88, February, 1976

Croley, T.E., "Sequential Stochastic Optimization for Reservoir System", Journal of Hydraulic Div., ASCE, Vol. 100, HY1, Pages 201-218, 1974

Croley, T.E., Sequential Stochastic Optimization in Water Resources, Ph.D Dissertation, Colorado State University, Fort-Collins, Fall, 1977

Curry, C.L., J.C. Helm and R.A. Clark, "Chance-Constrained Model of System of Reservoirs", Journal of Hydraulic Div., ASCE, Vol. 99, HY12, Pages 2353-2361, December, 1973

Dias, N.L.C., M.V.F. Pereira, and J. Kelman, "Optimization of Flood Control and Power Generation Requirements in a Multipurpose Reservoir", Paper Presented at Symposium on Planning and Operation of Electric Energy Systems, Int. Fed. of Autom. Control, Rio de Janeiro, Brazil, 1985

Druce, J.D., "Incorporating Daily Flood Control Objectives Into a Monthly Stochastic Dynamic Programming Model for a Hydroelectric Comple", Water Resource Research, Vol. 26, No. 1, Pages 5-11, January, 1990

Dupacova, J. "Water Resources System Modeling Using Stochastic Programming Models", in Recent Results in Stochastic Programming, Kall, P. and Prekops, A. (eds.), Springer-Verlag, New York, 1980

Eisel, L.M., "Chance-Constrained Reservoir Model", Water Resources Research, Vol. 8, No. 2, Pages 339-347, 1972

Ermoliev, Y., "Stochastic Quasigradient Methods and Their Application to Systems Optimization", Stochastics, Vol. 9,Pages 1-36, 1983

Gal, S., "Optimal Management of Stochastic Multireservoir Water Supply System", Water Resources Research, Vol. 15, No. 4, Pages 737-749, 1979

Gates, T.K., R.J-B. Wets and M.E. Grismer, "Stochastic Approximation Applied to Optimal Irrigation and Drainage Planning", Journal of Irrigation and Drainage Engineering, ASCE, Vol. 115, No. 3, Pages 489-503, June, 1989

Georgakakos, A., "Optimal Operation Planning of Uncertain Multireservoir Systems", Ralph M. Parsons Lab. for Water Res. and Hydrology., Dept. of Civil Eng., MIT., Cambridge, Mass., 1983

Georgakakos, A.P. and D.H. Marks, "Real Time Control of Reservoir Systems", Rep. TR 301, Ralph M. Parsons Lab. for Water Res. and Hydrology., Dep. of Civil Eng., MIT., Cambridge, Mass., May, 1985

Georgakakos, A.P. and D.H. Marks, "A New Method for the Real-Time Operation of Reservoir Systems", Water Resources Research, Vol. 23, No. 7, Pages 1376-1390, July, 1987

Georgakakos, A.P., "Extended Linear Quadratic Gaussian Control: Further Extensions", Water Resources Research, Vol. 25, No. 2, Pages 191-201, February, 1989

Gessford, J. and S. Karlin, Optimal Policy for Hydroelectric Operations, In Studies in the Mathematical Theory of Inventory and Production, ed. by Arrow, K.J., S. Karlin and H. Scarf, Pages 179-200, Stanford University Press, Stanford, California, 1958

Gjelsvik, A., "Stochastic Seasonal Planning and Multireservoir Hydroelectric Power Systems by Differential Dynamic Programming", Model. Ident. Contr., Vol. 3, No. 3, Pages 131-149, 1982

Hall, W. and D. Howell, "The Optimization of Single-Purpose Reservoir Design with the Application of Dynamic Programming to Synthetic Hydrology Samples", J. Hydrology, 1,

Pages 355-363, 1963

Halliburton, T.S. and H.R. Sirisena, "Development of a Stochastic Optimization for Multireservoir Scheduling", IEEE Transactions on Automatic Control, Vol. AC-29, No. 1, Pages 82-84, January, 1984

Hinks, R.W., Rliability-Constrained Problems in Water Resources Management, Ph.D Dissertation, Princeton University, Ptenceton, New Jersey, Fall, 1977

Houck, M.H., "A Chance Constrained Optimization Model for Reservoir Design and Operation", Water Resources Research, Vol. 15, No. 5, Pages 1011-1016, October, 1979

Houck, M.H., J.L. Colorn and C.S. Revelle, "Linear Decision Rules in Reservoir Design and Management, 6, Incorporation of Economic Efficiency Benefits and Hydroelectric Energy Generation", Water Resources Research, Vol. 16, No. 1, Pages 196-200, February, 1980

Howard, R.A., Dynamic Programming and Markov Processes, MIT Press, Cambridge, Mass., 1960

Hufschmidt, M.M. and M.B. Fiering, "Simulation techniques for Design of Water Resources Systems", Harvard University Press, Cambridge, Mass., 1966

Kelman, J., J.R. Stedinger, L.A. Cooper, E. Hsu and S-Q. Yuan, "Sampling Stochastic Dynamic Programming Applied to Reservoir Operation", Water Resources Research, Vol. 26, No. 3, Pages 447-454, March, 1990

Khuri, A.I. and J.A. Cornell, Response Surfaces, Marcel Dekker Inc., New York, 1987

Kleijnen, J.P.C., Statistical Tools for Simulation Practitioners, Marcel Dekker Inc., New York, 1987

Kushner, H. and F. Schweppe, " A Maximum Principle for Stochastic Control Systems",J. Math. Anal. Appl., Vol. 8, Pages 287-302, 1964

Labadie, J.W., ed., Scripts for Videotape Course on: Management of Water Resources - A Systems Approach, Colorado State University, Fort-Collins, CO, 1978

Labadie, J.W., R.C. Lazaro and D.M. Morrow, "Worth of Short-Term Rainfall Forecasting for Combined Sewer Overflow Control", Water Resources Research, Vol. 17, No. 5, Pages 1489-1497, October, 1981

LeClerc, G. and D.H. Marks, "Determination of the Discharge Policy for Existing Reservoir Networks Under Differing Objectives", Water Resources Research, Vol. 9, No. 5, Pages 1155-1165, 1973

Little, J.D.C., "The Use of Storage Water in a Hydroelectric System", Operations Research, 3, Pages 187-197, 1955

Loaiciga, H.A. and M.A. Marino, "An Approach to Parameter Estimation and Stochastic Control in Water Resources With an Application to Reservoir Operation", Water Resources Research, Vol. 21, No. 11, Pages 1574-1584, November, 1985

Loucks, D.P. and P.J. Dorfman, "An Evaluation of Some Linear Decision Rules in Chance-Constrained Models for Reservoir Planning and Operation", Water Resources Research, Vol. 11, No. 6, Pages 777-782, December, 1975

Loucks, D.P., "Some Comments on Linear Decision Rules and Chance-Constraints", Water Resources Research, Water Resources Research, Vol. 6, No. 2, Pages 668-671, April, 1970

Manne, A.S., "Product Mix Alternatives: Flood Control, Electric Power and Irrigation", International Economic Review, Vol. 8, No. 1, pp. 30-54, January, 1962

Marino, M.A. and B. Mohammadi, "Reservoir Management: A Reliability Programming Approach", Water Resources Research, Vol. 19, No. 3, Pages 613-620, June, 1983

Massé, P., Les Reserves et la Regulation de L'Avenir dans la Vie Economique, Vol. I&II, Paris, Hermann and Cie., eds, 1946

Mawer, P.A. and D. Thorn, "Improved Dynamic Programming Procedures and their Practical Application to Water Resources Systems", Water Resources Research, Vol. 10, No. 2, Pages 183-190, April, 1974

McLaughlin, D. and H.L. Velasco, "Real-Time Control of a System of Large Hydropower Reservoirs", Water Resource Research, Vol.26, No. 4, Pages 623-635, April, 1990

McShane, E.J., "On Multipliers for LaGrange Problems", Amer. J. Math., LXI, Pages 809-819, 1939

Myers, R.H., Response Surface Methodology, Allyn and Bacon, Boston, Mass., 1971

Nayak, S.C. and S.R. Arora, "Optimal Capacities for a Reservoir System Using the Linear Decision Rule", Water Resources Research, Vol. 7, No. 3, Pages 485-498, 1971

North, H.A., T.E. Unny, R. Thompstone, D.K. Sen and R.J. Silver, "Long-Term Optimum Operating Policies for a Multi-Reservoir Hydrosystem", Joint National ORSA/TIMS Meeting, San Francisco, California, May, 1977

Ouarda, T. B.M.J., Stochastic Optimal Operation of Large-Scale Hydropower Systems, Ph.D. Dissertation, Colorado State University, Fort-Collins, Spring, 1991

Pagageorgiou, M., "Certainty Equivalent Open-Loop Feedback Control Applied to Multireservoir Networks", IEEE Trans. Automat. Contr., Vol. 33, No. 4, Pages 392-399, April, 1988

Pontryagin, L.S., V.G. Boltyanskii, R.V. Gamkrelidze and E.F. Mishchenko, The Mathematical Theory of Optimal Process, Interscience, New York, 1962

ReVelle, C., E. Joeres and W. Kirby, "The Linear Decision Rule in Reservoir Management and Design 1: Development of the Stochastic Model", Water Resources Research, Vol. 5, No. 4, Pages 767-777, August, 1969

ReVelle, C.and W. Kirby, "Linear Decision Rule in Reservoir Management and Design 2: Performance Optimization", Water Resouces Research, Vol. 6, No. 4, Pages 1033-1044, August, 1970

Rockafellar, R.T. and R.J-B. Wets, "Scenarios and Policy Aggregation in Optimization Under Uncertainty", Working Paper WP 87-119, Int. Inst. for Appl. Systems Analysis, Laxenburg, Austria, 1987

Roefs, T.G. and L.D. Bodin, "Multireservoir Operation Studies", Water Resources Research, Vol. 6, No. 2, Pages 410-420, April, 1970

Roefs, T.G. and R.A. Guitron, "Stochastic Reservoir Models: Relative Computational Effort", Water Resources Research, Vol. 11, No. 6, Pages 801-804, December, 1975

Rossman, L.A., "Reliability-Constrained Dynamic Programming and Randomized Release Rules in Reservoir Management", Water Resources Research, Vol. 13, No. 2, Pages 247-255, April, 1977

Rubinstein, R.Y., Monte Carlo Optimization - Simulation and Sensitivity of Queuing Networks, John Wiley and Sons, New York, 1986

Russell, C.B., "An Optimal Policy for Operating a Multipurpose Reservoir", Operations Research, Vol. 20, No. 6, Pages 1181-1189, 1972

Saad, M. and A. Turgeon, "Application of Principal Component Analysis to Long-Term Reservoir Management", Water Resources Research, Vol. 24, No. 7, Pages 907-912, July, 1988

Sengupta, J.K., Stochastic Programming; Methods and Applications, North-Holland Publishing Co., Amsterdam, 1972

Simonovic, S.P. and M.A. Marino, "Reliability Programming in Reservoir Management, 2, Risk-Loss Functions", Water Resources Research, Vol. 17, No. 4, Pages 822-826, August, 1981a

Simonovic, S.P. and M.a. Marino, "The Role of Unconditioned and Conditioned Cumulative Distribution Functions in Reservoir Reliability Programming", J. Hydrology, 52(3/4), Pages 219-228, 1981b

Simonovic, S.P. and M.A. Marino, "Reliability Programming in Reservoir Management, 3, Systems of Multipurpose Reservoirs", Water Resources Research, Vol. 18, No. 4, Pages 735-743, August, 1982

Sniedovich, M. and D.R. Davis, "Comment on "Chance-Constrained Dynamic Programming and Optimization of Water Resources Systems", by A.J. Askew", Water Resources Research, Vol. 11, No. 6, Pages 1037-1038, December, 1975

Sniedovich, M., "A Variance-Constrained Reservoir Control Problem", Water Resources Research, Vol. 16, No. 2, Pages 271-274, April, 1980

Stedinger, J.R., "The Performance of LDR Models for Preliminary Design and Reservoir Operation", Water Resources Research, Vol. 20, No. 2, Pages 215-224, February, 1984

Stengel, R.F., Stochastic Optimal Control, John Wiley and Sons, New York, 1986

Su, S.Y. and R.A. Deininger, "Modeling the Regulation of Lake Superior Under Uncertainty of Future Water Supplies", Water Resources Research, 10, Pages 11-25, 1974

Trezos, T. and W. W-G. Yeh, "Use of Stochastic Dynamic programming for Reservoir Management ", Water Resources Research, Vol. 23, No. 6, Pages 983-996, June, 1987

Wasimi, S.A. and P.K. Kitanidis, "Real-Time Forecast and Daily Operation of a Multireservoir System During Floods by Linear Quadratic Gaussian Control", Water Resources Research, Vol. 19, No. 6, Pages 1511-1522, December, 1983

Yakowitz, S., "Dynamic Programming Applications in Water Resources", Water Resources Research, Vol. 18, No. 4, Pages 673-696, August, 1982

Yeh, W.W-G., "State of the Art Review: Theories and Applications of Systems Analysis Techniques to the Optimal Management and Operation of a Resevoir System", Rep. UCLA Eng. 82-52, Engineering Foundation, Univ. of California Los Angeles, June, 1982

Yeh, W.W-G., "Reservoir Management and Operations Models: A State-of-the-Art Review", Water Resources Research, Vol. 21, No. 12, Pages 1797-1818, December, 1985

Young, G.K., "Finding Reservoir Operating Rules", Journal of Hydraulic Div., ASCE, Vol. 93, HY6, Pages 297-321, 1967

AN OPTIMIZATION MODEL FOR MULTIPURPO L MULTIRESERVOIR, SCREENING, WITH EXPLICIT SYSTEM YIELD RELIABILITY CONSIDERATION

Amit Kumar Sinha and B. V. Rao
Department of Civil Engineering
Indian Institute of Technology, Bombay

ABSTRACT

An optimization algorithm is presented extending the work of *Lall and Miller* model [1988] for selecting among and sizing potential reservoir located at sites spread over basins located in Maharastra and Gujarat, in India. The basic objective of the model is to aid the state and country water planners to develop plans to meet projected water demand. Hence the decision variables are constructed to be physically interpretable and meaningful to the decision maker. The provision of water to meet the irrigation and hydropower demands are considered. A hybrid simulation- optimization strategy is used to consider the monthly operation of the reservoir in an annual yield framework. The model has a feel of simulation model with the usual parametric simulations replaced by a powerful optimization algorithm. Reservoir storage capacities are computed functionally using a sequent trough algorithm, this scheme is compared with *Lall and Miller*'s [1988] and *Lall*'s [1993] modified sequent peak algorithm and shown that the sequent trough algorithm performs better in terms of efficiency and computational time. Optimal generator capacity is also determined using the implicit simulation model. Applications of the model developed with data from sites on the Par, Ambica, Auranga and Purna basins are presented.

V. P. Singh and B. Kumar (eds.), Water Resources Planning and Management, 161–176.
© 1996 *Kluwer Academic Publishers. Printed in the Netherlands.*

INTRODUCTION

Digital simulation models have often been used by various government agencies to aid the planning process, wherein different candidate design scenarios (with specified sizes of reservoir and hydropower generator) may be simulated to identify efficient design and operation policies. Identification of an optimal model in such a way is usually difficult and concerns regarding water quantity, water quality may lead to complicated and large optimization model. Hence a compact and computationally efficient model capable of circumventing a variety of problems associated with the conventional models is desirable. *Lall and Miller* [1988] present a compact non-linear model with such abilities. *Lall* [1993] further extends his work by considering a conjunctive use of surface and ground water and thus demonstrates the linked simulation-optimization model to be a powerful tool for screening purposes. An application of the algorithm in an entirely new context with some modifications is presented here.

MODELING CRITERIA

The objective is to minimize the cumulative storage capacity of the reservoirs in the system subject to a set of constraints. The principal constraints considered are (1) limits on the total reservoir storage capacity for each site, (2) filling each reservoir at least once after the critical period, (3) maintaining prescribed ratios of conservation storage to total storage at each site, (4) upper bound on the submergence for each site, (5) meeting irrigation demands for each zone with the specified system reliability, (6) meeting system firm energy demand.

The state agencies in India specify a 75% operational year reliability, meaning that out of 100 years of operation their should be at least 75 success years. A success year is defined as one in which the system annual yield is at least equal to the specified target annual yield for a normal year by the decision maker. A minimum acceptable release is specified for any year through the minimum release or minimum reservoir flow requirement. This lower bound is set at 50% of the target system yield for each of the years of operation.

MODEL FORMULATION

For the model presented here, the yield definitions correspond to those used by *Loucks et al* [1981, pp 339-355] for their "yield model". Provisions for conservation for recreation. flood storage and municipal and industrial yield have not been made, although they are relatively easy to incorporate. The decision variables are symbolically defined as:

D_s dead (conservation) storage at site s (reservoir site index);

I_{si} firm annual irrigation yield from site s to irrigation demand zone i;

qf_{ys} degree (fraction) of failure of the firm irrigation yield at a given site s, in the failure year y

H_s firm annual release for hydropower from site s;

a_{st} monthly distribution factor for the firm annual release for hydropower from s:

Active Storage Capacity Determination

Modified Sequent Peak Algorithm (MSP)

Lall and Miller [1988] present a modification of sequent peak algorithm wherein they incorporate evaporation recursively. Here the first estimate of the reservoir capacity is computed using the sequent peak algorithm i.e. without incorporating evaporation. The storage state for each month is defined as the difference of active storage capacity and the accumulated deficit followed by incorporation of evaporation in the usual way. The maximum of the storage state is the storage capacity for the next iteration. The recursion continues till the convergence and termination criteria are satisfied. In yet another modification to the sequent peak algorithm *Lall* [1993] exploits the fact that reservoir capacity converges within a finite segment of time defined over the operation period. After the determination of the first estimate of the storage capacity, a backsweep in time is made from m1 to m2. m2 is the month of maximum accumulated deficit (in which reservoir is last empty during the critical period) and m1 is the first month of the critical period. Since the storage at the end of month m2 is known to be zero, *Lall* argues that the storage required at the beginning of the month m2 is the difference of the release and inflow in the month m2. Now the evaporation can be computed for the month m2 as the state is known at the beginning and end of the relevant month. This evaporation thus determined is added to the storage state at beginning of m2, thereby giving the correct estimate of state at the beginning of m2. This process is continued till month (m1-1) which gives the corrected estimate of storage capacity.

Lall [1993] makes a backsweep on the critical period defined by the first estimate of storage capacity. This is significant as it means that he implicitly assumes the critical period to be invariant even with incorporation of evaporation. *Loucks et al* [1981] say that sequent peak algorithm implicitly assumes that the reservoir is full at the beginning of the operation. An observation is made at this point. Critical period is the time segment with maximum accumulated deficit occurring at m2 defined above. Hence it is imperative that the reservoir must be full at the beginning of the critical period. This is a very important observation as the implicit assumption of sequent peak algorithm to begin the operation at reservoir full status turns out to be a virtual extension of the observation. In light of these discussions the assumption associated with the sequent peak algorithm is modified by stating that the reservoir must be full at the beginning of the critical period.

Storage capacity determined using the backsweep technique is a function of the critical period. It may be noted that the sequent peak algorithm doesn't give any due consideration to length of the run of critical period. Let us consider two apparent critical segments A and B with accumulated deficits differing infinitesimally, say Δ. Let segment B be longer than segment A. Sequent peak algorithm identifies A as the "real" critical period as the accumulated deficit associated with A is larger B, though infinitesimally. Hence one determines the "correct" estimate of the storage capacity by sweeping back in time on the segment A. Let the capacity be S_a. Since A is shorter than B, it implies that number of months for which evaporation has been accounted for are less than what would have been with the segment B. Since accumulated deficit associated with segment B is only Δ less than that for segment A, hence if the evaporation accounted for the segment B is more than Δ, then clearly storage capacity determined by sweeping back on B. say S_b will be greater than S_a. This implies that now the segment B becomes the real critical period. This phenomena is described as "critical period shift". Thus it is observed that the critical period as determined in first estimate is not invariant. It is again possible that B may not be the correct critical period and one may end up finding yet another critical period C with S_c greater than S_b. Since $S_a > S_b > S_c$ for segment A,B and C, it may be concluded that storage so computed is a monotonic function of iterations defined by A, B and C. Since it is known that there exist a finite value of "correct" storage capacity, it may be stated that this scheme will always converge and terminate after a finite number of iterations. Our experience show that the scheme never took more than 3–5 iterations to converge, though the number of iterations may also be greater (depending upon the data set) but will never be greater than the number of years of operation assuming that the seasonal cycle repeats itself each year.

Occurrence of critical period shift is primarily due to the incorrect determination of critical period, and hence incorrect determination of the storage capacity. Thus even when one begins the operation of the reservoir at full level, the active storage may not be adequate for unaccounted critical period. This may force the active storage in a time period, not an element of the critical period segment identified, to take a negative value. This necessitates updating the location of critical period and estimate of storage capacity. This forces the critical period to shift towards its correct location defined over the operation period. Hence a scheme, which is efficient and computationally faster than those discussed, is desirable. Such an An algorithm that correctly nails the critical period is proposed now.

Sequent Trough Algorithm

It would have been more appropriate if one described this algorithm as Rippl's algorithm but to sustain continuity in the usage of terminology it is preferred to use this description. Application of the mass-curve technique to storage reservoir analysis was introduced by Rippl in 1883. In the system analytic jargon, the method represents a "backward–moving, forward–looking recursive sequential maximization" which is the basic mode of the dynamic programming technique. For Rippl's method can be formally written as [*Klemes*, 1979]

$$K_t = max(K_{i-1}, C_i) \quad i = 1, 2, \ldots, n \tag{1}$$

the result being given as:

$$K = K_n \tag{2}$$

where the subscript runs backward in time, C_i is the minimal fillup necessary for the ith dry season, and K_t is the minimal storage capacity necessary for the period from the beginning of the ith dry season to the end of the whole period of operation. n is an index to represent the end of the operation period.

The appeal of the Rippl's algorithm lies in the fact that what one computes here are the values of storage state, and hence is realistic in contrast to the sequent peak algorithm wherein one does the computation in terms of deficit. Thus for a correct critical period the storage states are true. An option of determining storage capacity estimate for all critical periods is considered. Such a strategy will obviate the need to conduct checks to establish if the current critical period is correct. Even when the iterative scheme computes different estimates for different iterations, the correct storage capacity is the maximum of storage states defined by these iterations. The procedure is simply extended from a finite of iterations (sufficient for iterative scheme to converge) to maximum number of iterations (worst case convergence for the iterative scheme), albeit without any check specification. Thus the max of storage states with respect to all critical periods (equal to the number of iterations in a worst case convergence for the iterative scheme) is the required minimum storage capacity adequate to satisfy the demands imposed.

Thus for a direct determination of the correct critical period, hence the correct storage. the entire process is run in a backward sweep mode. This shifts (*Lall*, e-mail communication) the sensitivity to the end of operation period rather than the beginning of the operation. This should not be a source of error as the optimal release at any time depends not only on the past inflows but also on the future ones [*Klemes*, 1979]. The statement is also consistent with spirit to the stretched thread method. Rippl assumes that the reservoir is empty at the beginning of the operation. Thus to ensure that the reservoir is full at the beginning of the critical period, the reservoir must get an opportunity to fill before the onset of critical period. To circumvent this difficulty, as their is no directly available means to examine that the reservoir will fill before the beginning of critical period, Rippl's assumption is replaced with sequent peak algorithm's assumption that the reservoir is full at the beginning of operation. This assures that the reservoir is also full at the beginning of critical period.

Thus the sequent trough algorithm turns out to be the Rippl's algorithm coupled with the assumption that the reservoir is full at the beginning of the operation. Since the critical period and storage capacity is nailed in one sweep only, this procedure is more efficient and computationally faster than its predecessors. The algorithm has been shown in Figure 2. K_st is the critical period fillup; $SA_{s,t}$ is the storage in month t, at reservoir s: $EV_{s,t}$ is the estimate of monthly evaporation for month t , at site s; $e_{s,t}$ is net evaporation rate in month t, for reservoir s, T_s is the estimated storage capacity for reservoir s, SM_s is the active storage capacity for reservoir s and f_1s is the functional relationship between reservoir surface area and storage at site s.

Reservoir Simulation

The general scenario modeled may be illustrated using Figure 1 , the configuration of the system application reported on in this paper. It is considered that the candidate reservoir sites may exist on the stream system in parallel or series and that each is intended to serve specific demand areas. Local inflows into each site are identified as the difference between the gage flow record and all relevant upstream reservoir site gages. Their are no off stream reservoirs with primary inflows from a diversion from a upstream reservoir. Irrigation releases are routed through hydroplants to the irrigation fields. For lack of data irrigation or hydropower loss coefficients have not been used. Reservoirs are numbered sequentially from upstream to downstream. This allows the algorithm to proceed

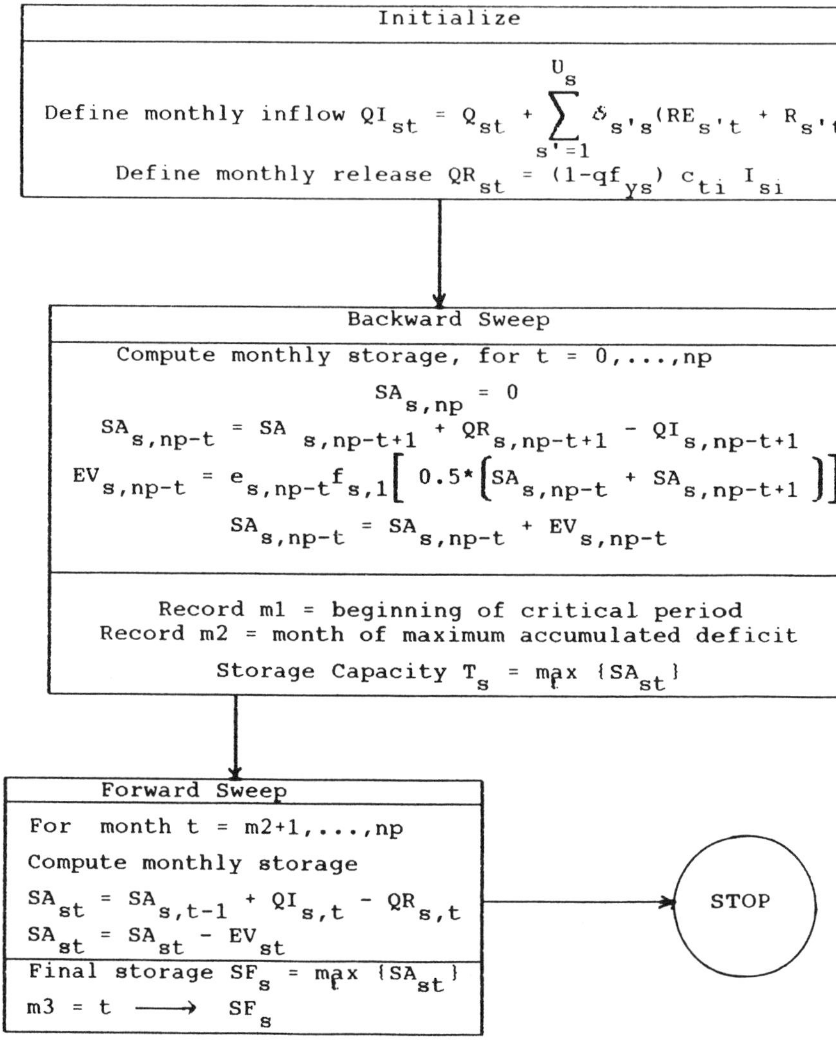

Figure 2 : Sequent Trough Algorithm

sequentially from upstream sites to downstream sites, evaluating the total release from an upstream reservoir and using it to determine the total inflow at the downstream reservoir. The model formulated thus leads to a sequential reservoir screening algorithm, with trade-offs in reservoir operation satisfied by a simultaneous examination of the reduced gradients with respect to all decision variables and constraints.

The simulation algorithm then initiates with the first reservoir and proceeds down stream. The inputs to STA are inflows and releases from the reservoirs. The symbols used are explained below:

Q_{st} local inflow into reservoir s in time period t (defined as the incremental reach gain/loss between an immediately upstream reservoir s' and reservoir;

U_s number of reservoirs upstream of s;

$\delta_{s,s'}$ fraction of the upstream releases reaching a reservoir s from a reservoir s';

RE_{st} spill or excess release from reservoir s in month t;

I total number of irrigation demand areas;

c_{ti} monthly demand fraction (monthly demand divided by annual demand);

e_{st} evaporation rate per unit area;

A_{st} surface area of reservoir s in month t;

Release is the sum of irrigation and hydro release. c_{ti} is pre-specified for all the demand zones. H_s, I_{si} and qf_{ys} are known at the current iteration of optimization model. Once these parameters are known, computation of release is trivial. The optimal hydropower generator size is determined by first computing the total release possible through the hydropower plant at site s in month t and then the corresponding possible power production treated as candidate generator sizes,and the one with highest net annual revenue for hydropower generation is selected as the optimal generator size. This is done at the completion of sequent trough procedure. The linked simulation-optimization formulation developed is generically illustrated in Figure 3.

OBJECTIVE FUNCTION

The objective is to minimize the cumulative storage capacity of the reservoirs in the system for meeting a specified target firm yield TY for the demand area.

$$MinZ \quad = \quad \sum_{s=1}^{S} T_s \tag{3}$$

CONSTRAINTS

Total Storage Capacity (TSC)

$$T_{min} \leq T_s \leq T_{max} \quad s = 1, \ldots, S \tag{4}$$

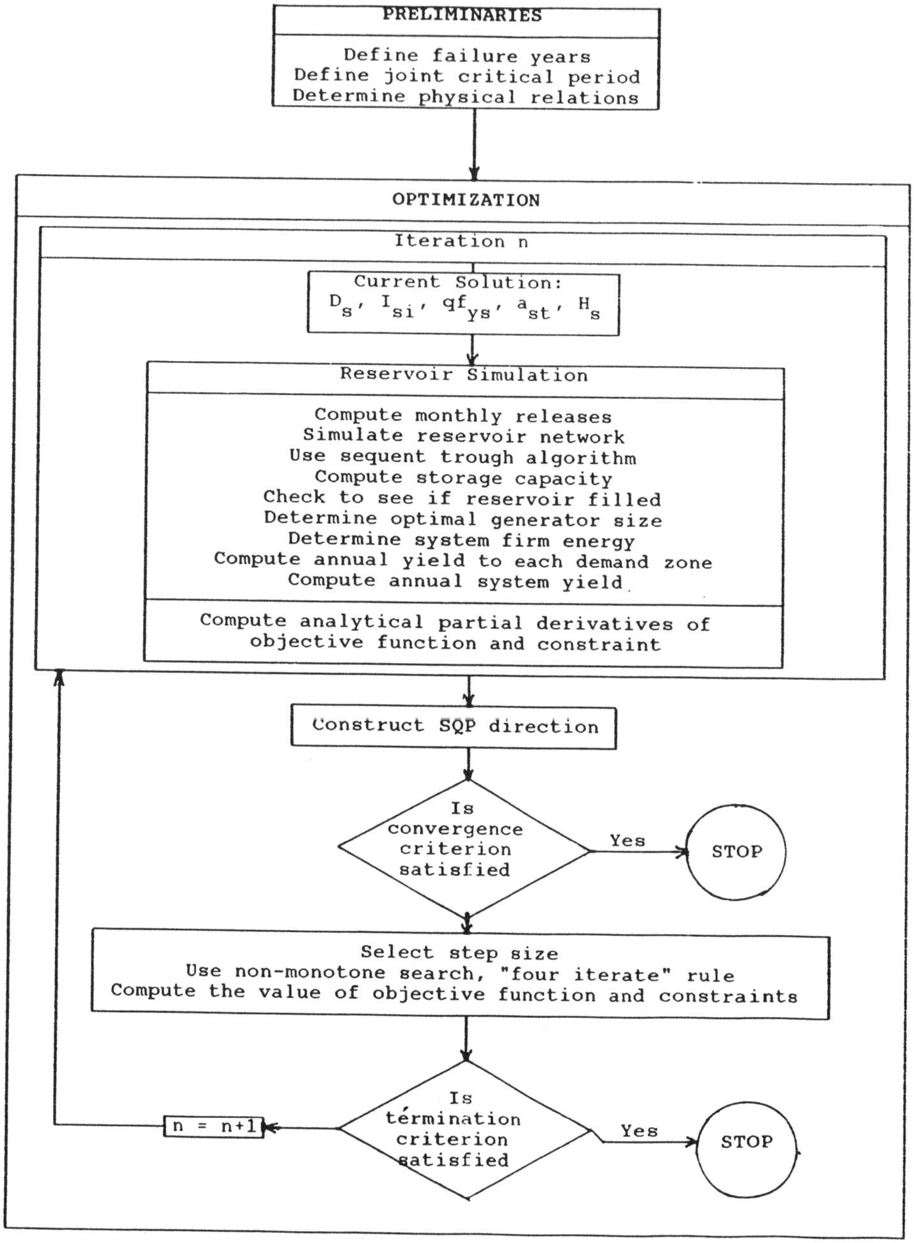

Figure 3: Generic representation of linked simulation-optimization model

The total reservoir capacity at site s should lie between some specified upper (T_{max}) and lower (T_{min}) bounds. T_{max} recognizes the physical limitation on size and also can be assigned a value that defines the upper limit on submergence. T_{min} is the minimum storage. If the system is being designed fresh then it may be arbitrarily set to zero else if the system is an extension to an existing one, with the provision that height of the dam can be increased in future, it is set to the existing storage level.

Reservoir Refill (RRC)

$$SF_s - T_s \geq 0 \quad s = 1, \ldots, S \tag{5}$$

Here S is the total number of reservoir in the system. Each reservoir must fill at least once during the simulation subsequent to a reservoir empty condition. Most of the conventional models never impose such kind of constraint. Since such a check is missing. there are quite a few reservoirs that never filled up. *Lall* (e-mail communication) describes it as "blindness" i.e. writing a matrix representation of the problem in LP or other form without ever explicitly checking to see if the reservoir is capable of filling at least once in the cycle. Hence one may view it as a "hard" constraint, which must be satisfied by the optimization algorithm.

Dead Storage (DSC)

$$DR_{s,min} \leq \frac{D_s}{T_s} \leq DR_{s,max} \quad s = 1, \ldots, S \tag{6}$$

The dead storage must fall within prescribed bounds. A lower bound is set on storage necessary to take care of silting, sedimentation etc. The values are specified by NWDA. New Delhi [1991].

System Reliability (QFC)

As discussed before, the state agency in India prescribe that 75% of the operation years should be success years. A "min" function approach is used to take care of this kind of reliability. Since at a given iteration of the optimization algorithm the values of decision variables are known, the system can be simulated to compute the system annual yield SAY_y. The set of annual system yield is then converted into an ordered set SAY_O such that $SAY_1 \leq SAY_2 \leq \ldots \leq SAY_T$. Let NF be the maximum number of years that can be failed to achieve a system reliability, as defined, of 75%. The years are selected, corresponding to $SAY_1, SAY_2, \ldots, SAY_{NF}$, as the potential failure years and apply a lower bound to the annual yield for these years (eq. 7). These are set at 50% of the firm target yield. The other years are success years and to them lower bound at the firm target yield is applicable (eq 8). This constraint may be expressed as,

$$(1, 2, \ldots, NF) min_y \{SAY_y\} \geq DY_{1,2,\ldots,NF} \tag{7}$$

$$(NF + 1, \ldots, T) SAY_y \geq TY \tag{8}$$

DY_y is the lower bound set for the annual yields in year from 1 to NF in the ordered set. TY is the target annual yield set for years $NF+1$ to NY in the ordered set defined above. Our experiments with this formulation revealed that the "min" function was well behaved and continuous in terms of decision variables.

Total Irrigation Demands (DMC)

$$TI_{i,min} \leq \sum_{s=1}^{S} I_{si} \leq TI_{i,max} \quad i = 1,\ldots,I \tag{9}$$

The irrigation demands for each demand area i must be satisfied at the specified reliability. $TI_{i,min}$ and $TI_{i,max}$ are the minimum and maximum requirements of irrigation water for demand area i.

Firm Energy (FEC)

Firm energy is also computed as part of reservoir simulation. This is done at the completion of determination of optimal size of generator at all sites. Let potential power produced in each month (i.e. without any consideration for

generator size) be P_{st} for $t = 1,\ldots,T$. Then the actual power produced, with an optimal generator size, is given by,

$$P_{st} = min\{P_{st}, P_{so^*}\} \tag{10}$$

P_{so^*} is the optimal generator capacity at a given iteration of the optimization model. After the computation of power with the optimal generator capacity at each site, system firm energy is defined as

$$FE = \frac{min \quad PS_t}{max \quad d_t} \tag{11}$$

where PS_t is the total power produced by the system in each month expressed as

$$PS_t = \sum_{s=1}^{S} P_{st} \tag{12}$$

d_t is the energy demand distribution factor for hydroenergy demand zone. Only one such zone is considered, although it is also possible to consider multiple power demand zones with different demand patterns. The system firm energy thus computed is subject to a lower bound set at the target firm energy (TFE),

$$FE \geq TFE \tag{13}$$

MODEL SOLUTION

Feasible Sequential Quadratic Programming (FSQP) algorithm of *Zhou* and Tits [1993] is used to solve the non-linear problem. The algorithm proceeds by forming a penalty function and then minimizing it subject to linear and non-linear inequality constraint. A quadratic subproblem is solved to establish a SQP direction feasible at Kuhn-Tucker points. It is followed by the construction of a local feasible direction and at this trial point objective function is tested and feasibility checked. If this direction is not accepted then a global descent direction is constructed and non-monotone search is performed. The step size is computed to satisfy the "four iterate" rule. A new symmetric positive approximation to the Hessian of the Lagrangian is computed which is used to compute the SQP direction at next iteration. The process continues till convergence or termination criteria is satisfied.

The model formulated has $(S + NI + S * NY + S + 12 * NS = S(14 + NY) + NI)$ number of decision variables (`nparam`) and $(S+S+S+NY+I = 3S+NY+I)$ number of constraints (`ncnstr`). Here I is the total number of arcs that define flow from each reservoir to each demand zone. The formulation results in 367 decision variables and 60 constraints. All the constraints are determined by reservoir simulation using the values of decision variables at the end of each iteration of optimization model. By comparison the linear programming based yield model of *Loucks et al* [1981, pp.368-371] would lead to greater than 18000 decision variables and 3000 constraints for a comparable problem. The actual number of variables an constraints in the linear programming formulation would be even greater than indicated here, once piecewise linearization to accommodate the non-linear functions is implemented.

Computation of Analytical Derivative

It is significant to observe here that computation of the derivatives proceeds sequentially exactly similar to the reservoir simulation. Indeed they are also obtained as part of simulation. *Lall* [1993] also presents expression to compute gradient but he neglects evaporation. This scheme was tried and compared with those incorporating evaporation and concluded that the error is too large to be neglected. Hence gradient computation with evaporation incorporation is desirable. The release can be expressed as:

$$QR_{st} = \sum_{s=1}^{S}(1 - qf_{ys}) * c_{it} * I_{si} + a_{st} * H_s \tag{14}$$

$$\frac{\partial QR_{st}}{\partial I_{si}} = (1 - qf_{ys}) * c_{it} \tag{15}$$

$$\frac{\partial QR_{st}}{\partial qf_{ys}} = -\sum_{s=1}^{S}(I_{si} * c_{it}) \tag{16}$$

$$\frac{\partial QR_{st}}{\partial H_s} = a_{st} \tag{17}$$

$$\frac{\partial QR_{st}}{\partial a_{st}} = H_s \tag{18}$$

Since inflow is a constant its derivative is zero. The following expressions are given without derivation (It is relatively easy to derive it, interested reader may do it) to compute the derivative of storage status (SB_{st}) during the backsweep spanning critical period.

$$SB_{s,t} = SB_{s,t+1} + QR_{st} - SP_{u/s,t} - Q_{st} \tag{19}$$

$$\frac{\partial SB_{s,t}}{\partial x} = UB * \frac{\partial SB_{s,t+1}}{\partial x} + VB * \frac{\partial QR_{s,t+1}}{\partial x} + XB * \frac{\partial SP_{u/s,t+1}}{\partial x} + YB * \frac{\partial D_s}{\partial x} \tag{20}$$

where,

$$UB = 1.0 + CE * (A + B) \tag{21}$$

$$VB = 1.0 + CE * B \tag{22}$$

$$XB = -1.0 - CE * B \tag{23}$$

$$YB = CE * (A + B) \tag{24}$$

$$CE = 0.5 * 10^{-5} * e_{st} \tag{25}$$

$$A = C_s E_s (SB_{s,t+1} + D_s)^{E_s - 1} \tag{26}$$

$$B = C_{1s} E_{1s} (SB_{s,t+1} + QR_{s,t+1} - Q_{s,t+1} - SP_{u/t+1} + D_s)^{E_{1s} - 1} \tag{27}$$

C_{1s} is the coefficient and E_{1s} is the exponent of storage area relation for a given reservoir s. Now the algorithm for gradient of storage capacity for each reservoir is readily developed as,

For s = $1, \ldots, S$

Sweep backwards for t = $m2_{s-1}, m1_{s-1}$

For x = $1, \ldots,$ nparam

$$SB_{s,t} := SB_{s,t+1} + QR_{st} - SP_{u/s,t} - Q_{st} \tag{28}$$

The gradient of storage capacity SM_s is expressed as

$$\frac{\partial SM_s}{\partial x} := \frac{\partial SB_{s,m1-1}}{\partial x} \tag{29}$$

A similar expression can be written for computation of gradient of correct storage state for each reservoir after the completion of computation of gradient of storage capacity for each reservoir. The storage state is determined by sweeping forward in time from $1, \ldots, T$.

$$SA_{s,t} = SA_{s,t-1} - QR_{st} + SP_{u/s,t} + Q_{st} \tag{30}$$

$$\frac{\partial SA_{s,t}}{\partial x} = UA * \frac{\partial SA_{s,t-1}}{\partial x} + VA * \frac{\partial QR_{s,t}}{\partial x} + XA * \frac{\partial SP_{u/s,t}}{\partial x} + YA * \frac{\partial D_s}{\partial x} \tag{31}$$

where,

$$UA = 1.0 - CE * (A' + B') \quad (32)$$

$$VA = -1.0 - CE * B' \quad (33)$$

$$XA = 1.0 - CE * B' \quad (34)$$

$$YA = -CE * (A' + B') \quad (35)$$

$$A' = C_s E_s (SA_{s,t-1} + D_s)^{E_s - 1} \quad (36)$$

$$B' = C_s E_s (SA_{s,t-1} - QR_{s,t} + Q_{s,t} + SP_{u/s,t} + D_s)^{E_s - 1} \quad (37)$$

Gradient of spill is computed in a similar way. Power is a function of storage state and the head available. Computation of gradient of power follows the same procedure as that of storage state except the fact that here the functional relation between storage and head is used. As before this computation is done sequentially as defined by the computation of power in reservoir simulation.

In summary, the non-linear optimization formulation is solved using FSQP, with non-linear functions evaluated via simulation, and with the derivatives of these functions evaluated as part of the simulation. The mathematical structure of the model is largely decoupled and is suitable for parallel computation (*Lall*, 1993)

MODEL APPLICATION AND DISCUSSION

The general configuration of the system is shown in Figure 1. Decisions for failure are applied to all the years. The results of application are shown in the Table-1. All the reservoirs are constructed and none of them are screened out. All the reservoirs have significant capacity with respect to the annual inflows, signifying efficient utilization of the resources with respect to the target demand and associated reliability. All the reservoirs are within bounds prescribed for their capacity. The model chooses to fail year 2,8,11,12,13,15,23 and 25 producing a system reliability of 76.47%. Year 25 produces minimum annual yield. The generator size at Mohankavchali reservoir is relatively small even though the storage capacity is large, as the head available is small. This is due to a very high dead storage provided at this site to ensure a gravity flow in the off take link canal. Paikhed reservoir has the largest generator as the head available for power production is largest. Generators at other sites also show a similar behavior. This is expected also since same economic parameters associated with hydropower generation are used. Significantly, firm annual hydro release is zero at the optimal solution. It implies that irrigation releases with the specified demand fractions at irrigation zones were sufficient to satisfy the system firm energy constraint. Hence no exclusive release for power generation to satisfy the FE constraint is required. This also keeps the dead storage at its lower bound. Possibly more power development is possible in the system. At a greater value of lower bound prescribed to FE, the dead storage is likely to increase to provide larger head. No experimentations have been conducted with it. Figure 4 shows the system annual yield for each year of operation. Target level is shown by a dashed line. The failure years are invariably under the target level and success years above or at the target level defined. As can be observed the inflow

Table 1: Summary Results for Optimal Solution of the Reservoir System

	Reservoir							
Category	Jheri	Mohan	Paikhed	Chasma	Chikkr	Dabdar	Kelwan	System
DS	015.00	192.00	011.00	003.00	012.00	017.00	027.00	277.00
AS	160.30	216.33	195.84	068.67	103.75	153.09	184.01	1081.9
TS	175.30	408.33	206.84	071.67	115.75	170.09	211.01	1358.9
IY	270.75	288.90	307.12	118.77	161.17	250.21	297.65	1694.6
HR	000.00	000.00	000.00	000.00	000.00	000.00	000.00	000.00
GS	7.8058	1.9595	10.212	2.1517	3.7327	5.6801	6.0648	37.606
FE								13699

DS Dead Storage in MCM

AS Active Storage in MCM

TS Total Storage in MCM

IY Irrigation Yield in MCM

HR Hydro Release in MCM

GS Generator Capacity in MW

FE System Firm Energy in MWH

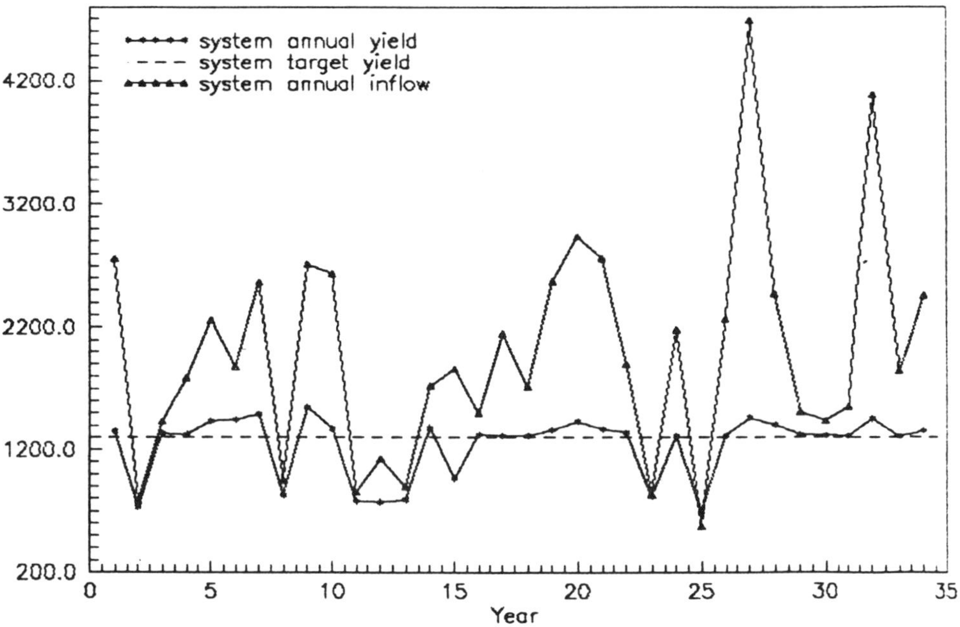

Fig. 4: Surface Water Yield for the system.

data is highly skewed. This gives an impression that a large volume of water is not being utilized. Probably a higher system reliability specification will result in better utilization of water. But it is likely to blowup the storage capacities of reservoirs as the inflows are highly erratic. The inflow reaches larger value only in some years and to satisfy a stronger reliability condition a high carry over storage may be required.

Following extensions have been made to *Lall's* [1993] algorithm. The modified sequent peak algorithm is replaced by a computationally faster and efficient procedure called sequent trough procedure. System firm energy demand is considered as constraint and finally gradient computation is done analytically with evaporation incorporation. The model has been applied in an entirely new context with different kind of reliability prescribed by the state agencies in India. As the reader might have noticed the model has feel of simulation that makes complicated policies and constraints to be incorporated relatively easily. The decision variables are directly interpretable to the decision makers. It is expected that such a model will further its acceptance to practitioners and aid in planning process.

ACKNOWLEDGEMENT

We are indebted to Dr. Upmanu Lall, Associate Member, ASCE, Assoc. Professor. UWRL and Department of Civil & Environmental Engineering, Utah State University. for a stimulating discussion throughout the span of work. His criticisms and comments are highly appreciated. We are extremely thankful to Prof. A. L. Tits and Prof. J. Zhou, System Research Centre, University of Maryland, with whom the communications regarding implementation of the FSQP algorithm were most fruitful.

REFERENCES

Klemes, V., Storage Mass-Curve Analysis in a System Analytic Perspective, Water Resources Research, Vol. 15, No. 2, pp. 359-370, April 1979.

Lall, U. and C. W. Miller, An Optimization Model for Screening Multipurpose Reservoir Systems, Water Resources Research, Vol. 24, No. 7, pp.953-968, July 1988.

Lall, U., Pre-Print, A Yield Model for Screening Surface and Ground Water Development, To Appear in ASCE Journal of Water Resources Planning and Management 1993.

Loucks, D. P., J. R. Stedinger and D. A. Haith, Water Resource Systems Analysis. Prentice Hall, Englewood Cliffs, N. J., 1981.

National Water Development Agency (NWDA), New Delhi, Pre-Feasibility Report on Par-Tapi-Narmada Link Analysis, 1991.

Zhou, J. L. and A. L. Tits, Nonmonotone Line Search for Minimax Problems. Pre-Print, TR 91-65, System Research Centre, University of Maryland, College Park, To Appear in Mathematical Programming 1993.

OPTIMAL OPERATING RULE FOR RESERVOIR WITH HYDROPOWER GENERATION AND IRRIGATION

U.Ratnayake and R.Harboe
Division of Water Resources Engineering,
Asian Institute of Technology,
Bangkok, Thailand.

ABSTRACT: A methodology is suggested to optimize on-peak firm energy generation from a reservoir while satisfying downstream water requirements for irrigation and other constraints. The model can be used in the case where the power grid is also fed by other sources of energy like thermal energy. An objective function which maximizes the minimum moving summation of on-peak energy generation within twelve months period is used to determine the most critical period. The optimal monthly generation coefficients as applied to the reservoir are determined using this twelve month period. The model uses the monthly historical inflow and irrigation demand series. A stochastic model to maximize the on-peak firm energy is then used to determine the annual target according to the above energy generation distribution. The results showed this model can improve the firm energy generation from a multipurpose reservoir and in turn reduce the on-peak energy generation cost. However the application require additional thermal power generation capacity. The model is applied to Sirikit Reservoir on the Chao Phraya river basin in Thailand as an example.

1. Introduction

Most of the reservoirs are usually constructed to have more than one purposes in order to be economically efficient and also to satisfy environmental and social objectives. Two of the most common purposes are the hydropower generation and water supply for irrigation. Both these purposes demand water supply of very high reliability. Therefore, in order to achieve the maximum possible returns with such reliability, it is very important to operate the reservoir system at the highest efficiency it can reach. This paper presents a model which can be used to optimize a single reservoir which mainly serves purposes like irrigation, hydropower generation, flood control, mandatory releases for navigation and salt water intrusion control.

Generally the purposes of a reservoir are in conflict with each other. For hydropower generation and irrigation purposes the water demands from a reservoir are usually not in phase. It is, therefore required to know the amount of energy generation in a month in such way as not to cause irrigation failures in the future. Such questions can be answered by the use of optimization models. One of the techniques used in optimization models is Stochastic Dynamic Programming

V. P. Singh and B. Kumar (eds.), Water Resources Planning and Management, 177–185.
© 1996 *Kluwer Academic Publishers. Printed in the Netherlands.*

(SDP) which has the advantage of incorporating explicitly stochastic variables and non-linear functions. The SDP models as proposed by WANG and ADAMS (1986), BRAGA et al. (1991) are used to optimize the operation of reservoirs with hydropower generation. The models use objective functions that maximize the benefits. VEDULA and MOHAN (1990) used a SDP optimization with streamflow forecasting and real time simulation to improve operation of a reservoir with irrigation and hydropower generation. The objective function used was to maximize the total energy generation. MOHAN and RAIPURE (1992) formulated a model for multiobjective analysis of a reservoir system with purposes such as hydropower generation, water supply for irrigation etc. This model used Linear Programming (LP) technique for solution, thus requiring a linear objective function and linear constraints.

A different requirement for optimization of such reservoirs, would be to maximize the on-peak firm energy generation while satisfying the irrigation and other demands. One way to do this is by using a max-min type objective function which maximizes the minimum energy generation subjected to a given set of constraints as formulated by HARBOE (1986 and 1990). The model is solved by deterministic dynamic programming technique using the historical inflow series and monthly time steps. Important parameters used in this model are the monthly energy distribution coefficients. These are the proportion of monthly on-peak firm energy generation to the annual total. These distribution parameters are an input to the model which uses them to convert each monthly generation to an annual value to facilitate comparison between different generations in different months. This distribution is calculated to follow the same pattern as the total energy demand on the grid. This resulted in energy generation proportional to the energy demand in each month

Reservoir releases which have to supply considerable quantity of water for irrigation, which has an annual water demand pattern, may often be in conflict with the above energy generation pattern. Performance of such reservoirs, in generating on-peak firm energy, can be improved only in the case where the reservoir power plant is connected to a local or national grid, which is also fed by other sources of energy such as thermal or other hydro power plants. It is assumed here that if one powerplant produces less on-peak energy relative to the demand, the other energy sources have the capability to compensate for the deficit firming up the energy production. Therefore, the reservoir can be optimized to yield the maximum possible on-peak firm energy following its own generation pattern and subjected to relevant constraints.

This paper describes a methodology that can be used to determine the energy distribution and optimize a multipurpose reservoir for on-peak firm energy generation. The methodology is presented in three steps. Step one determines the optimal energy distribution, step two optimizes the system using SDP and in third step evaluate the system performance using simulation. Energy distribution in proportion to the energy demand on the grid, is compared with the energy generation distribution computed by the model. The model is applied to a reservoir in Thailand - namely Sirikit Reservoir - as an example.

2. System Description

Chao Phraya river basin is the most important river basin in Thailand as it feeds largest and the best irrigated lands of which most is used for rice cultivation. The Chao Phraya river originates at the confluence of two rivers in the upper basin known as Ping River and Nan River as shown schematically in Fig.1. The Sirikit Reservoir is located in the Nan River Basin and is one of the largest reservoirs in the country. The releases from the reservoir are sent through the turbines for power generation and discharged into the original river. Part of the flow is diverted to Pitsanulok irrigation project and return flows are added to the river flow downstream of the project. This flow then feeds the Chao Phraya river. The Ping River basin has one major reservoir known as the Bhumibol Reservoir. In this study the common demand of water in Chao Phraya river is allocated to the two reservoirs in proportion to the average annual inflows to the respective reservoir.

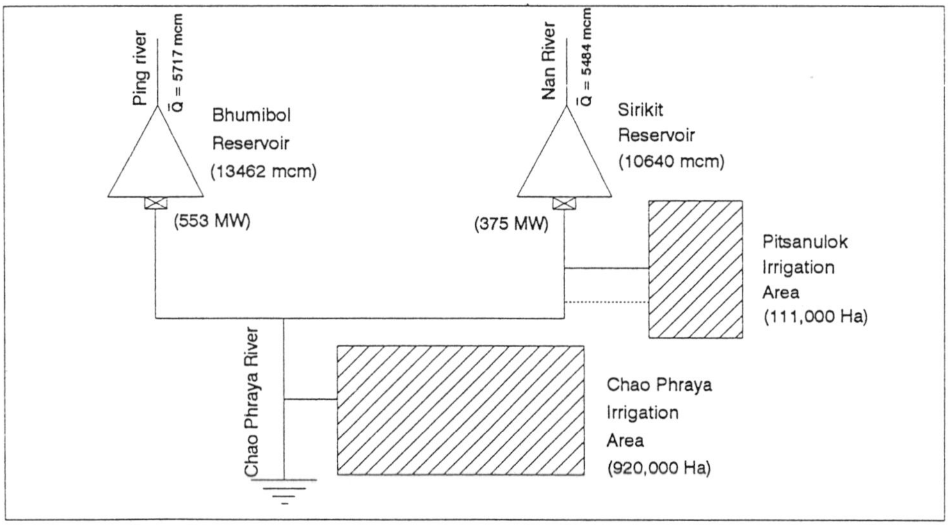

Fig. 1 Schematic Diagram of the Chao Phraya River System

3. Mathematical Model

The model used to find the optimal on-peak energy generation pattern maximizes the minimum moving summation of the energy generated within a period of 12 months. Therefore the model can find the maximized firm on-peak energy generation in each month of the critical 12 months period enabling the calculation of the energy generation proportions. These energy generations are due to releases required to meet all constraints as well as to maximize the energy generation. The model takes the following form,

3.1 OBJECTIVE FUNCTION:

$$\max Z = \max_{R_t} \left\{ \min_{t} \left[\sum_{t=t}^{t+11} EG_t \right] \right\} \qquad t = 1,2,3,\ldots(T\text{-}11) \tag{1}$$

where,

EG_t = Energy generation in month t in GWh.

R_t = Release through turbines during month t in mcm. (decision variable)

T = Length of the historical series.

The above objective function is to be maximized subjected to the following constraints,

3.1.1 Mass Balance in the reservoir:

$$S_{t+1} = S_t + Q_t - R_t - LOSS_t - SPILL_t \tag{2}$$

where,

Q_t = Inflow during month t in mcm.

S_t = Storage at the beginning of month t in mcm.

$LOSS_t$ = Evaporation loss during month t in mcm.

$SPILL_t$ = Spill during month t in mcm.

3.1.2 Storage Constraint:

$$D \le S_t \le C - FC_t \tag{3}$$

where,

C = Capacity of the reservoir in mcm.

D = Dead storage of the reservoir in mcm.

FC_t = Flood control reservation in month t in mcm.

3.1.3 Release constraint:

$$ID_t \le R_t \le TC \tag{4}$$

where,

ID_t = Downstream water demands in month t in mcm.

TC = Turbine capacity in mcm.

The above mathematical model is solved by Deterministic Dynamic Programming using the following recursive equation (backwards recursion).

3.2 RECURSIVE EQUATIONS

$$f_t(s_t) = \max_{R_t}\left\{\min\left[\sum_{t'=t}^{t+11} EG_{t'}(s_{t'}), f_{t+1}(s_{t+1})\right]\right\}, \quad t=(T-12),(T-13),..,1 \tag{5}$$

where,

$f_t(S_t)$ = Optimal moving summation of power generation from time period t to T if the initial state is s_t.

An alternative SDP model for a single reservoir optimization is formulated to maximize the expected annual on-peak firm energy production. Formulation of the model and the solution is as given in RATNAYAKE & HARBOE (1992) and HARBOE & RATNAYAKE (1993). The objective function of this SDP model is as follows,

$$\max Z = \max_{R_t}\left[\min_t E\left(\frac{EG_t}{\alpha_t}\right)\right] \tag{6}$$

The constraints are similar to the deterministic case, with inflows considered explicitly as a stochastic variable. The recursive equations are,

$$f_t(S_t,Q_{t-1}) = \max_{R_t}\left\{\min\left[\sum_{Q_t} P(Q_t/Q_{t-1})\cdot\frac{EG_t}{\alpha_t}, \sum_{Q_t} P(Q_t/Q_{t-1})\cdot f_{t+1}(S_{t+1},Q_t)\right]\right\} \tag{7}$$

Where,

$P(Q_t/Q_{t-1})$ = Conditional probability of inflow Q_t given that inflow Q_{t-1} occurred in the previous month.

$f_t(S_t,Q_{t-1})$ = Optimal expected annual firm energy if initial state is S_t and inflow in the previous month was Q_{t-1}, from stage t through the last stage.

4. Results and Discussion

In the first step of application of the model the on-peak firm energy generation distribution is found by maximizing the moving summation of monthly on-peak energy of 12 months. This optimization is carried out with the historical series of monthly inflows to the reservoir and monthly water demands. The water demands in the model are the totals of all net downstream demands arising from irrigation, water supply, navigation and salinity intrusion control. The result of the backward recursion, $f_1(S_1)$ gives the maximized minimum twelve month summations over the historical series. The objective function, had a maximum constant value above 60% of the reservoir active storage. Using this 60% storage as the starting value, monthly distribution of the energy generation is calculated as the proportions in the critical 12 months to the total of the same period. Results are given in Table I. Table I also gives the proportions of the energy demands on the grid. These two distributions are shown in Fig. 2.

Optimization of the system with each energy distributions are performed using the SDP model to obtain the annual expected firm energy target. This model maximizes the expected on-peak energy generation. The annual expected firm energy generation is later given as a target to a simulation model. These targets of annual on-peak energy generations, as obtained from the SDP model, are given in Table II. The higher target for the on-peak energy generation is achieved with the proportions calculated from the above model.

Table I On-peak energy generation distribution coefficient for 12 months.

Month	On-peak energy generation coefficients	
	Optimum From the Model	In Proportion to Energy Demand Distribution
April	0.168677	0.080415
May	0.085176	0.083623
June	0.059020	0.082297
July	0.185662	0.085075
August	0.090598	0.086060
September	0.005000	0.084734
October	0.001587	0.088605
November	0.070425	0.087172
December	0.066705	0.090165
January	0.055084	0.077763
February	0.066801	0.071795
March	0.145265	0.082297

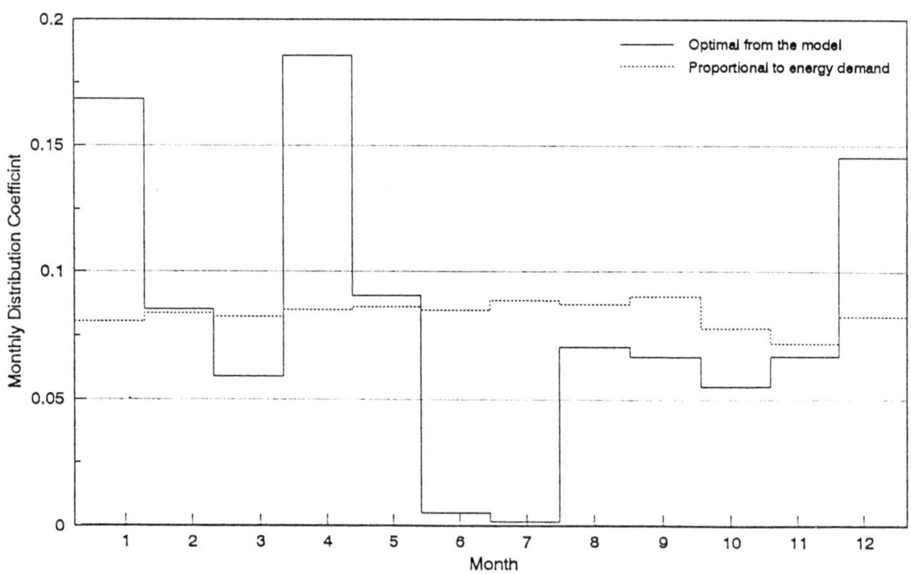

Fig. 2 Monthly Energy Distribution Coefficient

Table II Results of SDP optimization of the system.

	Annual on-peak energy target (GWh)
Optimal distribution	730.0
Proportional distribution using energy demand	436.0

Simulation of the system is performed using the historical series and a hundred year synthetic series. The results of the simulations are given in Table III. The reliability is defined as the ratio of number of successful irrigation seasons to the total number of irrigation seasons (two seasons per year). The application of the model improves the irrigation reliability, on-peak energy generation and reduces the average annual dump energy production. The reduction in the average annual on-peak energy generation (from the given target shown in Table II) is due to the failure to achieve the targets in some months.

Table III Results of simulations of the system using synthetic series.

	Average Annual On-peak Energy (GWh)	Average Annual Dump Energy (GWh)	Irrigation Reliability %
Optimal distribution	728.0	267.0	95.0
Proportional distribution using energy demand	430.0	564.0	92.0

An economic analysis is performed as an example to emphasis the applicability of the model in case where the thermal energy and the hydropower from the above reservoir is used to meet a given on-peak energy requirement. It is assumed that the annual on-peak energy requirement is 2000 GWh. and the daily on-peak energy demand has a triangular distribution of 13 hour duration. The cost of generation of thermal energy is 0.80 Baht/kWh (1 US$ = 25 Baht). The hydropower is taken to be generated at zero cost. Table IV shows cost of generation of this on-peak energy when the hydropower is generated according the two different distributions together with thermal energy. Also it shows the thermal generation capacities needed in addition to utilized hydropower capacity in each case to meet the demand. It is seen that the operation according to the model causes significant saving in the cost of generation of on-peak energy but in turn required higher installed thermal capacity. The example considered shows the additional thermal capacity requirement is about 332 MW which will cost 6224 Million Baht while the annual saving on the cost of generation is about 643 Million Baht.

Table IV Cost of generation of 2000 GWh/yr of on-peak energy using both hydro energy and thermal energy

	Cost of generation of 2000 GWh/yr on-peak energy (10^6 Baht)	Thermal capacity required for on-peak energy generation (MW)	Cost of thermal powerplant (@750US$/kW) ($10^6$ Baht)
Optimal distribution	3294	825	15468
Proportional distribution using energy demand	3937	493	9244

5. Conclusion

The model described here is an effective way of calculating on-peak firm energy distribution coefficients, to be followed in hydropower generation, if thermal energy is also available to feed the power grid. The model resulted higher on-peak energy generation and lower dump energy generation from the reservoir and also improved the irrigation reliability. The economic analysis shows that the model results in lower cost of generation for on-peak energy but in turn the application requires a higher installed capacity for thermal power plants.

ACKNOWLEDGMENTS: Research leading to this paper was made possible by the provision of financial support by the German Agency for Technical Cooperation (GTZ) through the Division of Water Resources Engineering, Asian Institute of Technology. The authors also thank the Royal Irrigation Department (RID) and Electricity Generating Authority of Thailand (EGAT) officials in Thailand for their kind support.

6. References

Harboe, R. (1986) 'Optimization, Simulation and Multiobjective Analysis of Operating Rules for Reservoir Systems', Proceedings, Water Forum '86: World Water Issues in Evolution (ASCE), Long Beach, California, August 1986, Vol. 1, pp 1026-1032.

Harboe, R. (1990) 'Optimal Power Output for Operation of a Reservoir System', Proceedings, VII Congress of the Asian and Pacific Division (IAHR), Beijing, China, Vol II, pp 171-174.

Ratnayake, U. & R.Harboe (1992) 'Optimal Operating Rule for a Reservoir with Stochastic Inflows', Proceedings, VIII Congress of the Asian and Pacific Division (IAHR), Pune, India.

Harboe, R. & U.Ratnayake (1993) 'Simulation of a Reservoir with Standard Operating Rule', Proceedings, IAHS/IAMAP Symposium J4-2, Yokohama, Japan.

Vedula, S. & S.Mohan (1990) 'Real-Time Multipurpose Reservoir Operation: a Case Study', Hydrological Sciences, Vol **35**,4 No. 8, pp 447-462.

Braga, B.P.F, W.W-G.Yeh, L.Becker & M.T.L.Barros (1991) 'Stochastic Optimization of Multiple-Reservoir-System Operation', Journal of Water Resources Planning and Management, Vol **117**, No 4, pp 471-481.

Wang, D. & B.J.Adams (1986) 'Optimization of Real-Time Reservoir Operations with Markov Decision Process', Water Resources Research, Vol **22**, No. 2, pp 345-352.

Mohan, S. & M.Raipure (1992) 'Multiobjective Analysis of Multireservoir System', Journal of Water Resources Planning and Management, Vol **118**, No. 4, pp 356-369.

APPLICATION OF STOCHASTIC DYNAMIC PROGRAMMING FOR RESERVOIR OPERATION - A CASE STUDY OF NARMADA SAGAR RESERVOIR

Dr. K. Venugopal
Asst. Professor,
Centre for Water Resources,
Anna University,
Madras.

MSV Subhakar
Scientist,
National Remote Sensing Agency,
Dept.of Space,
Hyderabad.

ABSTRACT

The need for an adequate institutional framework for water resources development and use, including planning operating policies, administration and law, has become evident, with the continuous increase in the demand for water. Linear and dynamic programming approaches, are mainly used for optimal allocation of available resources. This paper describes the application of stochastic dynamic programming to Narmadasagar reservoir to be taken up in Madhya Pradesh to assess the firm power generation. Even though Narmadasagar reservoir is meant to meet both irrigation and firm power requirements, in this, irrigation requirements are ignored for modelling. Thirtytwo years of flow data at Narmadasgar is used to arrive at the transitional probability matrix. For each run, a trial value of firm power target is assumed. Benefits, and penalties for short fall are given. Optimal operation policy is arrived at and the results are discussed.

1. INTRODUCTION

With continuous increase in the demand for water in all regions, the need for planning for water resources development and management, has become more important. The investments in water resources development are influenced by economic, social and political considerations along with the engineering aspects. Water under uncertain streamflows and multiobjective demands allocation process becomes more complex.

V. P. Singh and B. Kumar (eds.), Water Resources Planning and Management, 187–198.
© 1996 *Kluwer Academic Publishers. Printed in the Netherlands.*

Conventionally, engineers based on their ingenuity, experience and best judgement were operating the systems that were mostly meant for a specific purpose, isolated and simple. The operation of a system of multipurpose reservoirs present many problems due to conflicting nature of different uses, more so, when one considers the stochastic behaviour of inflows. Reservoir operation is a complex function. The operating policy adopted should lead to the optimal utilization of the available water. Decision making in this field is difficult in view of the number of parameters involved that are conflicting and complementary. Development of new methodologies and fast growth in computational capability have come to help in solving complex problems in water resources engineering.

This paper brings out the application of stochastic dynamic programming to evaluate the performance of Narmadasagar reservoir to be built in Madhya Pradesh in Narmada basin (Subhakar, 1987). Eventhough the reservoir is meant to meet both irrigation and firm power requirements, presently only firm power generation is considered for the purpose of modelling as the demand for irrigation is very meagure. Thirty two years of historic flow available at Narmadasagar is utilised for this study.

2. Mathematical Formulation

Mathematical formulation for the stochastic D.P through backward algorithm is discussed below.

Let the total number of stages be n'. Let the X_n & $X1_n$ be the starting and end states of the reservoir for n^{th} stage. Benefits are to be used in deciding the optimality in operating the system. Then the maximum value of objective function obtained from various combinations of X_n & $X1_n$ is to be considered. That is,

$$f\ (X_n,\ X1_n) = \text{MAX} \left[\sum_{q=q_n(1)}^{q=q_n(r)} \{P(q_n/q_{n-1}) * B(U_n)\} \right]$$

where

$f(X_n,\ X1_n)$ is the expected return form the optimal operation of the reservoir in n^{th} stage.

$P(q_n/q_{n-1})$ is the probability of occurence of q_n in n^{th} stage conditioned by q_{n-1} in$(n-1)^{th}$ stage.

q is the inflow, $q = q_n(1) \ldots\ldots\ldots, q_n(r)$ where r' represents number of class intervals used in probability calculations.

$B(U_n)$ is the benefits obtained from the release U' in the nth stage

The recursive equation for $(n-1)^{th}$ stage is given by,

$$f_{n-1}(X_{n-1}, X1_{n-1}) = \text{MAX}\left[\sum_{q=q_{n-1}(1)}^{q=q_{n-1}(r)} P(q_{n-1}/q_{n-2})\{B(U_{n-1}) + f_n^*(x_n, X1_n)\}\right]$$

The above equation states that for all possible values of X_{n-1} & $X1_{n-1}$, there is a value of release `$U_{n-1}$' which makes the right hand side of the above equation a maximum. The right hand side of the equation expresses that `U' shall be chosen such that return from the release of water in the current stage will be a maximum.

Generalizing the above equation, it can be written as

$$f_i(X_i, X1_i) = \text{MAX}\left[\sum_{q=q_i(1)}^{q=q_i(r)} P(q_i/q_{i-1})\{B(U_i) + f_{i+1}^*(x_{i+1}, X1_{i+1})\}\right]$$

Figure 1 shows the general flow chart of the Dynamic programme, CSUDP which was used in this study.

3. Reservoir Details

The Narmadasagar reservoir having a contemplated live storage of 10855 MCM (corresponding to the full reservoir level of 262.13 m) is the largest in the basin and is the last major control point in Madhya Pradesh portion of the Narmada River. The full reservoir storage is 12, 212 MCM at Narmadasagar, the general policy specifies first priority to downstream release through turbines required for Sardar Sarovar followed by its own irrigation requirement. Only after these releases are met, additional power releases are made, subject to the specified storage being built up. The power house will have an installed capacity of 1000 MW.

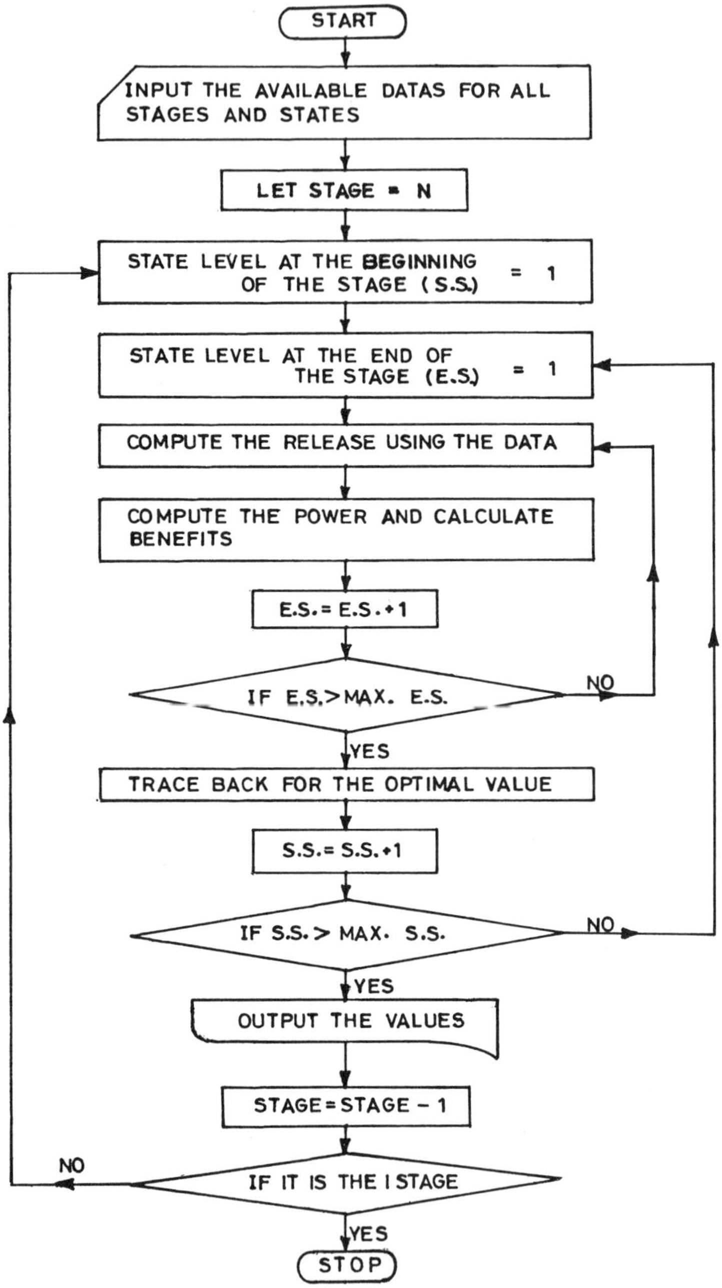

FIGURE 1 GENERAL FLOWCHART OF CSUDP

4. Modelling Details

For the purpose of analysing Narmadasagar reservoir using Stochastic Dynamic Programming, the inflow data of Narmadasagar reservoir is obtained from the results of System studies of Narmadasagar Complex (IIM Bangalore 1985).

Water year is taken as July to June of the ensuing year. Each month is considered as a stage in the dynamic programming formulation. The active storage i.e., the difference in storage at full reservoir level (FRL) and minimum draw down level (MDDL) 10855 MCM is divided into 20 equal parts and considered as states.

The inflow data available for 32 years is used to arrive at the steady state transitional probability matrices. The water year (July to June) is split into three groups. The inflows from July to September lie in one range i.e, 2,000-17000 MCM, from October to December lie in another range i.e., 500-3000 MCM and inflows from January to June fall in yet another range, i.e., 200-1,500 MCM. Major transitions take place during the months from September to October, December to January and from June to July. Each range of values are divided into five class intervals as shown in Table.1. The middle values of the class intervals are taken as the representative values.

The observed flows for each of the months are converted into the corresponding states. From these values, for any given state in the preceding month the frequency of occurence of various states in the succeeding month are determined. From this, the transitional probability matrix is obtained for each of the months. The transitional probability matrix obtained for July is shown in Table 2.

4.1 Steady State Probabilities

Each of the transitional probability matrices are used to arrive at the steady state transitional probability matrix. If Q denotes the matrix for a given month, powers of Q (Q^n) when computed leads to a matrix with identical rows. Each row is equivalent to a vector π, and the elements of each sum to unity. The steady State tranitional probability vectors (π) so obtainted are presented in Table 3.

TABLE 1 : DETAILS OF CLASS INTERVALS OF FLOW FOR THE THREE GROUPS

States	July - September		October - December		January - June	
	Class Interval	Middle Value	Class Interval	Middle Value	Class Interval	Middle value
1	2000 - 5000	3500	500 - 1000	750	200 - 460	330
2	5000 - 8000	6500	1000 - 1500	1250	460 - 710	590
3	8000 - 11000	9500	1500 - 2000	1750	720 - 980	850
4	11000 - 14000	12500	2000 - 2500	2250	980 - 1240	1110
5	14000 - 17000	15500	2500 - 3000	2750	1240 - 1500	1370

TABLE 2 : TRANSITIONAL PROBABILITY MATRIX OF FLOW FOR THE MONTH OF JULY

Current Stage (July)

States	1	2	3	4	5
1	0.75	0.25	0.0	0.0	0.0
2	0.7	0.15	0.15	0.0	0.0
3	0.5	0.35	0.0	0.0	0.1
4	0.65	0.35	0.0	0.0	0.0
5	0.8	0.0	0.2	0.0	0.0

TABLE 3 : π VECTOR OF FLOW FOR EACH MONTH (STAGE)

Stages	States Vector				
	1	2	3	4	5
July	0.77160	0.19390	0.02978	0.0	0.00472
August	0.0	0.33485	0.66515	0.0	0.0
September	0.38870	0.17409	0.20178	0.14095	0.9448
October	0.33503	0.11377	0.12458	0.19082	0.23580
November	0.0	0.0	0.0	0.15789	0.84211
December	0.77870	0.15574	0.03998	0.02558	0.0
January	0.0	0.0	0.0	0.00001	0.99999
February	0.83334	0.16666	0.0	0.0	0.0
March	0.01	0.99	0.0	0.0	0.0
April	0.00001	0.99999	0.0	0.0	0.0
May	0.77779	0.22221	0.0	0.0	0.0
June	0.37428	0.16043	0.15243	0.0	0.31287

TABLE 4 : OPTIMAL BENEFIT FOR DIFFERENT LEVELS OF FIRM POWER GENERATION - PENALTY 15

Reservoir States	Targets(MW)				
	250	275	300	325	350
1	3000	2908	2529	2585	2164
2	3000	3190	2895	2703	2676
3	3000	3300	3181	2734	2680
4	3000	3300	3457	3169	2680
5	3000	3300	3600	3473	2925
6	3000	3300	3600	3736	3446
7	3000	3300	3600	3900	3853
8	3000	3300	3600	3900	3853
9	3000	3300	3600	3900	3853
10	3000	3300	3600	3900	3853
11	3000	3300	3600	3900	3853
12	3000	3300	3600	3900	3853
13	3000	3300	3600	3900	3853
14	3000	3300	3600	3900	3853
15	3000	3300	3600	3900	3853
16	3000	3300	3600	3900	3853
17	3000	3300	3600	3900	3853
18	3000	3300	3600	3900	3853
19	3000	3300	3600	3900	3853
20	3000	3300	3600	3900	3853

4.2 Objective Function

The objective is to maximize the benefits from the firm power produced. The power (P) produced in MW by the turbine is calculated using the relationship.

$$P = 0.0037816 * U * HEFF * EFF$$

$$HEFF = HAVE - TWL$$

$$HAVE = (HX + HX1)/2$$

Where P is the power produced in MW

U is the discharge in MCM/month

HEFF is the affective head in metres

HAVE is the average water level obtained from initial and final storage in a stage in metres.

TWL is the Tail Water Level in metres

EFF is the efficiency of the power plant

HX & HX1 are the water levels corresponding to the starting and final storages in a stage in metres.

The water levels corresponding to given storages are computed using the elevation-capacity relationship of Narmadasagar reservoir.

The objective function is formulated keeping in view that the firm power generation is to be maximized. This is achieved by fixing a target value of firm power. If the power generation goes below the target value, heavy penalty is levied. On the other hand, power generation in excess of the target value is neglected. A case giving a very low value for power in excess of target value is also considered.

5. Results and Discussions

The objective of this work is to assess the firm power that can be generated from Narmadasagar reservoir. The prices considered are relative in nature. The price of unit of firm power namely 1MW per period is taken as 1 unit. For each run, a target value of firm power is assigned. If the power generation in any stage falls below the assigned value, penalty is levied.

TABLE 5 : OPTIMAL POLICIES TARGET - 300MW

States	July	Agu	Sep	Oct	Nov	Dec	Jan	Feb	Mar	Apr	May	June
1	3	14	15	5	1	4	1	2	2	1	2	3
2	4	15	15	1	2	5	1	3	3	1	3	4
3	6	12	16	1	3	6	1	4	3	1	4	1
4	7	12	17	1	4	7	3	5	3	1	5	1
5	8	13	18	2	5	8	4	6	4	2	6	1
6	9	14	19	4	6	9	5	7	4	3	3	1
7	9	15	16	5	8	10	6	8	6	4	4	1
8	9	16	17	6	9	11	7	9	8	6	5	1
9	9	17	18	7	10	12	8	10	9	7	5	1
10	9	17	18	8	11	13	9	7	10	8	5	1
11	9	17	15	9	13	14	10	8	10	8	5	1
12	9	17	15	10	14	10	11	10	10	8	5	1
13	9	17	16	12	15	11	13	11	10	8	5	1
14	9	17	17	13	16	13	13	11	10	8	5	1
15	9	17	18	14	17	14	14	12	10	8	5	1
16	9	17	19	15	17	14	15	13	10	8	5	1
17	10	17	20	16	18	15	16	13	10	8	5	1
18	11	17	20	16	19	16	16	13	10	8	5	1
19	12	17	20	17	19	17	16	13	10	8	5	1
20	13	17	20	18	19	17	16	13	10	8	5	1

TABLE 6 : OPTIMAL BENEFIT FOR DIFFERENT LEVELS OF FIRM POWER GENERATION - PENALTY 100

Reservoir States	Targets(MW)				
	250	275	300	325	350
1	3000	2287	2495	2071	1271
2	3000	2607	2495	2703	1432
3	3000	3300	2495	2703	1500
4	3000	3300	2696	2703	1500
5	3000	3300	3600	2703	1500
6	3000	3300	3600	2865	1500
7	3000	3300	3600	3900	2009
8	3000	3300	3600	3900	2009
9	3000	3300	3600	3900	2009
10	3000	3300	3600	3900	2009
11	3000	3300	3600	3900	2009
12	3000	3300	3600	3900	2009
13	3000	3300	3600	3900	2009
14	3000	3300	3600	3900	2009
15	3000	3300	3600	3900	2009
16	3000	3300	3600	3900	2009
17	3000	3300	3600	3900	2009
18	3000	3300	3600	3900	2009
19	3000	3300	3600	3900	2009
20	3000	3300	3600	3900	2009

TABLE 7 : OPTIMAL BENEFIT FOR DIFFERENT LEVELS OF FIRM POWER GENERATION ACCOUNTING SECONDARY POWER

Reservoir States	Targets (MW)				
	250	275	300	325	350
1	3194	3066	2665	2678	2241
2	3212	3371	3043	2807	2761
3	3229	3495	3351	2878	2775
4	2257	3512	3639	3322	2784
5	3270	3531	3800	3639	3071
6	3281	3550	3816	3923	3602
7	3290	3560	3826	4106	4031
8	3299	3569	3836	4106	4031
9	3308	3578	3845	4114	3039
10	3316	3586	3853	4123	4048
11	3325	3595	3862	4131	4056
12	3334	3604	3870	4140	4065
13	3342	3612	3879	4148	4073
14	3349	3618	3885	4154	4079
15	3353	3623	3890	4160	4085
16	3358	3628	3895	4164	4089
17	3362	3632	3899	4168	4093
18	3366	3636	3903	4173	4098
19	3371	3641	3908	4177	4103
20	3374	3644	3911	4181	4106

Two different penalties are tried. One with a penalty of 15 times the benefit, namely 15 units and the other a penalty of 100 units. Power generated in excess of the target value (secondary power) is ignored for the purpose of maximizing the benefit with the same set of target values and penalty as 15 units. Runs are taken assigning a value of 0.1 unit for unit of secondary power. Programme is run on IBM 360/44 system with a memory of 384K. The computer time (CPU) taken is 7 min 50 seconds for each run.

Table 4 shows the maximum benefit for an year for assumed target values of 250 MW, 300 MW, 325 MW and 350 MW, depending on the starting state of the reservoir namely 1 to 20. It can be seen that a target value of 250 MW can be achieved irrespective of the starting state. Depending on the carry over storage of each year, the firm power generation for that year can be increased. To say, with a carryover of about 1150 MCM, the firm power can be raised to 275 MW. A typical set of operating policies for the case of a target set of operating policies for the case of a target power of 300 MW is shown in Table 5.

Table 6 shows the results obtained with a penalty of 100 units. It can be seen that the behaviour is exactly similar. Table 7 shows the results obtained for a penalty of 15 units and 0.1 unit as the benefit per unit of the secondary power generated. It can be seen by comparing with the corresponding figures in Table 4, that as the firm power target increases the secondary power decreases.

References

(1) System Studies for Planning Narmadasagar Complex - Madhya Pradesh Government., IIM., Bangalore, 1985.

(2) Subhakar, M.S.V., Stochastic dynamic programming - A Case study , M.E. thesis submitted to Anna University, 1987.

IRRIGATION SYSTEMS MANAGEMENT - NEEDS AND OPPORTUNITIES

V.V.N. Murty
Irrigation Engineering and Management Program
Asian Institute of Technology
Bangkok, Thailand

K. Takeuchi
Regional Office for Asia and the Pacific
Food and Agriculture Organization
Bangkok, Thailand

R. Loof
Irrigation Engineering and Management Program
Asian Institute of Technology
Bangkok, Thailand

ABSTRACT. The development of irrigation systems with special reference to the Asian region is briefly reviewed. The scope and benefits of an efficient irrigation system management together with the essential technical components are outlined. These consist of examining the water deliveries, canal operation procedures, forming water users associations, infrastructural development, mitigating environmental hazards etc. The role of computer applications, monitoring and evaluation of irrigation systems, the need for development of interdisciplinary educational programs and irrigation extension services are discussed. Various suggestions for improved irrigation system management are summarized in the form of conclusions and recommendations.

1. Introduction

The importance of irrigation in increasing agricultural production is well recognized and as a result several irrigation systems have been established all over the world. In the last two to three decades a considerable number of irrigation systems have been constructed in the Asian region. Depending upon the size of the area irrigated and the cost of construction involved, irrigation projects are broadly classified as large, medium or small projects. Among these projects, the relatively larger ones have received more criticism relating to their

V. P. Singh and B. Kumar (eds.), Water Resources Planning and Management, 199–213.
© 1996 *Kluwer Academic Publishers. Printed in the Netherlands.*

performance compared to the smaller and medium sized projects. In case of the larger irrigation systems, several adverse performance parameters are usually mentioned. These include economic factors like inadequate returns on investments, performance factors like low water use efficiencies, social factors like failing to achieve equitable water distribution and finally environmental factors like soil salanization, contamination of ground water resources and adverse public health effects.

Initially, when the large irrigation systems were planned and constructed, many of them were part of the multipurpose projects. In case of the irrigation aspects, the main goal was to deliver water at the farm level. It was presumed that once the water is delivered at the farm level, the farmers will be able to manage the water for agricultural production. However, in the actual operation and management of the large irrigation systems several problems and issues have arisen, needing the attention of the administrators, engineers and scientists.

Large irrigation systems extend over considerable land areas and provide the most important input to crop production viz., water. The water available at times is used for muncipal and industrial purposes as well. Irrigation systems not only provide sustainability to agricultural production but also provide employment to the people in the command areas. Well managed irrigation systems will not generally cause adverse environmental impacts. In this paper the different aspects of irrigation system management are outlined indicating the scope of each activity. The discussions are with special reference to the irrigation systems in the Asian region.

2. Irrigation Systems in Asia

In the last two to three decades, in the Asian region there has been a phenomenal increase in the irrigated areas. It is estimated that from about 84.9 million ha in 1961, the irrigated areas have increased to 131.7 million ha in 1985 and at present a higher figure can be expected. The trend in the growth of the irrigated areas will, however, be at a slower pace as in many countries not enough sites are available and further there is some self-sufficiency in food grain production. Further development of irrigation should therefore concentrate on improving the performance of existing systems in a sustainable manner.

The following table (Table 1) gives the figures relating to irrigation development in the countries of the Asian region.

3. Previous Work

Considerable efforts are being made by several national and international institutions towards improving irrigation system management. The U.S. Agency for International Development (USAID), through a Water Management Synthesis Project have published several reports and handbooks in the area of irrigation water management (Lattimore and Fowler, 1988). Sagardoy et al. (1982) from the Food and Agriculture Organization (Rome) published a comprehensive manual for the organization, operation and maintenance of irrigation schemes. The International Commission of Irrigation and Drainage compiled the state-of-the-art papers related to the improvement in irrigation management with special reference to developing

Table 1 Irrigation Development in Asia and the Pacific Region

(Unit : 1000 ha)

Developing Countries	1965	1970	1975	1980	1985	1989
1. Bangladesh	572	1,058	1,355	1,639	2,073	2,738
2. Bhutan						34
3. China	*38,250	40,478	42,665	45,388	44,461	45,349
4. Cook Islands						
5. Dem. Kampuchea	753	89	89	89	90	92
6. DPR. Korea	500	500	900	900	1,070	1,400
7. Fiji			1	1	1	1
8. India	26,510	30,420	33,590	39,350	43,150	43,049
9. Indonesia	4,150	4,280	4,855	5,418	7,059	7,550
10. Iran	4,900	5,200	5,913	4,948	5,740	5,750
11. Laos	15	17	42	115	119	120
12. Malaysia	245	255	307	370	344	342
13. Maldives						
14. Mongolia			23	35	42	77
15. Myanmar	753	839	977	999	1,085	1,018
16. Nepal	86	181	232	230	650	943
17. Pakistan	12,043	12,958	13,601	14,680	15,620	16,220
18. Papua New Guinea						
19. Philippines	958	1,150	1,098	1,300	1,430	1,620
20. Rep. Korea	702	993	1,061	1,150	1,220	1,353
21. Samoa, West						
22. Solomon Islands						
23. Sri Lanka	341	465	480	525	583	560
24. Thailand	1,768	1,960	2,415	3,015	3,822	4,230
25. Tonga						
26. Vanuatu						
27. Vietnam	500	680	1,060	1,542	1,770	1,830
28. Australia	1,274	1,476	1,472	1,500	1,620	1,880
29. Japan	3,123	2,836	3,282	3,250	2,931	2,868
30. New Zealand	93	111	150	166	256	280
Asia-Pacific Total	97,536	105,946	115,568	126,760	135,126	139,294
Rest of World	59,701	66,243	72,906	83,566	89,094	93,534
World	157,237	172,189	188,474	210,326	224,220	232,828

Source: FAO Production Year Book

Remark: * = adjusted

countries (Framji, 1987). The American Society of Civil Engineers (ASCE) also compiled a manual containing recommendations for irrigation systems management (Johnston and Robertson, 1991). FAO in 1990, organised a regional workshop on Improved Irrigation System Performance for Sustainable Agriculture (FAO, 1991). This was held in Bangkok and was attended by participants from developing countries and variouse international organizations. In 1991, the Asian Productivity Organization (Tokyo, Japan) organised a study mission relating to the management of irrigation facilities in Asia and the pacific. The International Food Policy Research Institute (Washington, D.C.) is placing considerable emphasis on irrigation system management and conducting studies relating to the performance of irrigation systems in Asia (Svendsen, 1992).

The Central Board of Irrigation and Power (India) in their publication no. 164, outlined the multi-disciplinary perspective needed for utilization of canal waters (Singh, 1983). The Central Soil Salinity Research Institute (Karnal, India) have organised a National Symposium on Management of Irrigation Systems in 1988 (Tyagi et al., 1988). A similiar effort have been made in Sri Lanka (Gunawardena, 1990). Chambers (1988) reports in detail the issues involved in managing canal irrigation with special reference to the Indian irrigation systems. Lenton (1992) suggests a set of irrigation management stratagies applicable essentially to the Asian irrigation systems. The works cited above emphasize the need for efficient management policies for irrigation systems and indicate various steps to be implemented. In the context of the Asian irrigation systems, at present there is a need to identify and understand the policy options and adopt them at the field level.

4. Irrigation System Management

An irrigation system is planned and constructed to suit a particular site and agro-climatic condition and its management rules are framed accordingly. After an irrigation system is in operation for some years, the system management procedures need to be reviewed and updated continuously. In an irrigation system, it is likely that the original cropping patterns envisaged may not be existing, the available water supplies are not the same every year and there could be a change in the extent of the irrigation command. The environmental conditions in the command areas could undergo certain changes as large areas are irrigated year after year. It is desirable that the system operation is done in such a way that the available supplies are equitably distributed, the delivery systems are properly maintained and the environmental hazards in the irrigated areas are kept to a minimum. The irrigation systems should plan for drought situations i.e. when the water supplies are below normal as well as for situations of excess water in the form of floods. The irrigation systems could differ in the design and operational policies within a country or within a region. Uniform operating policies for all irrigation systems will not be possible, but certain guiding principles which are applicable for all irrigation systems could be formulated. For large irrigation systems, generally the source of water is a multipurpose reservoir constructed for hydropower generation, flood control and water supply. Water is released from the reservoir through spillways and through turbines for power generation. Downstream weirs and barrages divert the water into the irrigation system. The operation of the reservoir is guided by a set of operating rules which are primarily designed for power generation. Irrigation aspects, though considered may not get priority. Irrigation systems in such situations should be able to operate in phase with the water releases from the reservoir.

Drawing upon previous studies and experiences, the following items that need to be considered and understood in an irrigation system management are outlined.

5. Flow Control in Main and Secondary Canals

Flow control in irrigation canals is required to divide the total available discharge into smaller portions to meet the requirements of the users. In the large irrigation systems water is delivered by the main system to the tertiary units or outlets where the water users take over the responsibility. At times the main system operators fix rules for the distribution of water beyond the outlet (e.g. warabandi system). Eventhough flow control systems could be classified into several groups, four types are important. These are (i) upstream control, (ii) downstream control, (iii) proportional flow control, and (iv) mixed type. The flow control structures can be broadly grouped as (i) discharge regulators, (ii) water level regulators and (iii) discharge measurement structures.

Almost all the large irrigation systems in the Asian region are operated under upstream control. In these systems flow rates out of each turnout, lateral, and canal are estimated or computed in advance and the anticipated flows are added to determine the required source flow rate. Irrigation systems under upstream control require a central management in the operation of the system. The central management determines the distribution of flow to meet the demand at the tertiary offtakes.

In a given system, in order to achieve the required water deliveries the hydraulics of the system need to be fully understood. The canal capacities, travel times and the discharge characteristics of all the control structures should be carefully determined and documented. These need to be incorporated in procedures for determining the delivery patterns under different operating conditions.

6. Water Delivery to the Tertiary Units

One of the important aspects of the irrigation system management is to effect the delivery of water to the tertiary units within a prescribed policy. In case of the larger systems, the delivery of water is influenced by the operation of the main and secondary canals. Water to the individual farms is ultimately delivered through the tertiary units. The area covered by a tertiary unit and the water allowance provided varied widely across the projects and regions. For example, in the Bhakra system (India) a water allowance of 0.19 l/s/ha provided where as in Gandak (India) project it was 0.70 l/s/ha. In the Mae Klong project of Thailand, a system designed primarily for paddy, water allowance of 1.70 l/s/ha was considered.

The two main issues relating to the water deliveries to the tertiary units are the rate of flow and the duration for which the flow is provided. It is desirable that the water delivered matches the crop water requirements in the command area of the tertiary units. A mismatch between the deliveries and requirements (Fig. 1) will result either in excess or deficit supply.

Agrometeorological observatories are routinly maintained by most of the large irrigation systems. The data collected from these stations can be used for estimating crop water requirements and the information in turn can be used for determining water deliveries. In rainy season allowance for rainfall could be given in scheduling the water deliveries.

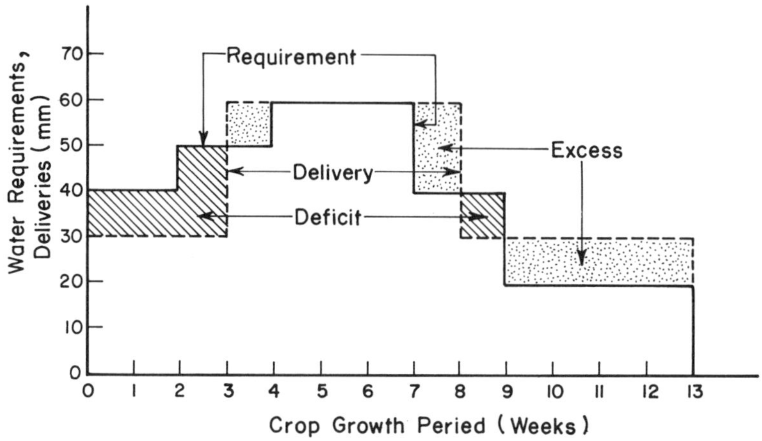

Fig. 1 Irrigation Deliveries and Crop Water Requirements

In case of water deliveries to the tertiary units, the following aspects steps need to be considered in each irrigation system.

i. Using the knowledge of crop water requirements and other components of the water balance equation, assess the water requirements and required water deliveries at secondary and tertiary levels.
ii. Establish procedures on the timing and amount of water to be delivered at various levels in the irrigation system. These precedures should consider the situation when the water is adequate and also the situation when water is in short supply.
iii. The officials responsible for the water deliveries and the farmers should be made aware of the operating policies.

External influences (like political) on water deliveries are likely to occur particularly during water shortages. Such situations are to be handled with tact and fairplay on the part of the system management.

7. Farm Irrigation Systems

The performance of an irrigation system considerably depends on the on-farm development and water management at the farm level. At present there is a recognition of this aspect and in most of the irrigation systems on-farm development works are initiated.

On-farm development should consider the following aspects.

i. Consolidation of land holdings wherever possible and applicable.
ii. Land grading and field layout for surface drainage.
iii. Lining of field water courses along with ancilliary structures.
iv. Surface irrigation design.

In many irrigation systems, on-farm development work is not the responsibility of the agency which deals with the construction and operation of the main system. Harmony between both the aspects of works is required. Introduction of irrigation and on-farm development should be closely followed by introduction of other improved agricultural practices like improved crop varieties, use of fertilizers and control of crop pests and diseases. For this prupose, in many situations agricultural extension services are established. However, these extension services are limited to the agricultural aspects only. In irrigation systems, farmers for some years need some guidence and assistance in water management and irrigated agriculture. Considering the situation in most of the irrigation systems in the Asian region, an irrigation extension service responsible to the system management appears to be desirable.

8. Computer Applications

Microcomputers can be effectively used for a wide variety of applications in irrigation system management. There is a need to understand and appreciate the utility of computer applications. Implementation of some of the management approaches based on mathematical models is possible only through computers. With the increasing availability of microcomputers at affordable costs the possibility of their applications will increase.

As and when computers are available, the first step could be to develop the basic microcomputer skills for the purpose of data processing and analysis. For this purpose, commercially available softwares could be used. There are a wide variety of softwares available for several aspects of irrigation system management. The use of such softwares should be done with caution. It is desirable to use such software with understanding so that the results obtained are properly interpreted. Use of software for technical problems on a "black box" approach basis is not desirable as the end results could be misinterpreted.

Broadly, microcomputers can be useful for the following purposes:

i. Data base management: hydrological, meteorological and any other information could be stored and retrieved as necessary.
ii. Statistical analysis of hydrological and meteorological data.
iii. Reservoir operation.
iv. Planning irrigation deliveries: Water balance models and models for estimating crop water requirements could be used for planning irrigation deliveries at the tertiary and secondary levels.
v. Surface irrigation system design and evaluation.
vi. Canal operation: Complex hydraulics of the canal system could be analysed with appropriate computer models to determine the canal flows at different sections.
vii. Geographical information system: Geographical information system for the command area could be useful for mapping soils, cropping patterns and locating areas of environmental concern.

Several other applications in drainage, ground water and water quality problems are possible.

9. Drainage and Conjunctive Use

In spite of the general awareness of the requirement of drainage in irrigation systems, the implementation of the drainage measures has not progressed to the desired extent. The problems of salanisation and rising ground water levels are being reported in some irrigation systems. As an initial step, in all irrigation systems, the surface drainage systems should be given attention and these should be installed as per requirement. Large scale subsurface drainage systems have not been installed in the Asian region except in the Indus basin in Pakistan. Some efforts relating to the installation of tile drain systems are reported from India. It is reported that subsurface drainage appear to require further investigation for large scale adoption in the irrigation systems in the Asian region.

In all large irrigation systems, with the passage of time ground water tables have a tendency to rise. Wherever ground water aquifers are available for pumping, tubewells could be installed to supplement surface water. Such conjunctive use of surface and ground water help to stabilise the ground water tables and also increase productivity as water from tubewells can be provided to suit crop needs.

Utilization of ground water for crop production using tubewells is practiced all over the region. However, the agencies responsible for ground water management and irrigation system operation have to function in an integrated manner.

10. Water Users Associations

The role and importance of farmers' participation in irrigation is realised and it is accepted that farmers participation is not only useful but at times becomes essential for an effective management of the system. In organising water users associations, the sociological conditions and farmers attitudes need to be considered. The irrigation system management should encourage the formation of the water users associations and assist them for viable functioning. The assistance of the water users associations could be taken for distribution of water in the tertiary units, settling minor disputes in water allocation and wherever possible in the maintenance of the system at the tertiary levels. In respect of the water users associations the following items need to be considered.

i. Type of water users association, formal or informal organization.
ii. Methodology for introducing and strengthening the association.
iii. Structuring of the association according to it functionality, according to village structure or hydrological unit.
iv. Arrangements for membership and leadership.
v. Support for water users association from government agencies and others.
vi. Activities undertaken by the water users association in irrigation management.
vii. Financial structure of the water users association, membership fees and O&M costs, government subsidies.
viii. Participation of water users association in the planning, design and maintenance of

irrigation improvement works.

11. Monitoring and Evaluation

A regular and systematic monitoring and evaluation relating to the performance of an irrigation system provides the necessary feedback for improvement. Monitoring should consist of both physical and economic parameters and should include cropped areas, crop yields, cost of cultivation, water deliveries, condition of the water delivery system, ground water levels, drainage conditions etc.

Evaluation of the irrigation system should consist of a set of performance indicators. These may include the following:

i. Area of land irrigated in relation to water supply
ii. Crop yields in the command area
iii. Profitability - actual income for the farmer
iv. Volume of complaints against the agency
v. Illegal tapoffs in the system and other administrative problems
vi. Departure from planned cropping pattern
vii. Comparison between canal head and tail performance
viii. Responsiveness of agency to farmers needs

It is difficult to develop an uniform procedure for evaluation of all irrigation systems as the physical conditions under which each of the system operate are different. An attempt should therefore be made to define or develop a set of performance parameters for a particular irrigation system and evaluate these parameters either seasonally or annually.

Abernethy (1986) discusses the issues involved in the performance measurement in canal irrigation. Molden and Gates (1990) present a set of parameters for evaluating irrigation water delivery systems. These parameters consist of adequacy, efficiency, dependability and equity of the water delivery system. Procedures for numerically calculating these parameters have been indicated. This is one example and there could be other ways for evaluating the water deliveries and the overall performance of the system. With the available procedures and information, each system should have some procedure for its evaluation, the results of which could be useful for improving the system performance.

12. Communications

Irrigation systems as they come into operation, change the cropping patterns and contribute to increased production of several agricultural commodities. The farmers in the irrigation command areas should be assisted by way of providing marketing facilities, roads, etc. Quite often these activities do not fall under the purview of the irrigation system manager, but he has to negotiate with other organizations to get such activities initiated. The project should also have good communication facilities for operating the water delivery system and for emergency maintenance operations.

13. Irrigation Service Fees

There appears to be no uniform policy regarding the charges for irrigation supplies effected through the irrigation systems. In some large systems, no charges are levied (as in Thailand) while in many of the systems nominal charges are imposed on the farmers based upon the extention of the irrigated area.

The water pricing in irrigation systems in an issue for consideration. Suitable charges levied from the beneficiaries could meet the maintenance cost of the system and may also contribute to the farmers' awarness and participation in the water management aspects in the irrigation systems.

14. Diversified Use of Irrigation Systems

Eventhough irrigation systems are designed essentially for providing water for crop production, in some instances irrigation water is being used for other purpose like muncipal and industrial uses. Such uses need not necessarily be banned but can be allowed taking into consideration the water quality aspects. Also muncipal or industrial activities should not result in the contamination of water is the irrigation system.

The irrigation system manager should be aware of the diversified uses of the system. Crop diversification and possibilities for aquaculture need to be investigated wherever possible and suitably supported.

15. Watershed Management

The irrigation system is supplied either from a reservoir or a diversion structure and they in turn receive the water from the catchment area. The erosion process in the catchment both by geological processes and human activities brings in considerable quantities of sediment, a part of which is deposited in the reservoir. This could lead to loss of the storage capacity of the reservoir. Excessive sediment in irrigation water is also responsible for silting of the irrigation channels.

The methods for controlling reservoir sedimentation can be broadly divided into three categories. The first consists of methods that reduce sediment inflow into the reservoirs. These are: control of sediment generation through watershed management; retention of sediment in debris basins before the river enters the reservoir and, bypassing sediment. The second category consists of methods that use hydraulics of flow to reduce that accumulation of sediment load that has entered the reservoir. Sediment flushing operations, sediment sluicing through specially designed reservoir operation policies and release of density currents belong to this category. The third category consists of hydraulic dredging of existing sediment deposits in the reservoir. All of these methods have been tried to some extent with limited success. Mahmood (1987) reviews the relative effectiveness of these measures and in respect of watershed management concludes that it cannot be very useful in alleviating sedimentation of large reservoirs. The irrigation system management should, however, be aware of the watershed management practices as in case of small reservoirs intensive conservation measures could effect the inflows into the reservoir.

16. Environmental Considerations

It has been noticed that major irrigtaion systems are responsible for several environmental hazards and steps to mitigate these hazards are necessary. Introduction of irrigation on a large scale can bring both favourable and unfavorable changes in the irrigated areas. With the introduction of irrigation, there is a change in the cropping patterns. In irrigated areas, changes in ground water levels and in some cases changes in the quality of ground water are reported. Because of a rise in ground water levels, waterlogging and salinity problems appeared in several projects. Drainage waters in many cases are of poor quality and the drainage channels became source of aquatic weeds.

A major aspect which has not received adequate attention is the effects on health of the habitants in the command area. In the Asian region both water-based and water related vector-borne diseases have been reported from irrigated areas. Oomen et al., (1991) examine the irrigation health problem in detail and offer suggestions for mitigating some of the hazards. The International Rice Research Institute (IRRI, 1988) highlight the problems involved in vector-borne disease control in humans through rice agroecosystem management. Lowland paddy being one of the major crops in several large irrigation systems, could be responsible for human health problems.

The environmental hazards that are likely to occur in an irrigation system could be listed as follows:

i. Soil degradation, salinity and alkalinity
ii. Deterioration of ground water quality
iii. Deterioration of surface water quality
iv. Waterlogging, formation of stagnent water pools and vector breeding places
v. Stagnent drainage channels with aquatic weeds
vi. Changes in cropping patterns resulting in cropping systems injurious to human health
vii. Uptake of undesirable elements by plants causing long term health problems
viii. Spread of certain water borne diseases
ix. Irrigation return flows causing water quality problems for aquatic habitat
x. Contamination of irrigation water with urben wastes

The irrigation system management should be aware of these and any other possible hazards in the system. In the monitoring program of the system, it is necessary that the likely hazards are included. In order to mitigate several of the hazards listed, co-operation and assistance from other agencies might be required. The irrigation system management may not be able to implement the preventive measures but they should act as a catalyst in getting the required measures carried out.

17. System Maintenance, Rehabilitation and Betterment

With the passage of time, the structures originally constructed in an irrigation system could deteriorate effecting the water deliveries. Depending upon the materials of construction originally used and the local conditions, the problems of maintenance in each irrigation system could be different. In case of lined canals, the lining could get damaged at various locations. Siltation and weeds are perennial problems. Irrigation systems being public property could be subject to vandalism.

Maintenance involves the main structures, structures along the canals, outlets, structures in tertiary units and finally the drainage systems. Budgetary provisions for maintenance are a sensitive issue. It could be ideal if the maintenance of the installations in the tertiary units could be attended to, by the farmers' associations. Maintenance guidelines for each project may have to be developed depending upon the type of the structures, materials of construction, location conditions, etc.

In some irrigation systems, projects have been initiated for rehabilitation and betterment. Rehabilitation and betterment (R&B) involves a joint process that combines the elements of maintenance and changes for upgrading system performance in light of current needs and opportunities. The efforts in R and B projects are directed towards correcting errors in the original planning process and respond to new informations. It has been reported that most R and B projects have experienced the problems encountered in new irrigation projects and have also posed new constraints and challenges to the project designers and implementors (Lattimore, 1988).

The following guidelines and recommendations were proposed (Lattimore, 1988) to obtain more effective rehabilitation and betterment projects:

i. Due to dynamic nature of the irrigation system and its environment, the irrigation system generally need not be restored to its original design specifications.
ii. The current construction orientation of most agencies should be replaced by an approach that improves management as well as structures.
iii. R&B projects should be flexible, allowing adjustments during project implementation to correct errors in planning and to respond to new information.
iv. The institutional capacity for actively involving farmers should be developed.
v. Greater emphasis should be placed on project design. Planning for design should invove consultation with relevant national, regional and local officials, and with operation and maintenance personnel and farmers.
vi. Diagnostic analysis of the irrigation system should be conducted to provide input into the planning process.
vii. Roles, responsibilities, sharing of funds, and training opportunities should be clearly defined for all participating agencies before a project begins. The role that each agency plays should be complemented by specific incentives.
viii. New policies should be formulated and communicated to water users and other involved in the process prior to project implementation and not during and after.
ix. Farmers should be involved in collecting and using irrigation fees to improve the

collection and use of these fees.

x. The monitoring and evaluation unit should be linked with the project management office to effectively make revisions in project implementation and operation.

18. Education and Training

Educational and training programs need to be planned to meet both the short term and long-term goals of irrigation system management.

Training programs in irrigation system management have been initiated by several organisations and institutions. These need to be continued and intensified. In all the countries in the region irrigation systems are being managed by the engineers who constructed these systems. It has been noticed that in general, in the undergraduate curricula of the students not enough emphasis is being placed on the irrigation system management aspects and the interdisciplinary requirements of the irrigation system management. It is suggested that in the undergraduate program where irrigation engineering is covered, irrigation system management should also be included.

Several countries have initiated educational programs covering the engineering aspects of agriculture. Such professionals could be utilised in the irrigation system management. Interdisciplinary educational programs involving the traditional civil engineers, agricultural engineers etc. could be initiated.

19. Conclusions/Recommendations

Considering the irrigation systems in the Asian region and the information related to their management, the following conclusions and recommendations are made.

(1) In the Asian region, particularly the large irrigation systems need an overall management concept for their efficient operation. Close coordination between the system operation, agricultural and environmental aspects and the farmers is essential.

(2) The water deliveries in relation to cropping patterns in each irrigation system should be carefully reviewed. Water deliveries from the main system to the tertiary units should as far as possible match with the crop water requirements. Low water duties in some situations may encourage ground water exploitation by the farmers. Conjunctive use of surface and ground water supplies need to be encouraged.

(3) Improvement in the management of the farm irrigation systems is possible by improving the surface irrigation methods by proper land grading and field layout. For the purpose of design, evaluation and improvement of surface irrigation systems including guiding the farmers for proper irrigation scheduling, there appears a need for suitable extension services.

(4) In irrigation systems, farmers participation and water users associations need to be responsible for water distribution as well as maintenance of the system at the tertiary

levels.

(5) The environmental hazards that are occurring or likely to develop in an irrigation system need to be understood. Necessary steps to mitigate these hazards need to be initiated. Several of the environmental hazards could be reduced by providing adequate drainage facilities in the system.

(6) Microcomputers are useful in several aspects of irrigation system management. Computer applications could be from simple data-base management to application of complex mathematical models. Use of available softwares for system management should be done after suitable adoption for the particular system under consideration.

(7) There could be demands for water from the irrigation system for muncipal and industrial uses. Such uses could be allowed depending on the situation and the water quality standards. At times when irrigation canals pass through urban areas, contamination from such urban areas is possible and this should be avoided. Aquaculture in irrigation systems need to be suitably supported.

(8) The irrigation system should have a program of monitoring and evaluation which should consider crop production, environmental and economic factors. The feedback from such programs should help in improving the management of the system.

(9) Educational programs particularly at the relevant undergraduate engineering level should have topics of courses related to irrigation system management. Keeping in view the multidisciplinary nature of the irrigation system management, there is a scope to introduce joint educational programs for civil and agricultural engineering students. Training programs in irrigation system management for in-service personnel need to be continued.

(10) With the passage of time, irrigation structures could deteriorate effecting the performance of the system. In such cases, suitable rehabilitation programs for the system improvement need to be initiated.

In conclusion it can be stated that in the Asian context, irrigation systems offer an excellent opportunity for sustainable agricultural development, rural employment and improvement. However, progressive methodologies are needed for the management of the irrigation systems.

20. References

Abernethy, C.L., (1986) *Performance Measurement in Canal Irrigation*, IIMI/ODI Irrigation Management Network, Paper 86/2d

Asian Productivity Organization, (1991) *Management of Irrigation Facilities in Asia and Pacific*, APO, Tokyo (Japan)

Chambers, R., (1988) *Managing Canal Irrigation*, Oxford & IBH Publishing Co., Ltd., New Delhi (India)

FAO, (1991) *Improved Irrigation System Performance for Sustainable Agriculture*, Publication No. AGL/MISC/18/91, Food and Agriculture Organization, Rome (Italy)

Framji, K.K. (1987) *Improvement in Irrigation Management with Special Reference to Developing Countries*, International Commission on Irrigation and Drainage, New Delhi (India)

Gunawardena, E.R.N., (1990) *Irrigation and Water Resources*, (Symposium Proceedings) University of Peradeniya (Sri Lanka)

Johnston, W.R. and J.B. Robertson (Editors), (1991) *Management, Operation and Maintenance of Irrigation and Drainage Systems*, American Society of Civil Engineers, New York (USA)

International Rice Research Institute (IRRI), (1988) *Vector-borne Disease Control in Humans Through Rice Agroecosystem Management*, Workshop Proceedings

Lattimore, D. and D. Fowler, (1988) *Developing Irrigated Agriculture : A Socio-Technical Approach*, U.S. Agency for International Development, Washington, D.C. (USA)

Lenton, R., (1992) *Irrigation Management Strategies for the 21st Century*, Canadian Journal of Development Studies, Special Issue

Mahmood, K., (1987) *Reservoir Sedimentation, Impact, Extent and Mitigation*, World Bank Technical Paper No. 71, The World Bank, Washington, D.C. (USA)

Molden, D.J. and T.K. Gates, (1990) *Performance Measures for Evaluation of Irrigation Water Delivery Systems*, Journal of Irrigation and Drainage Engineering, American Society of Civil Engineers, 116(6):804-823

Oomen, J.M.V., J. de Woft and W.R. Jobin, (1990) *Health and Irrigation Vol. 1 and 2*, International Institute for Land Reclamation and Improvement, The Netherlands

Sagardoy, J.A., A. Bottrall and G.O. Uittenbogaard, (1982) *Organization, Operation and Maintenance of Irrigation Schemes*, FAO Irrigation and Drainage Paper No. 40, Rome (Italy)

Svendsen, M., (1992) *Assessing Effects of Policy Change on Philippine Irrigation Performance*, International Food Policy Research Institute, Washington D.C. (USA)

Singh, K.K., (1983) *Utilization of Canal Waters : A Multi-disciplinary Perspective on Irrigation*, Publication No. 164, Central Board of Irrigation and Power, New Delhi (India)

Tyagi, N.K., P.K. Joshi, R.K. Gupta and N.T. Singh, (1990) *Management of Irrigation System*, Central Soil Salinity Research Institute, Karnal (India)

IRRIGATION SYSTEM PERFORMANCE : CONCEPTS, INDICATORS, STANDARDS AND APPLICATION

N.K.Tyagi*

ABSTRACT

A state of the art review of the irrigation system performance concepts, measures and specifications of performance is presented. On the basis of critical review, suitable indicators for evaluating performance at different hydraulic levels of irrigation system are identified. The equity, adequacy (relative water supply, RWS) and reliability (timeliness) at watercourse level and the application, storage and distribution efficiencies at the farm level were selected as the performance indicators. Using the established indicators, the performance was evaluated in part of Bhakra Canal Command in Haryana. The efficiencies of the on-farm system were influenced most by the stream size. Though application efficiencies were quite high, the storage efficiency representing adequacy and distribution uniformity were low. In irrigation with fixed frequency schedules, the inadequate application are likely to affect crop yields adversely.

The equity in water distribution decreased with size of the watercourse (flow rate) and the average value of interquartile ratio (water supplied in the most favored quarter/water supplied in the least favored quarter) was 1.85. There were locational and intraseasonal variations in adequacy which improved form tail to head and was higher during rainy season as compared to winter season. With average value of RWS at 0.64 across watercourses and seasons, the system had a highly deficient water supply. Water productivity which reflects both adequacy and timeliness had an average value of only 0.51 across watercourses and seasons. Lower values of water productivity as compared to seasonal adequacy reflect time mismatch between supply and demand.

The major drawbacks of the system seem to be volumetric inequity, time mismatch between demand and supply and non-uniform water application. Improvement in performance is possible through provision of proper unit command areas, variable warabandi time, auxiliary storage and adjustment in water delivery schedules.

INTRODUCTION

Talk of irrigation system performances now a days generates considerable interest in government agencies associated with irrigation development, funding agencies like World Bank and the water users. The rising cost of irrigation development, dissatisfaction with current level of performance and greater awareness of environmental issues, are mainly responsible for undertaking performance audits. These efforts are being made on

* Principal Scientist, CSSRI KARMAL - 132101, India

V. P. Singh and B. Kumar (eds.), Water Resources Planning and Management, 215–233.
© 1996 *Kluwer Academic Publishers. Printed in the Netherlands.*

the premise that it is possible to improve the performance through modern technology and management techniques. The knowledge of existing levels of performance and the projections of performance levels in post improvement phase is essential for suggesting improvements. Meaningful performance appraisal is possible only with a clear understanding of how we define an irrigation system, its management and the objectives for which it has been created. The vastness of the subject makes it incumbent to delineate the boundaries alongwith anticipated outputs and users of the recommendations.

1.0 PERFORMANCE OBJECTIVES

The performance of a management system can only be measured in terms of its objectives. Irrigation schemes have multiple and sometimes conflicting objectives and the perception about performance will vary depending upon the value attached to the different goals by the person or organization making the assessment. There are different aspects of project performance which may relate to financial, institutional, design construction and operation of the system. Smith (1990) gives detailed list of the characteristics of the project performance in respect of financial institutions and socio-economic agricultural operational factors. Chambers (1984) has listed some of the perception of the good performance by different disciplines and sections of the society. An indepth critique of the objectives of irrigation is given in Small and Svendsen (1990) who stated that the goals were crucial to performance assessment and their clear specifications and classification as to whether they were related to inputs, outputs or efficiency were needed. They conceptualized irrigation purposes within a nested means and end framework in which a narrow purpose is seen as a means of achieving some specified end. In the hierarchical order of objectives, the end of the first level of objective becomes the means of next higher level of objective (Table 1). At each level of assessment a whole system view is required because the constraints at different levels influence the performance of other levels.

If we limit our discussion to technical performance at the hydraulic level of water distribution system, then the success of an irrigation system can be measured by how it supplies the required quantity of water at the right time in an equitable manner to users served by the system.

The water delivery system can have many objectives but there is a broad agreement that most of them can be included in adequacy, equity and timeliness. In cases where water quality is a problem, a fourth dimension of quality will have to be added. Water delivery management is not an end in itself but is a means of increasing agricultural productivity in a sustainable manner and ultimately improving the quality of life. It is therefore, necessary to establish linkages between the secondary or intermediate objectives and the ultimate objectives. Abernethy (1987) has shown how all the objectives can be combined into a single objective of productivity.

Table 1 Irrigation purposes as nested means and ends.

Level of end	Means	End
Proximate	Operation of irrigation facilities	Supplying water to crops
Intermediate-1	Supplying water to crops	Sustained increase in agricultural productivity
Intermediate-2	Sustained increase in agricultural productivity	Increased incomes in rural sector
Intermediate-3	Increased in rural sector	Rural economic development
Ultimate	Rural economic development	(1) Improved livelihoods of rural people (2) Sustained socio-economic development for entire economy

Source: Small and Svendsen (1990)

2.0 PERFORMANCE INDICATORS

The performance objectives have to be expressed in terms of measurable quantities which we call the state variables. These state variables are combined in various forms to develop indicators of performance. Nonexhaustive lists of these indicators can be found in Eisel (1988), Smith (1990) and Svendsen and Small (1990). Clyma and Lowdermilk (1988) identified the parameters that need to be monitored for diagnosing the irrigation system performance with a management focus for system outcomes (Table 2) Important indicators of the various performance objectives are briefly reviewed.

2.1 Adequacy

Adequate water supply to irrigate crops is the primary concern of a water supply system. The water requirement for a given agricultural policy is a function of the land area, crop consumptive use, application losses and leaching requirement etc. The adequacy of water supply is controlled by available supply for distribution, water delivery schedule, capacity of the hydraulic structure to supply water according to schedule and operation and maintenance of the system. Few important measures of adequacy are discussed.

2.1.1 Water duty: The early measures of adequacy revolved around duty of water which is defined as the ratio of irrigated area and water discharge rate (e.g. litres/ha). Abernethy (1990) mentioned that the concept of duty originated as a planning and design parameter and is especially suited to water deficit environment.

Table 2 Framework for diagnostic analysis of irrigation systems with a management focus for systems outcomes.

Key Management Outcomes	Susbsystem	Performance Parameters
Productivity of agriculture	Field	Yield, cropping intensity, and variance in yield, cropped area
	Farm	Same as field subsystem
	System	Same as field subsystem
	National	Same as field subsystem
Resource conservation	Field	Salinity level, waterlogging erosion, sedimentation,fertility structural degradation seawater intrusion
	Farm	Same as field subsystem
	System	Same as field subsystem
	National	Same as field subsystem; Permanent (sustained agricuture)
Return on investment	Farm	Net returns
	Cooperative	Same as farm
	System	Return on investment
	National	Return on investment; social returns
Water control for -equity -adequacy	Main	Volume; flow and time, variance of outlet (Qs); ratio, dependability - actual water delivered/scheduled delivery; ratio of actual supply designed supply
	Farm	
	-delivery	Same as main subsystem
	-application	Defined efficiencies
	-water use	Stress index
	-water removal	Water content of soil and water level for optimum yield
	National	All of above; return on investment

Source: Clyma and Lowdermilk (1988)

It is essentially a crude parameter and is related to water requirement only approximately.

2.1.2 Relative water supply (RWS): In the physical sense, relative water supply (RWS) is the ratio of water supplied to an irrigation unit and the demand for water in that unit over a period of time. The RWS is inverse of water use efficiency and is computed as follows:

$$RWS = \frac{(IR + RN)}{(ET + S\&P)} \tag{1}$$

where,

IR = Mean irrigation application mm/day
RN = Mean rainfall mm/day
ET = Mean evapotranspiration mm/day
S&P = Seepage and deep percolation mm/day

The concept of RWS is related to available water supply, water demand and the management intensity in an irrigation system. Oad and Podmore (1989) suggested that the knowledge of RWS can be used to specify characteristics, irrigation organization and system management. The capacity of the system to control water is determined by both physical resources and the institutions. The degree to which the capability is actually realised is called the intensity of management. By varying the intensity it is possible to match supply with the demand. When RWS is low it represents water scarcity giving rise to higher tensions.

Abernethy (1990) pointed the possible objections to the inclusion of percolation in the crop demand for water. He argued that inclusion of percolation losses into the demand affected the meanings of the relative water supply because according to the definition if the losses were more (may be due to poor water management practices), the demand would be more and hence the RWS would be low indicating a better management.

2.2 Contribution of Structural and Management Factors to Adequacy Performance

It is possible to define the water delivery performance in terms of the amount of water which may refer to either rate, volume or frequency or duration of water delivery. Molden and Gates (1990) developed a procedure to determine the relative contribution of hydraulic structures and management procedures to water delivery adequacy. They defined the following 4 water quantities over time (t) and space (x) to derive performance measures. $Q_D(x,t)$ as the amount of water actually delivered, $Q_R(x,t)$ the amount of water required to be delivered, $Q_S(x,t)$ the amount of water scheduled to be delivered and $Q_D(x,t)$ the deliverable amount of water.

The performance relative to adequacy due to structural characteristics Pas was defined as the ratio of amount deliverable to the amount of water scheduled to be delivered over the region and period of interest.

$$P_{as} = \frac{1}{T} \sum_T \left(\frac{1}{R} \sum_R p_{as} \right) \tag{2}$$

where,

p_{as} = Qd/Qs when Qd ≤ Qs

p_{as} = 1 otherwise

P_{as} is temporal and spatial average of adequacy measure relative to hydraulic structure characteristics. A value of less than unity indicates the amount that cannot be delivered due to structural limitations, p_{as} is the point performance function of adequacy relative to structural characteristics. R is the region or sub-regions within the water delivery system, T is the time period and t is the time.

Similarly, the performance due to management characteristics can be written as the ratio of the amount of water delivered to the amount deliverable.

$$P_{am} = \frac{1}{T} \sum_{T} \left(\frac{1}{R} \sum_{R} p_{am} \right) \tag{2}$$

where,

$$p_{am} = QD/Qd \quad \text{when } QD \leq Qd$$

$$p_{am} = 1 \quad \text{otherwise}$$

P_{am} and p_{am} are the average and point values, respectively of the adequacy performance due to management characteristics. If the delivered amount is less than deliverable amount, it represents a break in operation policy. This approach does not handle the case of more than adequate supply.

2.3 Other Statistical Measures of Adequacy

The performance of water delivery schedule can be evaluated by looking at the ratio of intended to required water (flow rate, volume, duration and frequency). Clemmens and Bos (1990) divide the overall performance of water delivery into two components, the delivery schedule and the operations. They compute the overall performance as the product of these two ratios actual divided by the required water. Number of standard deviations between unity and mean value of actual to intended flow rate (or volume/duration/frequency/pressure) ratio during the delivery suggested to be used as a measure of adequacy.

3.0 EQUITY

Equity refers to the quality of distribution of a given output and is one of the main objective of irrigation management. Equity as related to water delivery may be defined as the fair share of water to users at different points in the system. The fair share may be based on legal right for water as in a prior appropriation system or may be set as a fixed proportion of water supply as is done in irrigation systems in Northwest India. The concept of 'fair share' is complicated and is often interpreted subjectively. A critique on equity concept is given in Sampath (1990) who argued that equity should not be treated synonymous with equality, though in most of the case studies, the case is otherwise.

3.1 Theil Information Index

Sampath (1990) analyzed several possible measures of equity including, coefficient of variation, standard deviation of logarithm, relative mean deviation G in coefficient and Theil's information measures and decided in favour of the last measure. Mathematically Theil's information measure (T) can be defined as

$$T = \frac{1}{\log n} \sum_{i=1}^{n} Y_i \log(nY_i) \qquad (4)$$

where,

n = Number of units under consideration

Y_i = Fraction of the total available water delivered to ith unit

3.2 Inter-Quartile Ratio

Abernethy (1990) proposed the use of inter-quartile ratio defined as the ratio of the volumes delivered respectively, to the most favoured upper 25% and the least favoured lowest 25% of the units. This indicator has a better communication value than several others.

3.3 Relative Equity Ratio

As reported by Abernethy (1990), Rao (1987) proposed a relative equity ratio on the basis of middle range (units getting within ±10% of their proportionally fair amount). The ratio of amounts of water received by these units which get more and less than the middle range, is defined as relative equity ratio.

3.4 Spatial Coefficient Of Variation (CVR)

The relative spatial variability of the amount of water (Q_D) delivered to the amount required over the time period of interest was used to describe the degree of variability in water delivery from point to point over the region by Molden and Gates (1990). The corresponding measures associated with structural and management contributions to performance relating to equity are.

$$P_{es} = \frac{1}{T} \sum_{R} CVR \left(\frac{Qd}{Qs}\right) \qquad (6)$$

$$P_{em} = \frac{1}{T} \sum_{T} CVR \left(\frac{QD}{Qd}\right) \qquad (7)$$

where,

P_{es} and P_{em} are the average values of spatial variation due to structure and management components respectively.

CVR = Spatial coefficient of variation of the ratio of QD/QR over region R

3.5 Some Other Statistical Measures Of Equity

Christiansen(1942) uniformity coefficient (Cu) has since long been used a a measure of uniformity of water distribution. While comparing spatial uniformities with several other methods Tyagi (1988) found that Cu did not represent spatial uniformity very well. Clemmens and Bos (1990) have suggested standard deviation of the ratio of actual over intended flow rate (or volume/frequency/duration/pressure) at a given time as a measure of spatial equity.

4.0 DEPENDABILITY, RELIABILITY AND TIMELINESS

Dependability is defined as the delivery of a relatively uniform amount of water over time (Molden, 1988). There are two aspects of dependability: reliability which relates to a fairly constant rate of supply and the predictability which relates to the timing of water supply. The concept of reliability means the degree to which the irrigation system and its water deliveries conform to the prior expectation of the users. Timeliness means correspondence of water supply to crop water requirements and it relates to distribution of water supply across the season to some utility based standard. The standard is derived either from crop requirements or contractual obligations. Timeliness and reliability both carry the implication of some external demand or need to be fulfilled. Svendsen and Small (1990) argued that in the absence of any readily identifiable distinction between reliability and timeliness, the single concept of timeliness would be more useful.

The response of the water users is greatly influenced by the reliability but it has not been systematically studied and for that reason no numerical standards have been developed for comparing performance.

4.1 Temporal Coefficient Of Variation (CV_T)

Molden and Gates (1990) use the temporal coefficient of variation (ratio of standard deviation to mean) of Q_D/Q_R over a period to represent the effect of variability in dependability. The expressions for structural and managerial contribution for dependability are given as:

$$P_{ds} = 1/R \sum_{R} CV_T\,(Qd/Qs) \qquad (7)$$

$$P_{dm} = 1/R \sum_{R} CV_T\,(Q_D/Qd) \qquad (8)$$

Where,

P_{ds} and P_{dm} are the average values of the dependability performance due to structural and management components, respectively.

As the values of P_{ds} or P_{dm} approach zero, the relative water delivery over time becomes more uniform indicating a more dependable supply.

4.2 Some Other Measures of Dependability

Clemmens and Bos (1990) proposed standard deviation of the ratio of the actual over intended flow rate (or volume/duration/frequency/pressure) at a given time (season) as measure of reliability. Mean prediction error methodology originally proposed by Theil (1966) has been used by Sharma et al (1989) to evaluate dependability for water deficient systems.

5.0 WATER QUALITY

The characteristics of irrigation water itself also have bearing on the performance of water delivery system. Temperature, sediment and dissolved salts affect the performance of irrigation system. Temperature of irrigation water through its effect on soil can affect the plant growth.

Sediment in water can have both positive and negative effects. The positive effect is in its soil building role in sandy soil and supply of nutrients for crops. The negative effects can be the sealing of soil surface, raised field elevations relative to canal outlets and thereby reduced water supplies. The standards for sediment load and will however have to be site specific.

Dissolved constituents in water particularly salts are important indicators of irrigation system performance. The effects of salt in irrigation water supplies have been extensively researched and standards are available for various salts for crops in a given soil and climate (Ayers and Westcot, 1988, Tyagi, 1992). Like water stress, the effect of salinity stress also varies with stage. The management can therefore vary concentration by mixing water to suit the physiological stage. The salinity can have both long-term and short-term effects.

6.0 PRODUCTIVITY

Irrigation systems are meant for providing water to crops for raising productivity of land and water. The excess or deficit of water supply has impact on crop yield though the effect varies with stage. The effect of water shortage may be reflected within a season whereas the excess application may not have immediate effect unless it creates conditions of oxygen deficiency. But in areas underlain by saline groundwaters it will affect sustainability of irrigation system. There are several approaches to simulate the effect of water supply regime (Bhirud et al 1990, Vijayaratna, 1988) but the one that is more elegant and has potential for application on large scale is due to Lenton

(1984).

6.1 Lenton's Water Delivery Performance Index

Lenton (1984) suggested a measure of irrigation system performance and termed it as water delivery performance (WDP). This performance measure takes into consideration of both the adequacy and the timelines of water supply and essentially represents productivity on 0-1 scale. The water delivery performance by Lenton's method is computed by the following expression;

$$WDP = \sum_{i=1}^{n} \frac{K_t V_t}{V^*_t} \qquad V(t) \leq V^*(t) \qquad (9)$$

where,

$$\sum_{t=1}^{n} K_t = 1$$

V^*_t = Target volume of water delivered to farm i during period 't' of cropping season, calculated for actual crops grown and other sources of water.

V_t = Actual volume of water delivered to the farm i during period 't'of cropping season.

K_t = Weighing factor indicating the relative importance of water at crop growth stage 't' of the cropping season.

n = Number of periods in the season

The detailed procedure to estimate values of K_t is given in Tyagi et al (1992).

The potential productivity index of Abernethy (1987) is also similar to Lenton's (1984) water delivery performance index. It also produces a number in the range 0 to 1 which would be obtainable under a given water supply regime. Abernethy (1990) suggests the possibility of developing a fully integrated indicator of the three major water supply characteristics of adequacy, equity and timeliness. Such an index would, in one value, sum up the system's performance in meeting crop demand fairly well if the relevant data could be collected.

7.0 STANDARDS OF PERFORMANCE

Performance indicators will be of little value unless they can be compared to targets set in implementation plan. Standards are important in formulating the indicators for assessing the performance. Svendsen and Small (1990) identify two categories of indicators, objective and normative. The objective indicators are derived directly from raw data and the application of standards to objective indicators results in normative indicators. The basic normative indicators of water supply system are adequacy, equity and timeliness and their specifications vary depending upon the standards used. Some targets range of values have been indicated as standards for some of the objective indicators.

Adequacy when expressed in terms of relative water supply may have value of 1.2 and 1.5. It may be pointed that if it is desired to ensure a RWS of 1.2 at farm level then the value has to be higher at watercourse or distributary level. Wade (1982) while working in Southern Indian irrigation projects, has shown that farmers willingness to organize into voluntary groupings, for performance and supervision of water management function is low if the RWS is more than 2.0 because they then perceive it a case of abundant water supply. The competition is killed again if RWS is less than 1.2 as it gives rise to high tension system. Standards of RWS can not be set without consideration for equity and vice versa. With high RWS, high equity will not be a valid objective and interquartile ratio of 3-4 may be acceptable, but with scarce water supply (low RWS) an interquartile ratio of 2 may also look to be high. Some of these performance indicators were monitored in a field study in Bhakra Canal Command.

8.0 IRRIGATION SYSTEM IN STUDY AREA

The area selected for performance analysis lies in the command of Fatehabad Branch Canal (FBC) which takes off from Bhakra Main Branch (BMB) canal. Eight watercourses in pairs of two (one lined and the other unlined) taking off from Gorakhpur and Khajuri distributaries in the head reach, Adampur in the middle reach and Kutiyana and Sheronwali in the tail reach (Fig 1), were selected for investigation. In addition, two watercourses on Fatehabad distributary in the tail reach near Fatehabad were also chosen. The salient hydraulic data for these watercourses are given in Table 3. The detailed procedure used in conducting the study is given in Tyagi (1993).

9.0 PERFORMANCE EVALUATION

The performance indicators chosen for evaluation were irrigation efficiencies at the farm level and adequacy, equity and productivity at the watercourse level.

9.1 On-Farm Irrigation System Performance

Only hydraulic performance of the field irrigation was evaluated. Graded borders are the most commonly used method of applying water to crops. The test borders were specified along the entire length of Fatehabad Branch Canal, in the command of selected watercourses. In general, three borders on each watercourse with locations in head, middle and tail reaches were selected. A detailed procedure described by Meriam (1978) was adopted and the variables that were monitored are given in Table 4.

9.2 Specifications of Irrigation Borders

Though the flow measurements were made in a limited number of borders, an extensive survey was conducted to establish the distribution of border dimensions (length and width) and the leveling index (Table 4). It is interesting to note that in most of the watercourse command areas, 60-70 percent of borders had leveling index less than 4 cm which indicates a reasonably good leveling.

Table 3 Salient hydraulic data for different watercourses

Distri-butary	Water-course No	Lined/ Unlined	Length (m)	Design discharge (lps)	GCA (ha)	CCA (ha)
Gorakhpur	5000L	Lined	1811	45.0	330	282
	3275L	Unlined	1980	45.0	296	288
	8738L	Lined	2010	46.4	305	288
Khajuri	9400L	Unlined	1440	28.0	177	169
	11260L	Unlined	1125	19.5	139	114
Fatehabad	179415L	Lined	3800	68.0	462	408
	204200L	Unlined	3300	45.6	392	273
Adampur	44000L	Lined	1500	29.7	178	185
Kutiyana	780L	Lined	3020	45.5	456	274
Sheronwali	2000R	Unlined	2850	34.6	290	195

GCA = Gross command area CCA = Culturable command area

Table 4 Variables measured for evaluation of field irrigation efficiencies

Sr. No.	Name of variable	Unit	Dimension
1.	Rate of flow	lps	L^3T^{-1}
2.	Duration of flow	minutes	T
3.	Advance time	minutes	T
4.	Recession time	minutes	T
5.	Border length	meter	L
6.	Border widths	meter	L
7.	Border slope	percent	1
8.	Leveling index	cm	L
9.	Field capacity	cm	L
10.	Soil moisture before irrigation	cm	L
11.	Infiltration rate*	cm/minute	LT^{-1}

* Computed from ring infiltrometer test data

9.2.1 Irrigation efficiencies

Measurements mentioned in Table 4 were used to compute the efficiencies of water application (water stored in the root zone/water applied), storage (water stored in the root zone/water needed to fulfill the moisture deficiency) and distribution (Table 5).

Table 5 Irrigation borders dimensions and leveling indices in different watercourses (% distribution)

Water-course	Length (m)			Width (m)				Leveling index(cm)		
	40-50	50-60	>60	4-6	6-8	8-10	>10	0-2	2-4	>4.0
				Gorakhpur Distributary						
5000L	14	68	18	8	27	50	15	5	75	20
3275L	12	77	11	11	25	45	9	7	58	35
8738L	16	74	10	8	30	54	6	9	62	20
				Khajuri Distributary						
9400L	10	75	15	16	35	42	7	9	59	32
11260R	14	71	15	30	44	22	4	6	57	37
				Fatehabad Distributary						
179415L	8	80	12	3	26	55	16	3	55	42
204200L	32	61	7	45	37	18	0	14	78	36
				Adampur Distributary						
27670L	19	75	6	22	29	36	13	18	61	31
				Kutiyana Distributary						
780L	29	63	8	32	36	24	0	12	52	36
				Sheronwali Distributary						
2000R	47	53	0	59	34	7	0	21	51	28

The application efficiency in most cases is quite high whereas the storage efficiency is low. Because of higher unit stream size, water spreads quickly and the irrigation is terminated before the required quantity has been diverted into the border. The distribution uniformity is also poor (less than 60% in majority of the cases). Thus the scope for improvement in design of water application practices exists.

9.3 Water Delivery Performance of Watercourses

Water supply to five watercourses on different distributaries was recorded. Delivery losses were measured along the watercourse to estimate the water reaching the head, middle and tail reaches of the watercourse. Cropping pattern and irrigated areas under different crops were taken from the records of irrigation department. Using data on climate, soil, crop and irrigation schedules, the daily values of evapotranspiration, effective rain and irrigation requirements and the crop yield during the season were determined. These outputs were then used to compute equity, adequacy, relative water supply and water productivity of the watercourse system in the head, middle and tail reaches(Table 6).

Table 6: Field irrigation efficiencies (%) in head, middle and tail reaches of watercourses.

Water-course	Location	Inflow rate (lps)	Stream size (lps/m)	Irrigation required (cm)	Land slope (%)	AE (%)	SE (%)	DU (%)
		Gorakhpur Distributary($Ic = 0.14t^{0.30}$)						
5000L	Head	50.8	4.064	5.9	0.15	85	49	70
	Middle	40.3	4.112	6.1	0.15	82	50	65
	Tail	29.6	2.846	6.5	0.15	87	57	68
3275L	Head	38.2	4.064	6.7	0.20	78	52	67
	Middle	23.1	2.852	7.4	0.16	75	56	72
	Tail	15.6	1.835	8.0	0.15	62	61	64
8738L	Head	36.1	3.539	5.6	0.20	65	52	36
	Middle	29.0	2.990	6.8	0.12	79	58	62
	Tail	21.5	2.529	7.5	0.15	80	64	58
		Adampur distributary ($Ic= 0.95t^{0.65}$)						
27670L	Head	39.5	5.640	6.9	0.25	84	55	42
	Middle	32.7	5.940	5.7	0.25	60	51	55
	Tail	26.4	3.120	6.5	0.30	62	59	35
		Fatehabad distributary ($Ic = 0.95t^{0.65}$)						
179415L	Head	65.0	9.630	7.0	0.30	80	42	90
	Middle	60.7	7.490	5.5	0.35	82	48	84
	Tail	51.5	6.870	6.2	0.35	80	55	59
204200L	Head	30.4	4.540	6.8	0.30	81	56	54
	Middle	20.3	2.550	7.1	0.30	76	71	19
	Tail	16.7	2.090	8.0	0.30	79	68	27
		Kutiyana distributary ($Ic = 1.2t^{0.5}$)						
780L	Head	39.5	5.640	5.4	0.30	92	65	30
	Middle	32.7	5.940	6.2	0.25	86	68	35
	Tail	26.4	3.120	4.9	0.25	96	75	61
		Sheronwali distributary ($Ic = 1.2\ t^{0.55}$)						
2000R	Head	30.5	4.760	7.5	0.20	72	78	65
	Middle	20.7	3.630	6.6	0.20	67	87	49
	Tail	15.8	3.200	5.8	0.20	58	74	58

AE = Application efficiency, S.E. = Storage efficiency,
DU = Distribution uniformity Ic = Cumulative infiltration, depth, cm,

9.3.1 Equity

Conveyance losses at different points were measured to compute the water being supplied to different farms and the equity as represented by Christiansen uniformity coefficient (C_u), coefficient of variation (C_v), modified interquartile ratio (IQR) and Theil index (T) were computed (Table 7).

Table 7 Values of equity measures in different watercourses

Distri-butary	Water-course No	Design discharge (lps)	Cu	Cv	T	IQR
Gorakhpur	5000L	45.0	0.81	0.22	0.63	1.69
	3275L	45.0	0.78	0.27	0.77	2.01
	8738L	46.4	0.88	0.14	0.25	1.65
Khajuri	9400L	28.0	0.95	0.06	0.66	1.48
	11260L	19.5	0.91	0.11	0.48	1.33
Fatehabad	179415L	68.0	0.82	0.12	0.40	2.58
	204200L	45.6	0.73	0.31	0.19	2.29
Adampur	44000L	29.7	0.94	0.12	0.19	1.36
Kutiyana	780L	45.5	0.89	0.18	0.17	1.60
Sheronwali	2000R	34.6	0.63	0.35	0.21	2.53

The values of Cu range from 0.63 to 0.95 for different watercourses with an average value of 0.8. If Cu were chosen as a measure of equity, the values of equity are apparently quite high, though there is a decrease in equity with increase in the size of the watercourse. The distribution looks more non-uniform when one computes IQR which represents the ratio of the average depth in the most favored quarter to the average depth in the least favored quarter. The IQR value at the lowest discharge of 19.5 lps is 1.33 and it increases to 2.58 at the highest discharge of 68 lps with an average value of 1.85. In other words the farms located in the head reaches of the watercourses were receiving nearly twice the water supply to the tail end farms. Kaushal et al (1992) also obtained similar results for distributary taking of from Sirsa Branch Canal.

9.3.2 Relative water supply (RWS)

Relative water supply of canal (Table 8) varied across the watercourses, season and reaches. Watercourse 9400L (Khajuri) and 179415L (Adampur) had relatively higher adequacy of about 80% as compared to watercourse 5000L(Gorakhpur) with 49% and 2000R (Sheronwali) with 58%. Watercourse 179415L was actually drawing more water than its designed discharge due to inaccurate installation of APM outlet. The sill level of the adjusted proportionate module (APM) was lower than provided in the design. RWS values during summer season, when crop water requirements are partly met by rainfall, were higher by 8-12% as compared to winter season. There was marked variation in RWS along the watercourses with values at the head exceeding those at the tail by 25%. For example, the average value at the head of watercourses in winter season was 0.65 as against 0.50 in the tail reach. Such large differences obviously call for immediate remedial measures.

Table 8 Relative water supply in different watercourses

Distributary	Watercourse	Head Summer	Head Winter	Middle Summer	Middle Winter	Tail Summer	Tail Winter	Average
Gorakhpur	5000L	0.58	0.54	0.52	0.47	0.45	0.38	0.49
Khajuri	9400L	0.88	0.77	0.86	0.87	0.72	0.70	0.80
Fatehabad	179415L	0.72	0.65	0.61	0.52	0.54	0.46	0.58
Adampur	780L	0.83	0.72	0.80	0.64	0.71	0.55	0.71
Sheronwali	2000R	0.61	0.56	0.56	0.52	0.47	0.43	0.62
Average		0.72	0.65	0.67	0.60	0.58	0.50	

9.3.3 Productivity of water delivery

The computed values of productivity are given in Table 9. There are some similarities as well as differences with the results obtained in RWS analysis. Like RWS, the productivity of water delivery was higher in the head reaches of all the watercourses, the average value during summer being 0.61 in head reach as against 0.48 in tail reach. But unlike RWS, the relative productivity potential is higher during winter season as compared to summer season. This may be due to occurrence of rain during supply periods in the summer season. In such cases, the supply is rendered surplus. Higher sensitivity of crops grown during summer to moisture stress could also be the reason for low water productivity. Another important difference is that in general the values of productivity are lower than seasonal RWS. It speaks of the mismatch between supply and demand with respect to time.

Table 9 Water delivery performance in terms of relative productivity potential

Distributary	Watercourse	Head Summer	Head Winter	Middle Summer	Middle Winter	Tail Summer	Tail Winter	Average
Gorakhpur	5000L	0.49	0.53	0.44	0.50	0.36	0.41	0.45
Khajuri	9400L	0.79	0.85	0.72	0.81	0.62	0.64	0.70
Fatehabad	179415	0.59	0.59	0.52	0.55	0.43	0.45	0.51
Adampur	780L	0.79	0.74	0.69	0.71	0.61	0.56	0.41
Sheronwali	2000R	0.51	0.55	0.47	0.50	0.44	0.46	0.49
Average		0.61	0.64	0.55	0.60	0.48	0.50	

10.0 SUMMARY AND CONCLUSIONS

The art and science of evaluating the performance of irrigation system is still in evolving stage. There is lack of field studies to correctly evaluate the usefulness of performance indicators. Further, no uniforms standards are available to judge the performance as good or bad. Evaluation of water delivery performance using the currently available measures indicated that major shortcomings in the system were:

(1) Lack of volumetric equity due to non-compensation of time for the water lost in conveyance and larger size unit command areas.

(2) Temporal mismatch between supply and demand, mainly due to system capacity constraint which makes it difficult to meet peak demand efficiently. Also fixed frequency schedules that are in vogue do not necessarily match with the demand for water.

These problems are amenable to solution with certain modifications in the existing system. Some of the suggested improvements are as follows.

(1) Provision of optimum unit command size

(2) Provision of variable Warabandi time schedule from head to tail to compensate for the conveyance losses.

(3) Provision of auxiliary storage at watercourse level during peak supply period for releasing the stored water during peak demand period.

(5) Intra-seasonal change in water delivery schedules to provide irrigation at more critical stages.

11.0 REFERENCES

Abernethy, C.L., 1987. The Objectives of Irrigation Water Management and Methods of Measuring their Achievement. In: Improvement in Irrigation Management with Special Reference to Developing Countries (ed K.K. Framji), I.C.I.D., New Delhi, State of the Art No. 4, pp.55-68.

Abernethy, C.L., 1990. Indicators of the Performance of Irrigation Water Distribution Systems. Memiographed Paper, International Irrigation Management Institute, Colombo, Sri Lanka, p 1-22.

Ayers, R.W. and Westcot, D.W. 1985. "Water Quality for Agriculture", FAO Irrigation and Drainage Paper 29 (Rev. 1), p. 74.

Bhirud, Sanjay, Tyagi, N.K. and Jaiswal, C.S., 1990. A Rational Approach for Modifying Rotational Water Delivery Schedule. Irrigation and Drainage Engineering, ASCE, 116(5):632-644.

Chambers, R., 1984. Irrigation Management: Ends Means and Opportunities. In: Productivity and Equity in Irrigation System (ed. Niranjan Pant), Ashish Publishing House, New Delhi.

Christiansen, J.E., 1942. Hydraulics of Sprinkler System for Irrigation. Trans. ASAE, 107:221-239.

Clemmens, A.J. and Bos, M.G., 1990. Statistical Methods for Irrigation System water Delivery Performance Evaluation. Irrigation and Drainage Systems, 4(4):345-365

Clyma, W. and Lowdermilk, M.K., 1988. Improving the Management of Irrigated Agriculture : A Methodology for Diagnostic Analysis. Water Management Synthesis Service Centre, Colorado State University, Fort Collins.

Eisel, H., 1988. Formulation and Evaluation of the USU Main System Allocation Model. Agricultural and Irrigation Engineering Department, Utah State University, Logan, Utah, WMS II Report 76.

Lenton, R.A., 1984. A Note on Monitoring Productivity and Equity in Irrigation Systems. Productivity and Equity in Irrigation Systems (ed N. Pant), Ashish Publishing House, New Delhi

Meriam, J.L. and Keller, J., 1978. Farm Irrigation System Evaluation: A Guide for Management. Utah State University, Logan, p.271.

Molden, D.J., 1988. Water Control as a Basis for Evaluation of Irrigation Water Delivery Systems. Unpublished Ph.D. Dissertation, Civil Engineering Department, Colorado State University, Fort Collins, USA.

Molden, D.J. and Gates, T.K., 1990. Performance Measures for Evaluation of Irrigation Water Delivery Systems. J. of Irrigation and Drainage Engineering, Proc. ASCE, 116(6):804-823.

Rao, P.S. 1987. "Relative Equity Ratio: Concept and Method of Computation", Unpublished Paper.

Sampath, R.K. 1990. "Measures of Inequity for Distribution Analysis of Large Public Surface Irrigation Systems: A Welfare Theoretic Approach." Chapter 5 In: Social, Economic and Institutional Issues in Third World Irrigation Management, (ed R.K. Sampath and R.A. Young), Studies in Water Policy and Management, Westview Press, Boulder, Colorado.

Small, L.E. and Svendsen, M., 1990. A Framework for Assessing Irrigation Performance. Irrigation and Drainage Systems, 4(4):283-312.

Svendsen and Small, L.E., 1990. Farmer's Perspective on Irrigation Performance. Irrigation and Drainage Systems, 4(9):385-402.

Smith, L.E.D., 1990. An Economist's Perspective on Irrigation Performance Assessment with Examples form Large Scale Irrigation. Morocco, Irrigation and Drainage Systems, 4:329-343.

Tyagi, N.K., 1988. Development of Management Systems of Irrigation Under various Sets of Conditions. Annual Report, Central Soil Salinity Research Institute, Karnal, pp. 52-55.

Tyagi, N.K. 1992. " Managing Irrigation with Saline Waters", Seminar on Irrigation Water Management, New Delhi, July 31 - August 2, 1992, Vol.2

Tyagi, N.K. 1993. Performance evaluation of an irrigation system: A case study , In: Sustainable Irrigation in Saline Environment (eds Tyagi et al), National Workshop on Sustainable Irrigation, CSSRI, Karnal, Feb. 17-19, 1993. 1:1-15.

Vijayaratna, C.M. 1986. Assessing Irrigation System performance; A Methodological Study with Application to Gal Oya Scheme. Sri Lanka, Ph.D. Dissertation, Cornell University, Ithaca, USA.

Wade, R., 1982. Employment Water Control and Irrigation Institution: India and South Korea. International Labour Organization and Institute of Development Studies, Discussion Paper No. 182, Institute of Development Studies, U.K.

CAN PROTECTIVE IRRIGATION BE MORE EFFICIENT?

M. Jurriëns[1]; W. Wolters[2]
[1] International Institute for Land Reclamation and Improvement (ILRI), P.O.Box 45, 6700 Wageningen, The Netherlands.
[2] Agricultural University, Nieuwe Kanaal 11, 6709 Wageningen, The Netherlands

ABSTRACT

In India, numerous irrigation systems have been designed for "protective irrigation". Since developments in the agricultural sector have lead to the wish for more productive irrigation, this has resulted in problems of inequitable water distribution and under-utilization of the irrigable area. The paper makes the point that problems now faced are directly related to the difference between actual water demands and available resources.
This issue is illustrated with detailed information from the Tungabhadra Left Bank System in Karnataka. First, the initial design is reviewed in terms of irrigation intensities and duties. Subsequently, the essentials of the current situation are depicted and it is shown how the present practices are conflicting with the initial design. In this situation it is often suggested that increasing the irrigation efficiencies would be (part of) the solution to combat water shortage. This is assessed in the second part of the paper.

After a brief review of the efficiency terminology and the problems involved, actual irrigation efficiencies realized for several years are discussed. Results show a wide variation and it is concluded that no definite values can be determined. Very high values point at serious under-irrigation and low values indicate water losses. Results in rabi are generally higher than in kharif, even when not counting rainfall. It is concluded that one should concentrate on identifying possibilities for reducing avoidable losses, based on field information. This is done in the remaining of the paper, distinguishing between kharif and rabi, between various system levels and considering the effects of rainfall.

In the Tungabhadra case the possibilities in the main system seem to be limited and most promising measure would be to apply a canal closure period during the kharif harvesting, when losses are highest. There seems to be some scope for reducing losses at chak and field level. Moreover, losses could be reduced when a more reliable outlet flow could be guaranteed. Although losses during rain can be considerable, a number of practical reasons is given why it is not likely that measures can be found to substantially reduce canal water losses during rainfall periods.
On a number of aspects, the Tungabhadra case may not be representative for all Indian irrigation. But at least one lesson will be more generally valid, namely that it is not realistic to expect, particularly with protective irrigation, that efficiencies could be drastically improved.

1. INTRODUCTION

Irrigation plays a major role in Indian agriculture. Since independence many millions of hectares have been brought under new irrigation systems. The past decade one began to realize, however, that more attention should

V. P. Singh and B. Kumar (eds.), Water Resources Planning and Management, 235–245.
© 1996 *Kluwer Academic Publishers. Printed in the Netherlands.*

be paid to the performance of existing schemes, which were generally showing problems and scope for improvements.
Logically in this situation, a wealth of literature has been produced on the problems in Indian irrigation, apart from the common numerous textbooks on technical design. This literature deals with nearly all aspects of irrigation in the fields of technology, management, agriculture, sociology, economics, politics, etc. Strangely enough, however, one crucial aspect is very scarcely discussed and analyzed: the relation between the amounts of required and available water, or in other words: the design of the systems in terms of water allocations and its implications for the present performance.
The paper concentrates on this water aspect and illustrates that important problems are directly related to the design objectives, which in principle involve the ratio of water availability and water requirements.

Increasing the irrigation efficiencies is often advocated as a solution to all problems in situations of water shortage. Therefore, actual efficiencies are reviewed and possibilities for improvement are analyzed, showing that increasing the efficiencies is not a realistic option to solve the problems. At the same time it is illustrated that the common efficiency definitions are not well workable in such situations of protective irrigation.

The paper is based on research in the Tungabhadra Left Bank System in South India. Because it is realized that this system is not completely representative for the entire Indian irrigation scene, further analysis is needed on similarities and differences with other situations, to assess the general validity of the conclusions of this paper.

2. IRRIGATION SYSTEM DESIGN PRINCIPLES

The design of an irrigation system is in principle based on three elements: the command area to be covered, the irrigation intensity to be achieved and the amount of water to be allocated per unit area, all this taking into account the amount of water (expected to be) available in the river or reservoir.
Irrigation design in large parts of India, as well as in Pakistan, is primarily based on the "duty" concept. The duty, expressed in acres per cusec, gives the area that is to be irrigated by a unit flow size. For a given (estimated) amount of available water it fixes the size of the area to be commanded. The reverse value is the "allowance", or the amount of water to be supplied to a unit area. A duty of 70 acres/cusec equals 1 l/s per ha.

In theory, the duty or allowance could be based on estimated crop water requirements, but the essence of the design of most Indian schemes is that this is not the case. Many systems have been designed for "protective" irrigation, meaning that a limited amount of water is to be spread thinly over as large an area as possible, to protect as many farmers as possible from total crop failure. This usually means large command areas, high duties and low intensities. Duties often are as high as 200-400 acres per cusec, which means a unit supply of about 0.15-0.30 l/s.ha (at main canal head)! Evidently, this will usually not suffice to adequately irrigate the entire command area. Consequently, when a reasonable allowance would be estimated for instance at three times the design value, the implication would be that the target irrigation intensity is about 30%. In India (and Pakistan), usually both duties and target intensities are explicitly given. Berkoff (1991) and Jurriëns (1993) give a number of examples, based on Indian literature and experience. Sometimes, however, design intensities are not explicitly given.

Even when the protective irrigation concept would have worked in the beginning, the problems may well have increased because over the past decades requirements in terms of water distribution in quantities and timing have increased due to the changed agricultural situation (more paddy, more high yielding varieties, more irrigation in both seasons and remunerative agriculture; see Jurriëns, 1993).

Evidently, irrigation has yielded impressive results in Indian agriculture. But at the same time it is known that there are serious problems in various large canal systems. Many problems are undoubtedly related to the above sketched implications of design elements and protective irrigation. This water aspect will now be illustrated in more detail for the Tungabhadra Left Bank System in Karnataka.

3. THE TUNGABHADRA CASE

Basic information on the Tungabhadra Project is given in Jurriëns et al. (1987) and in Satnarayan Singh (1989). Here we will only discuss the Left Bank Command Area (LBCA), which covers 244,000 ha and is supplied by one of the three main canals departing from a reservoir. This Tungabhadra scheme was designed for protective irrigation, both expressed in intensities and duties. The basic principle is depicted in Table 1.

Table 1. Design intensities and duties and estimated requirements

CROPS	Area (ha)	% of total	Duties (ac/cus)	Allowances (l/s.ha)	Theor. CWR (l/s.ha)
Kharif					
- paddy	21100		55	1.27	1.65
- light	89300	46	150	0.47	0.88
Two-seasonal					
- Sugar	8400		75	0.93	-
- Cotton	30000		120	0.58	1.18
- Garden	6300	18	100	0.70	-
Rabi					
- paddy	-		40	1.75	2.06
- light	88800	36	120	0.58	1.18
Total	243900	100			

The design irrigation intensity was 100%: the protective element is that every piece of land can only have one irrigated crop. Part of the command area (18%) is supposed to be irrigated under bi-seasonal crops. Of the remaining land, part is to be irrigated in the wet season ("kharif") only (46%), the other part (36%) in the dry season ("rabi") only.

The design duties further underline the protective nature. Different design duties were given for the various planned crops. These are also given in Table 1, together with their conversion into l/s.ha allowances and the theoretically calculated crop water requirements. Whatever may be said of the latter estimates, they support the general statements in the area that duties "cannot be realized", and they are in line with results from research stations and proposals made by various committees, all opining that duties do not cover crop water requirements.

Implementation of these design principles was thought to be facilitated by the typical South Indian practice of crop "localization". This prescribes which crop has to be cultivated on which land and in which season. Such details are given for the entire scheme, for each distributary, each "chak" (tertiary unit), and even for every survey number (a cadastral unit, varying from some 5 to 25 ha). Table 1 gives the localization for the LBCA, showing that large parts are planned under "light" (low water demanding) crops.

Farmers are not much interested in the above design ideas. They want to have an irrigated crop in both seasons. Irrigation canals are running throughout the year, because in each season and in each block there are areas to be served which are localized for that season, and thus, upstream farmers are largely succeeding in having two irrigated crops a year, though being localized for one season only.
Also, farmers do not care much about the officially localized crops. They have their own ideas on which crops to grow, resulting in much more paddy being cultivated instead of localized "light" crops (crops with low water requirements, including various grains, pulses and oil seeds); even in rabi, when paddy is prohibited at all. Moreover, one logically tries to give sufficient water to cover crop water requirements, which is more than according to the duties, and again the head-enders often succeed in doing so. Finally, some land under the command was, for unknown reasons, not localized at all, but appears to be perfectly irrigable in practice. Much of such land is being irrigated in reality, thus increasing the struggle for the scarce water.

The above clearly pictures a major problem in the scheme: in the upstream areas of scheme and distributaries there is widespread double cropping (often paddy!) with high yields, at the cost of large tail-ends without any irrigation at all. In essence this is the result of the fact that the initial objectives of protective irrigation and related design elements are in essential conflict with the farmers wishes for productive irrigation. For detailed figures see for instance Jurriëns and Landstra (1989) and Jurriëns and Ramaiah (1989), where it is shown that there is a problem related to the ratio of water demands and water availability.

4. IRRIGATED AREAS AND WATER USE

The nature of protective irrigation is most clearly understood when it is seen in quantitative terms of water. For the Tungabhadra case it can be illustrated as follows.
Initially, the Left Bank Main Canal would be allowed a total annual release from the reservoir of 102 TMC (thousand million cubic feet) or 2890 MCM (million cubic meters). For the command of 243,900 ha with a 100%

intensity, this would come down to 1185 mm per year. Even assuming an overall efficiency of 45 % (which seems rather high), this would make available about 530 mm for crop level evapotranspiration. Evidently, with most crops requiring more than that, this illustrates a rather limited water availability. Over the years, the situation has even become worse, because less water has become available due to siltation of the reservoir.

Detailed data have been recorded on main canal releases (from the reservoir) and on irrigated crops and areas. Processing of these data for the period 1975-1987 shows that on the average, 2280 MCM has been available from the reservoir for the LBC with a maximum year of 2465 MCM. Apparently not more water can be made available than this order of magnitude. The released water was used to irrigate an average area of 141,100 ha, coming down to a water supply of 1620 mm for the area actually irrigated. Evidently, this is much more than the 1185 mm according to planning, and certainly much more than the 935 mm which should be supplied when the available reservoir water (2280 MCM) would have been spread over the entire command area. Plan and reality are schematically depicted in Figure 1.

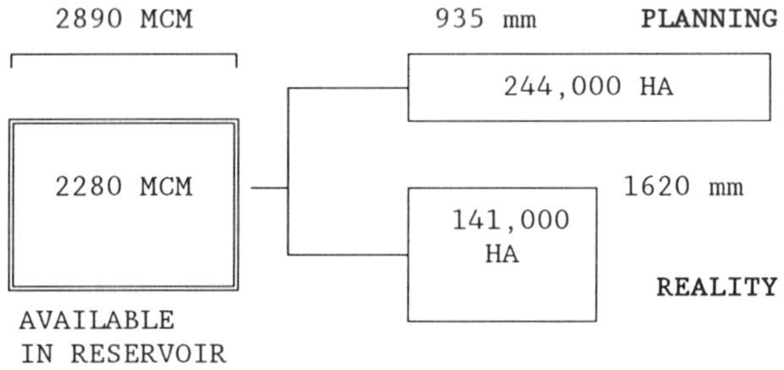

Figure 1. Plan and reality in water distribution and water use.

Farmers do not feel the official allocations sufficient to grow an acceptable crop. Therefore, they try to take more than allowed, leaving less for downstream areas and the result is that the land actually irrigated is much less than targeted. Data for the above mentioned period indicate that it has never been possible to irrigate the entire command area.

Summarizing, the simple, but nevertheless important factual observation from the above is that on the average about 60% of the area is irrigated with a water supply per irrigated ha of about 1.7 times the allocated amount. Under-irrigation in terms of area and over-irrigation in terms of water are of course directly related.

Some other aspects are involved, which need a brief mentioning. In reality the situation is even more problematic. Many farmers who succeed in getting sufficient water for a kharif crop are also in the position to irrigate a second crop in rabi. This means that a same piece of land is irrigated twice whereas the localization is for one crop only. Thus, the land actually irrigated, must have been much less than the above 141,000 ha. Detailed figures are not available, but there are indications that only about half of the land in the command area is actually getting some irrigation water. Finally it should be realized that in years with high irrigation intensities (and low unit supplies in terms of mm), yield reductions are common, which is not expressed in the figures.

The above shows that the low ratio between available and required water has lead to considerable problems in terms of not realizing target intensities and of inequitable water distribution. Figure 1 suggests that to irrigate the entire area, some 40% water should be saved to restore the balance between requirements and supplies. But even if the intensity would then be 100%, still not the entire land in the command would receive irrigation, because of the double irrigation on part of the area (Jurriëns, 1991). In this paper we emphasize on the possibilities to increase the efficiencies. This is the more interesting because "more efficient irrigation" is often suggested to be the solution to all problems.

5. THE EFFICIENCY CONCEPT AND IMPLICATIONS

Although one of the most discussed indicators, the notion of irrigation efficiencies still involves numerous complications. Because an understanding of this is necessary for the following considerations, some aspects will be briefly mentioned in this section.

First the definition itself. The commonly used definition is in the form of $E=(ET_p-R_e)/S$ with ET_p the potential crop evapotranspiration, R_e the effective rainfall and S the total irrigation supplies. Even with zero rainfall and data being collected on seasonal or annual supplies, the problem still is to determine ET_p. The real ET_p in a large scheme is always difficult to estimate (crop calendars, crop coefficients) and an error in ET_p is linearly translated into an error in the efficiency. But also, actual evapotranspiration may well be less than the potential, particularly when there are relative water shortages in part of the area. It is practically impossible to know even approximately how much the actual ET_a in a given situation is. When then the ET_p is used in the calculation this will result in an over-estimation of the efficiency (see Wolters, 1991, for an extensive discussion on this aspect).
A further complication is how to properly take rainfall into account. When water is running off unused during rainfall, nobody can say whether this canal water or rain water. The real R_e simply cannot be determined and any change in its assumed value directly affects the calculated efficiency value. Moreover, if rain is high compared to ET_p, the efficiency value can become very low or even negative!.

The above refers to efficiencies in which crop water use at field level is involved. For higher level efficiencies in the system, the efficiencies only include seepage and operational losses. But those as well cannot be determined in practice, and certainly not for the entire system.
Finally it is observed that existing efficiency figures usually are for the entire season. In a peak period they may well be higher. And conversely, it may not be important in a practice to strive for high efficiencies beyond the peak period.

6. ACTUAL EFFICIENCIES REALIZED

To estimate actual efficiencies for the Tungabhadra system we have to estimate the crop water requirements and areas actually irrigated. It has been indicated above that this is difficult. For instance, crop calendars vary widely over the vast scheme area and it is not known what a reasonable representative crop calendar for the different crops would be. Also, the real crop coefficients are not precisely known. Consequently, only indicative estimates can be made.

On the other hand, main canal releases from the reservoir have been recorded and monthly totals are available. Using these data and ET_p estimates, the scheme efficiencies have been estimated, with and without taking into account the rainfall. Three years have been taken, one with very high releases, one with very low releases and an average year. Considering the observations in the previous section, the formula given there has not been used. Instead, calculations have been made with the formula: $E = ET_p/(S+R)$, with S=canal supplies and R=total rain. The latter means that rainfall is considered as additional "irrigation" sypply, which complies with the protective irrigation principle. Results for the three years are given in Table 2. Canal releases are given in million m^3 (MCM).

The table shows that results are varying over the years and between the seasons. If rain is not counted for kharif, efficiencies roughly vary from 30-70% for this season, and from 35-80% for the year. In rabi, efficiencies are always higher than in kharif, roughly from 50-95%.

Table 2. Actual irrigation efficiencies

Year/ season	Irrigated area (ha)	Releases (MCM)	Eff. (no rain)	Rain (mm)	Eff. (incl. rain)
1980/81					
kharif	64920	1327	0.27		0.21
rabi		1138	0.47		
year	116350	2465	0.36	650	0.21
1985/86					
kharif	102940	1189	0.46		0.30
rabi		932	0.81		
year	161680	2121	0.61	600	0.33
1987/88					
kharif	117280	962	0.68		0.28
rabi		875	0.96		
year	202360	1837	0.81	1190	0.36

Not only are results varying, but it is also seen that the higher the irrigated area, the higher the efficiency. There could at least be two explanations for this. One is that when water management is more efficient, more area can be irrigated. The other is completely different: the more area one tries to cover, the less the unit water use per ha will be and consequently under-irrigation and yield reductions will occur. In that case, both irrigated areas and efficiencies are over-estimated. The high values for 1987-88 for instance must point at considerable under-irrigation (yield reductions were indeed recorded in reality) and the low values in 1980/81 may point at real losses and inefficient water use. This analysis is supported by the supply figures for the irrigated areas: 910 mm in 1987 and as high as 2120 mm in 1980!.

When kharif rain is taken into account the efficiencies are considerably lower, about 20-30% for kharif and 20-35% for the year.

7. SCOPE FOR IMPROVEMENT

Comparing the two results for 1980 and 1987 it may be concluded that the 1980 performance was not good because of too much losses, but the 1987 results were not good because of too much yield reduction. With a certain amount of water available there is a limit to the area that can be irrigated. Too much area irrigated can result in high efficiencies on paper, but this does not say anything on actual water losses or yields. It is therefore impossible to say what an attainable efficiency would be that at the same time would include acceptable yields. Probably, increasing the efficiencies would only mean even more under-irrigation and yield reductions. Consequently it is better not to talk too much about efficiencies, but rather to look at possibilities to reduce avoidable losses in the system.
At least three aspects are then to be investigated:
- difference between kharif and rabi;
- losses at various system levels;
- coping with rainfall.

* Difference between kharif and rabi

The above high efficiencies for rabi are partly due to under-irrigation. But also, irrigation is indeed more efficient in rabi because there is no rain and canal water is often less or less reliable. Field observations as well as flow measurements in drains in Tungabhadra confirmed that losses in rabi were much less than in kharif. Table 3 gives results for some drains in more upstream reaches of one distributary command. Drain 4 is the most upstream and drain 1 the most downstream (still in the middle of the command). Figures indicate that losses rapidly decrease in downstream direction, which would suggest that reducing losses will only be possible over a limited area.

Table 3. Measured losses in some tertiary drains (l/s.ha)

	Year 1; months			Year 2; months		
Drain no.	ASO	NDJ	FMA	ASO	NDJ	FMA
1	0.01	0.03	0.00	0.03	0.07	0.00
2	0.04	0.08	0.00	0.03	0.09	0.01
3	0.20	0.20	0.07	0.09	0.15	0.02
4	0.18	0.18	0.09	0.10	0.13	0.04

The above suggests that generally not much can be gained in rabi and that efforts to reduce the losses should concentrate on the kharif season and on parts of the area only.

The table also illustrates another interesting aspect. Average losses are highest in the months November, December and January, because this period includes the kharif harvesting period, roughly in November-December. Much less water is then needed for irrigation. But canals are still running and much of the canal water is running off. A well organized canal closure for some weeks could substantially reduce the losses.

* Losses at various system levels

Water losses are of different nature, related to the level of the system concerned. Seepage losses in the 227 km long, lined main canal have been estimated at 15% (standard design value) but are not really known. A few measurements carried out gave results in this order of magnitude. Seepage losses in the unlined distributaries are also not known. The canals are largely in black cotton soils and a few measurements here indicated that the design assumption of 10% losses may be reasonable. Lining of all distributaries does not seem to be justified, the more so because the real gains in water losses will be limited.
Operational losses in the main system are limited because the canals are running continuously. Only the above mentioned canal closure during harvesting could reduce the losses.

It is likely that most of the water losses occur within the chak. Again, nothing is known on the actual losses in watercourses and field channels. The impression from field experience is that there are considerable run-off losses from the fields, particularly of course in the upstream reaches of the canals. Table 3 above gives an illustration. This seems to be mainly due to three factors.

One is that because of the ever fluctuating outlet flows farmers try to take as much water as possible when it is there (see Jurriëns, 1992). Subsequently the excess water runs off. Another aspect is that farmers want to irrigate their paddy once every two to three days. When intervals would be prolonged this could lead to more efficient water use. And finally the sometimes large flows are inappropriate to efficiently irrigate dry-foot crops on the often sloping fields. Better land grading and more adequate flow sizes could possibly reduce the losses.
For the latter it will usually be needed to change the internal water distribution and a good irrigation service for assistance could be useful in that sense. But this does not mean that "introduction of warabandi", as is often advocated, would be a relevant recommendation. Some kind of organized rotation with rosters and rules exists in nearly all chaks (except where water is abundant or absent). "Introduction" is not necessary, improvement of existing schedules may sometimes be possible.

* Coping with rainfall

Table 2 indicates that total efficiencies are substantially reduced when rain is taken into account. However, it will be extremely difficult to use rainfall more efficiently. Rainfall is not only very much varying over the years, but also it can fall in a few heavy showers or many light showers. Moreover, on this vast command area it has a large spatial variability. It is hard to see how canal operation could be adjusted to the rainfall pattern. Moreover, design duties are in addition to rainfall, so that normally canals should indeed continue to run, which with the given structures can only be at full supply level. Subsequently, even if one would be able to save some water, often at that moment the reservoir is full and the water to be saved is flowing off over the reservoir spillway.

Finally, one must be aware of the following. When heavy rain is falling in some distributary command, and no canal water is needed for some period, outlets are closed which means that canal water reaches the tail end of the canal, where it normally does not reach. It has been observed that in such tail ends, close to the canal, farmers are able to produce a kharif light crop just because of one or two of such additional irrigations. Successful reduction of canals flows would even deprive them of this small benefit.

8. SUMMARY AND CONCLUSIONS

In the first part of this paper the essentials of protective irrigation are outlined, showing that it is crucial to look into the water aspect when analyzing canal irrigation performance. For the Tungabhadra case it is illustrated how the combination of high duties and low intensities is at the basis of the current problems of inequitable distribution and large areas not being irrigated at all.

To solve the Tungabhadra problem, it would at least be required to reduce the water losses by some 40%. Possibilities to reduce the losses in the main system seem to be limited and most promising measure would be to apply a canal closure period during the kharif harvesting, when losses are highest. There seems to be some scope for reducing losses at chak and field level. Improving the existing rotation schedules could facilitate longer paddy irrigation intervals and more appropriate flow sizes for field application. Better land levelling could also improve the field irrigation. Furthermore, losses could be reduced when a more reliable outlet flow could be guaranteed. Although losses during rain can be considerable, a number of practical reasons is given why it is not likely that measures can be found to substantially reduce canal water losses during rainfall periods.

All above cannot be quantified, but field information and data suggest that the 40% reduction of losses will by far not be possible. And even so, one would still face the distribution problem that an increase of available water would go to double irrigation on the same land instead of single season irrigation on new land.

It is fully realized that the described Tungabhadra case may not be representative for all Indian canal irrigation, and certainly not for large systems in the North. Yet, the analysis may illustrate a number of points of more general validity:

- When assessing protective canal irrigation, it is crucial to investigate the principle water elements such as intensities, duties and command area, on the basis of which one can compare the available water and the real requirements;
- In such situations, trying to use the limited resources more efficiently is a first (but not sufficient) requirement;

When discussing the efficiency definition and calculation results for some years it was seen that use of the common efficiency concept in this respect only yields indicative information:

- The real efficiencies cannot be determined, particularly not when rain or under-irrigation is involved.
- Yet, calculation results can be used to get a broad impression on what happened. In one year for instance the efficiency figure is very low,

pointing at high losses. In another year it is very high, indicating severe under-irrigation.
- Figures also indicate that efficiencies in rabi are generally higher than in kharif and that it is appropriate to concentrate on improvements in kharif.

To assess the scope for more efficient water use one should systematically analyze the possibilities to reduce losses at various levels in the system, on the basis of the real situation in the field, supported by measurements. Such information from Tungabhadra suggests that it may not be realistic to expect that the efficiency of protective irrigation can be significantly improved.

REFERENCES

- Berkoff, D.J.W. (1991). India- Irrigation sector review: irrigation management on the Indo- Gangatic plain. In: Improved Irrigation Systems Performance for Sustainable Agriculture; pp. 67-92. FAO, Rome.
- Jurriëns, M. (1991). Irrigation problems in India: a matter of water? Paper presented at ODI Lunch-time meeting, October 1991. ODI, London.
- Jurriëns, M. (1993). Protective irrigation: essence and implications. In: Proceedings of the 15th ICID Congress, The Hague, 1993.
- Jurriëns, M., V. Ramaiah and J.G. v.Alphen (1988). Irrigation water management in the Tungabhadra Scheme, India. In: ILRI Annual Report 1987. ILRI Reprint No 50
- Jurriëns, M. and V. Ramaiah (1989). Water distribution in a secondary irrigation canal: Results of a measurement program. In: ILRI Annual Report 1988. ILRI Reprint No 55
- Jurriëns, M. and W. Landstra (1989). Water distribution: Conflicting Objectives of Scheme Management and Farmers. In: Irrigation: Theory and Practice, Proceedings of the International Conference, Southampton, 1989. ILRI Reprint No 58
- Jurriëns, M. (1992). Interactions between main system and tertiary units. In: Irrigators and engineers. G. Diemer and J. Slabbers (eds); Thesis Publishers, Amsterdam, 1992. ILRI Reprint No. 76.
- Satnarayan Singh (1989). Issues in canal irrigation in South India; past, present and future. Wamana, July 1989. Bangalore, India.
- Wolters, W. (1991). Water shortage and irrigation efficiencies. In: ILRI Annual Report 1990.
- Wolters, W. (1992). Influences on the efficiency of irrigation water use. ILRI Publication 51. ILRI, Wageningen.

SECTION 4:

REMOTE SENSING IN WATER RESOURCES INVESTIGATIONS

POTENTIAL OF REMOTE SENSING IN APPLIED HYDROLOGY

IVAN MUZIK
Department of Civil Engineering
The University of Calgary
Calgary, Alberta, T2N 1N4
Canada

ABSTRACT. Population growth, extension of irrigated agriculture and intense industrialization are stressing the quantity and quality of water resources globally. Increasingly, the assessment, planning and rational management of water resources requires more and better information on the state of various components of hydrologic systems on a large scale, then ever before. Remote sensing offers a unique approach to fulfil this need. The purpose of the paper is to summarize first the physical principles of remote sensing to provide the basis for understanding the use of remotely sensed data, and secondly to review the most significant applications of remote sensing techniques in hydrology, including estimation of rainfall, runoff, snow cover, soil moisture, groundwater, and water quality.

1. Introduction

It is generally recognized that availability of water has played a crucial role in development of human societies. Even today, the amount and quality of usable water is a severe limiting factor for human development. According to Catley-Carlson (1988), "lack of clean water and adequate sanitation is a contributing factor in the death of 12.4 million people annually. That's a daily toll of 34,000 deaths, or the equivalent of fatal crashes of 100 jumbo jets each day - one every 15 minutes!"

Equally devastating can be the effects of catastrophic floods. Clearly, water resources management is an on-going process and much effort will be necessary to alleviate existing problems. One reason for inefficient management of water lies in inadequate knowledge of spatial and temporal distributions of hydrologic parameters. To overcome this inadequacy we usually apply one of the following remedies (Schultz, 1988): (a) collection of more, and more reliable conventional data; (b) application of more sophisticated mathematical techniques; and (c) use of new data acquisition techniques such as remote sensing. Relative merits of each of the above remedies become quite illuminated by considering the statement of Klemes (1981): "Improvement of an operational (hydrologic) model requires either more empirical data or more casual input, i.e. more information on the phenomenon being modelled. It cannot be achieved merely by more mathematics. The degree of mathematical sophistication has no point beyond the "carrying capacity" of the amount of the physical information available". Thus, while collection of conventional data must continue the benefits of this remedy will not be usually

V. P. Singh and B. Kumar (eds.), Water Resources Planning and Management, 249–260.
© 1996 *Kluwer Academic Publishers. Printed in the Netherlands.*

realized until many years or decades from now. For the present day water management needs remedy (c), specifically application of remote sensing carries the greatest potential for improving our ability of resolving conflicts and reducing uncertainty arising in water resources engineering and management.

For hydrological applications the most relevant characteristics of remote sensing data are: (1) relatively high resolution in space and/or time; (2) area measurements instead of point measurements; (3) coverage of remote areas where no measurements could be obtained otherwise; (4) data available in digital form; and (5) data collected and stored centrally. The objective of the paper is to provide an overview of the current state of applications of satellite remotely sensed data in hydrological modeling and water resources management. This includes a brief description of physical principles of remote sensing, data acquisition and analyses.

2. Basic Principles of Remote Sensing

Remote sensing may be broadly defined as the collection of information about a target (i.e. an object, are, or a phenomenon) by a measuring device (sensor) which is not in physical contact with the target. This definition would include techniques such as magnetometery, gravimetry, sonar and spectroscopy, however, the term "remote sensing" has acquired a more restrictive meaning in that it usually refers to the measurement of electromagnetic energy which is reflected and/or emitted by earth surface features.

A distinction can be made between active and passive remote sensing depending upon the source of electromagnetic energy sensed. An active remote sensing system artificially generates a signal which is transmitted towards a target. The return signal is detected and the difference between the transmitted and return signal is of interest. An example of an active remote sensing system is a radar. In passive remote sensing sensors detect the natural energy, either reflected or emitted, which arrives from the target. The most significant source of natural electromagnetic energy is the sun. However, all objects having temperature above the absolute zero emit electromagnetic radiation.

Application of remotely sensed data can in many cases benefit engineering and management analysis and decisions made in water resources. Efficient use of remote sensing requires some knowledge of basic elements of the data acquisition process, which can be listed as follows: energy source, propagation of energy through atmosphere, energy interaction with earth features, airborne/spaceborn sensors, sensor data (pictorial or numerical), data analysis and interpretation (visual or digital). A brief discussion of these elements is given in the next few sections, followed by a review of major applications in water resources.

2.1. ELECTROMAGNETIC RADIATION

According to the wave theory, electromagnetic energy propagates through space in a sinusoidal fashion such that

$$c = f \lambda \tag{1}$$

where c is the speed of light (299, 893 km/s), f is the frequency, and λ is the wavelength.

It is common in remote sensing to define electromagnetic radiation in terms of its wavelength. The electromagnetic spectrum has been divided for convenience into "bands", and each band given a name. Commonly used bands in remote sensing are listed in Table 1. It should be noted

TABLE 1. Electromagnetic Spectral Bands Used in Remote Sensing

BAND	WAVELENGTH
Ultraviolet, UV	3 nm to 0.4 μm
Visible	0.4 to 0.7 μm
Infrared, IR	0.7 to 300 μm
Reflected IR	0.7 to 3 μm
Thermal IR	3 to 5 μm 8 to 14 μm
Microwave (Radar)	0.3 to 300 cm

that the infrared radiation is divided into two bands, according to its origin. The first band, 0.7 to 3 μm, is the reflected sun's infrared radiation. The second band, between 3 and 14 μm, is the emitted radiation by the target and depends on the target's temperature. It is called the thermal infrared radiation. How much energy an ideal radiator (i.e. an object that totally absorbs and re-emits all energy incident upon it) is given by the Stefan-Boltzmann law

$$W = \sigma T^4 \tag{2}$$

where W is the total emitted radiation from the surface of the object in Wm^{-2}, σ is a constant (5.6697×10^{-8} W m^{-2} $°K^{-4}$), and T is the absolute temperature of the emitting surface in °K.

The dominant wavelength λ_m, or wavelength at which an ideal radiator emits maximum energy, is related to its temperature T by Wien's displacement law

$$\lambda_m = \frac{A}{T} \tag{3}$$

where A is a constant equal to 2898 μm °K.

One can easily verify, by means of the Wien's law, that the sun (average surface temperature 6000 °K) emits maximum energy in the visible range, whereas for the earth surface (average temperature 300°K) the maximum occurs in the thermal infrared band.

The amount of energy to be sensed and its wavelength are two very important parameters in remote sensing. In a broad sense, they relate to sensors design on one hand, and to the selection of the most appropriate band or bands in which to collect data for a given task, on the other hand. Electromagnetic energy interacts both with atmosphere, as it travels to and from a target, and the target itself. This interaction is due to the processes of scattering, absorption and transmission, which are all wavelength dependent. Atmosphere selectively attenuates radiation thus limiting the electromagnetic spectrum usable for remote sensing to the so called atmospheric windows, or spectrum bands for which atmospheric attenuation is negligible or small.

When electromagnetic energy reaches the earth's surface the proportions of energy reflected (scattered), absorbed, and transmitted will vary for different earth features. These differences (measured by sensors as reflected or emitted radiation) permit us to distinguish different features on an image. Moreover, the wavelength dependency means that, even for a given feature (target), the proportion of reflected, absorbed (and re-emitted) and transmitted energy will vary at different wavelengths. Typical spectral reflectance curves for vegetation, soil and water are shown in Fig. 1. This radiometric differentiation of different features is the essence of remote sensing. A particular application, such as land use classification for example, will influence the choice of wavelength region in which remote sensing data should be collected. Obviously a band will be selected which will provide the best reflectance differentiation between different land use classes of interest. Often, several bands, or multispectral sensing, are used simultaneously to achieve optimal results.

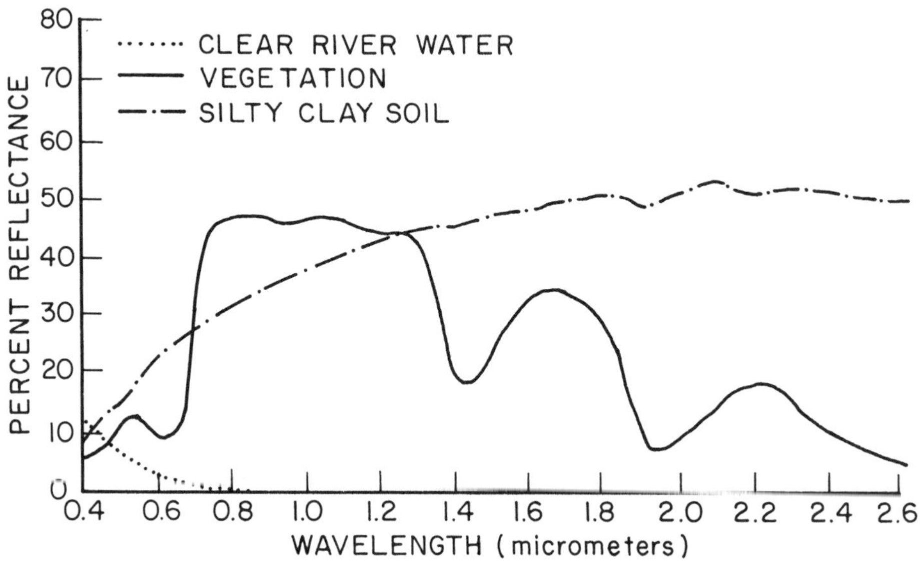

Figure 1. Typical spectral reflectance curves.

2.2. DATA COLLECTION SYSTEMS

There is a number of satellites providing data suitable for application in water resources. These satellites are either geostationary (i.e. their position relative to earth does not change with time) or sunsynchronous, which follow elliptical orbits crossing near the poles.

Two geostationary satellites of interest are GOES (USA) and METEOSAT (European Space Agency). They operate from a relatively high altitude of 36,000 km and thus, the best spatial resolution these satellites can provide is only between 0.9 and 5 km. However, they have 30 minute repeat coverage which makes the data suitable for precipitation and flood prediction and monitoring.

Sunsynchronous satellites most often used for water resources applications are the Landsat (USA) and SPOT (France) series of satellites. The orbits of Landsats 4 and 5 are repetitive and near-polar at a nominal altitude of 705 km at the equator. Each orbit takes nearly 99 minutes, and the spacecraft completes slightly more than 14 orbits a day, covering the entire earth every 16 days. The sensors on board the spacecraft obtain data along the ground track at a fixed width or "swath", which for Landsat satellites is 185 km, as shown in Fig. 2.

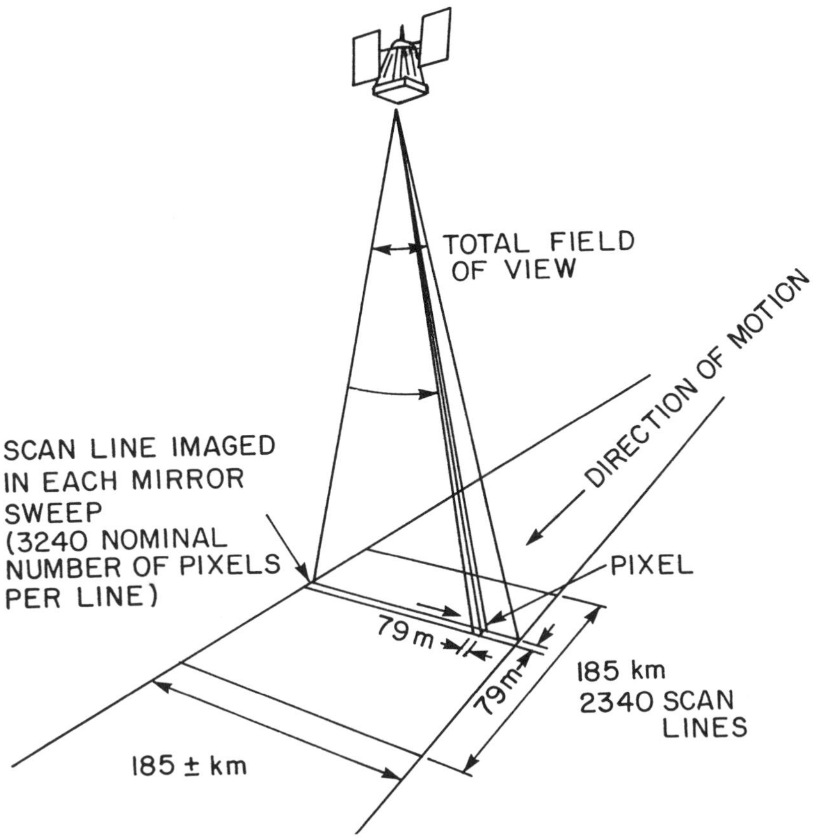

Figure 2. Landsat operating configuration.

The Multispectral Scanner (MSS) is the primary sensor system on aboard of all the Landsat satellites. It records the solar radiation reflected by the earth's surface in four distinct wavelength bands in the visible and near-infrared portion of the electromagnetic spectrum. The radiation coming from the earth's surface is reflected by an oscillating mirror to detectors which produce a voltage related to the amount of radiation. The voltage produced is an analog signal which is converted to digital values ranging from 0 to 63. The radiance, or brightness value of an area on the earth's surface is represented by these numbers, 0 for black and 63 for white. The MSS scans across track swaths of 185 km. The analogic signal is sampled during the scan at intervals which result in the picture elements (or pixels) on the imagery corresponding to a rectangular area on the ground with the 79 m side in the satellite movement direction and the 56 m side in the scan direction. The along-track scan is produced by the orbital velocity of the spacecraft itself. The sampled radiation signals are digitized, formatted into a serial data stream and stored on a magnetic tape for later transmission to earth when the satellite passes over a receiving station.

Landsats 4, 5 and 6 carry aboard also the Thematic Mapper, which is a seven-band multispectral high-resolution (pixel size 30 x 30 m) scanner. It quantizes radiation into 256 brightness values, rather than 64 for MSS, thus providing a better radiation differentiation in the image.

2.3. DATA ANALYSIS

Once data have been collected from a remote sensing system the raw data usually undergo geometric and radiometric corrections before further application. The main reasons for geometric distortions are changes of spacecraft attitude (roll, pitch, yaw), velocity and altitude, sensor characteristics, earth's curvature and rotation. These distortions are removed by mathematical operations and the image is then converted into one of the standard map projection systems, such as Universal Transverse Mercator.

Raw data also require radiometric correction to reduce errors in the data caused by detector noise and environmental variations.

Often, data enhancements are performed to improve the visual presentation of the information content of images. Examples of data enhancement are rationing two images, edge enhancement, principal component transformation, filtering, and contrast enhancement algorithms.

The final output of the evaluation of remote sensing data will be thematic information, presented as statistics or a thematic map. This is achieved by classifying all pixels into a number of distinct classes according to their brightness values. Each class represents a certain physical attribute, such as type of land use, land cover, soil type, crop type, level of soil moisture, range of temperature, etc. Such classification can be done visually but digital interpretation techniques have become increasingly important.

Two basic approaches to digital image classification have been developed: supervised and unsupervised classification. Whereas no or minimal interaction of the operator is required for unsupervised classification, the supervised algorithms require from the operator a set of statistical parameters describing the spectral characteristics of the classes to be identified. These "training" statistics are developed from areas of known types (ground truth).

3. Remote Sensing in Water Resources Engineering

Remote sensing techniques can provide valuable information for design and operation of water resources projects. However, it has to be born in mind that because remote sensing data do not match the accuracy of ground-based measurements, ground control measurements should always accompany the acquisition of remotely sensed data. Remote sensing data can then be used to extent the ground point measurements in space and time to obtain improved estimates of hydrological features which are notoriously variable in both time and space. The decision to use remote sensing techniques has to be based on the consideration of data requirements, and the availability and cost of conventional data versus the additional benefit and cost of remotely sensed data. The aim of this section is to summarize the use of remote sensing in those areas of water resources engineering where it can be expected to provide the most benefit at present.

3.1. RAINFALL

The use of radar for precipitation measurement is an example of active remote sensing. Radar emits electromagnetic energy in the microwave region which interacts with precipitation by means of absorption, scattering and reflection. Part of the reflected energy returns to the transmitter and can be measured. The average return power of the signal $\bar{P}_r$ is related to the raindrop diameter D by the relation

$$\bar{P}_r = \frac{C}{r^2} \sum D^6 \tag{4}$$

where C is a coefficient dependent on the radar type and r is the range. The difficulty in application of eq.(4) arises from the fact that the dropsize distribution in a storm is unknown. Thus, eq. (4) is rewritten as

$$\bar{P}_r = \frac{C}{r^2} Z \tag{5}$$

where the so called reflectivity factor Z is related to the rate of rainfall R by an empirical equation

$$Z = a R^b \tag{6}$$

in which parameters a and b must be evaluated by simultaneous radar and raingage network measurements. Unfortunately, the parameters a and b have the tendency to vary not only aerially but also from storm to storm in the same location. Thus, while radar measurements have marked advantages, such as digitally processed areal amounts of rainfall and the ability to track the storm, they always have to be related to ground measurements for corrections.

A number of satellite techniques to infer precipitation from the visible and/or infrared imagery of clouds have been developed (Schultz and Barrett, 1989; Engman and Gurney, 1991). All these techniques are physically indirect, i.e. satellites provide measurements which pertain to clouds as seen from above, rather than to the rainfall below the cloud. Typical measurements include cloud top brightness values which are converted into temperature values, areal extent of clouds, and the rate of change of cloud's areal extent. Various methods use these parameters in an empirically based procedure or equation to provide estimates of precipitation.

While these techniques rely on empirically developed relationships they nevertheless are all based on certain physical principles pertaining to cloud development.
This can be seen, for example from an equation for the accumulated rainfall ΔP during a time interval Δt derived by Ingraham (1980) as a simplified cloud development model (Brass, 1990)

$$\Delta P = -\frac{Ew\rho}{\Gamma} \ln \left[\frac{A(t+\Delta t)}{A(t)} \right] \Delta T \tag{7}$$

in which E is an efficiency factor (to account for the proportion of vapour that actually condenses in the cloud), w is the mixing ratio of air in the updraft supplying the cloud system, ρ is the density of moist air, Γ is the temperature lapse rate, A(t) and A(t+Δt) are the cloud top areas at times t and t+Δt, respectively, and ΔT is the difference in temperature between the level of zero divergence in the cloud and the cloud top.

It can be seen from eq.(7) that the rate of rainfall is related to the rate of growth of the cloud top, and the cloud top temperature (in ΔT), which both can be monitored by satellites. In general, faster cloud growth rates and cooler cloud top temperatures are associate with higher rainfall rates. The rest of the terms in eq.(7) are usually accounted for by empirically developed relations.

For details of operational satellite techniques the reader is referred to Schultz and Barrett (1989), and Engman and Gurney (1991).

3.2. RUNOFF AND CATCHMENT CHARACTERISTICS

Present satellites cannot measure runoff directly but the multispectral data they provide can be used to derive hydrologic parameters needed in runoff prediction, and to measure catchment geomorphological characteristics. Watershed boundary, channel length, drainage density, extent and time variation of flooded areas, lakes, rivers, swamps, etc. can be readily determined from satellite imagery.

The greatest success in estimating parameters for prediction of runoff by remote sensing appears to be in the area of land use and land cover classification. This is not surprising since classification is the primary objective of image analysis. Land use/cover classes are to be used in a hydrologic model to predict infiltration and excess rainfall rates, and eventually runoff. An example of a distributed model which uses remotely sensed data to simulate rainfall runoff conversion is Hydrotel (Fortin et al, 1990). Spatial information available from remote sensing is used in this model to define "homogeneous" hydrologic units with respect to topography, pedology and land use/cover. Such units are then assigned specific infiltration characteristics of the soil profile which is divided into three layers. The top layer is limited to about 10 cm in order to be compatible with measurements of soil moisture by remote sensing. Furthermore, data from Spot of Landsat are used to determine the length of a river.

For those preferring lumped models based on the Soil Conservation Service (SCS) runoff curve number method (1972), classification of remotely sensed data into land use/cover classes provides a new tool, supplementing field and map data, or quite often, providing the only source of current data. Rango et al (1983) compared conventional and Landsat-derived land-use data for use in SCS type hydrologic models, and the resulting discharge frequency curves were analyzed. Landsat classification accuracy on a cell-by-cell basis (size about 1.1 acres or 0.45 hectare) was only 64 percent, but, when the grid cells were aggregated into watersheds, the classification accuracy increased to 95 percent. When both conventional and Landsat land use data were input into the HEC-1 model for generating discharge frequency curve, the difference in calculated discharge were judged insignificant for subbasins as small as 2.6 km^2. For basins larger than 26 km^2, use of the Landsat approach was found to be more cost-effective than use of conventional methods. Similar results were found by Cermak et al (1977), and Still and Shih (1985).

3.3. SNOW MAPPING

Hydrologists generally want to know how much water is stored in a basin in the form of snow. This information, and subsequent prediction of snowmelt runoff, can be derived from observed areal extent of snow cover, and the depth and water equivalent of the snowpack. Traditionally, snow data have been obtained manually by means of snow courses, or by snow pillow and storage gage measurements. Remote sensing offers a new and valuable tool for obtaining snow data for predicting snowmelt runoff.

The most readily derived parameter from satellite data is the areal extent of snow cover. Until recently, satellite based snow mapping techniques have relied upon visible and/or infrared data, such as imagery supplied by NOAA from the Advanced Very High Resolution Radiometer (AVHR), and Landsat TM/MSS sensors. According to Rango (1990), the best satellite for providing highly repetitive coverage of a basin is NOAA with 2 coverages per 24 hour period. Landsat coverage is only once every 16 days. However, because of the poorer resolution of NOAA imagery (1.1 km) it is suitable only for basins larger than about 200 km^2. Based on NOAA data, the National Weather Service in the U.S. produces operational digital maps of areal extent of snow cover on approximately 4000 basins in North America on a weekly basis.

Two types of snowmelt runoff prediction models, utilizing satellite derived snow cover extent data, evolved. The first type relies on an empirically derived relationship between remaining snow cover as of a given date and the subsequent melt water runoff (Ostrem et al 1981). This method can be used only when snow has started to melt and parts of the basin have become snow-free. Such empirical relations can be developed for many basins, but indications are that the method is limited to essentially vegetation-free areas.

The second type is characterized by the Snowmelt Runoff Model (SRM) developed by Martinec (1975). SRM requires three input variables, namely, daily values of temperature, precipitation, and snow-covered area. SRM is a distributed model in which the snowmelt calculation is based on the daily degree-day value.

Mapping snow cover using visible and/or infrared data can be limited by the occurrence of clouds which cannot be penetrated by this radiation. Furthermore, the discrimination of snow and clouds, because of their similar reflectances, requires special techniques. Other confusion factors which may lead to misinterpretation are forest cover, shadows, and rocks (Engman and Gurney, 1991).

Passive microwave sensing of snow cover is a promising new technique offering a number of advantages. Microwaves penetrate clouds, thus microwave sensing is not hampered by cloudy conditions. Also, microwaves penetrate dry snow and can be used in theory to measure the depth and water equivalent of dry snow cover. Interaction of microwaves with snowpack is complex, depending on temperature, snow properties and other factors. Nevertheless, good results have been reported. In general, the microwave method requires the establishment of a relationship between microwave radiation and the snow water equivalent determined by a conventional method. Goodison (1989) assessed the utility of passive microwave data for determining snow water equivalent on Canadian prairies. The best algorithm for prediction of snow water equivalent resulted from the regression of the microwave radiation gradient between 37 and 18 Gttz [(37 - 18)/19] on snow water equivalent values determined by airborne gamma survey using U.S. National Weather Service procedure.

3.4. OTHER APPLICATIONS

It would not be possible to give in this paper a comprehensive review of all possible remote sensing applications in hydrology and water resources. Just a few of the more important applications shall be mentioned briefly. More information is available in the literature (Engman and Gurney, 1991; Kennie and Matthews, 1985).

3.4.1. *Soil Moisture*. Both visible and infrared data have been used to measure soil moisture, but the most promising approaches are based on passive and active microwave data. In the passive microwave approach, the sensor measures the so called brightness temperature T_B which is related to the soil temperature T_s by

$$T_B = e\ T_s \tag{8}$$

where e is the emissivity, $0 \leq e \leq 1$.

The microwave sensing of soil moisture utilizes the fact that the emissivity strongly depends on the dielectric constant of the soil k, which in turn exhibits a large contrast between water (k = 80) and dry soil (k = 3). For example, a change in k for a soil from 4 to 80 would result in a decrease of the soil emissivity from 0.95 to 0.6. A major problem, however, arises from the fact that the emissivity (or dielectric constant) depends on the air-soil interface. Thus, the microwave sensing provides information only on the top layer of the soil, not more than 10 cm thick. The problem, therefore, remains to correlate the top-layer soil moisture measurements with the moisture profile further below. For hydrological modelling purposes this means to estimate the soil moisture profiles down to about 2 m. Another complication is due to vegetation cover, which attenuates microwave emission from the soil.

Active microwave sensing (radar) measures the reflected energy of an emitted signal. Surface roughness effects the soil reflectivity, whereas vegetation cover is a less important factor in active microwave sensing. The depth of measurement is still limited to the top layer of soil.

Presently, the available satellite sensors have a rather poor spatial resolution (150 km). Good resolution sensors are still too big to be flown on satellites, therefore airplanes or ground based platforms are used presently.

3.4.2. *Evapotranspiration.* Evapotranspiration is one of the most important parameters in continuous hydrologic modeling. It is also one of the most difficult parameters to meausre even by conventional methods. There is no remote sensing method of evaporation measurement as yet operational. However, certain parameters of the energy and water balance equations, from which evapotranspiration can be computed, may be derived from remotely sensed data. These include surface temperature, surface soil moisture, surface reflectance, vegetative cover, and incoming radiation. For a review of recent research results the reader is referred to Engman and Gurney (1991).

3.4.3. *Groundwater.* At present there is no direct method of remote sensing which could detect deep groundwater aquifers. Only indirect inferential methods can be generally applied. Most of those methods rely either on the recognition of surface water or moisture anomalies and their related vegetational effects as indicators of possible groundwater presence or on the interpretation of geologic and topographic information and surface hydrology (Blyth, 1985). Satellite imagery enables the geologist to view very large areas which would not be possible to achieve from ground surveys. This broad perspective helps to infer the location of aquifers from surface features such as land-form, drainage patterns, vegetation characteristics, land use patterns, linear and curvilinear features, and image tones and texture. In general, the presence of groundwater is indicated by land forms such as alluvial fans and flood plains, dense vegetation, lineaments (it is assumed that lineaments mark the location of geologic joints and faults which may contain water), medium and coarse-textured drainage patterns (indicating coarse-grained material in the basin with high infiltration rates).

Infrared imagery may be used to identify shallow ground water and springs trough the difference in temperature between the seeping groundwater and its surroundings. Radar can be used to obtain images of topographic relief even under dense vegetation canopy.

3.4.4. *Water Quality.* Remote sensing can only measure reflected or emitted energy from the surface or near surface of water. Thus, there is a limit on the water quality characteristics that can be detected by remote sensing. In general, an empirical relationship between the water quality parameter and one or more spectral bands must be established for a given application. Water quality indicators such as colour, turbidity, chlorophyll, suspended solids, oil slicks and water temperature have been successfully determined from remotely sensed data and used in many water quality monitoring applications. The spatial and temporal nature of remotely sensed data is of great significance in water quality studies. In many situations monitoring large areas on a frequent basis can only be achieved economically with remote sensing.

4. Conclusions

Major areas of remote sensing application in hydrology and water management were briefly reviewed. Many important applications could not be discussed because of space limitation but should be at least listed. These include flood monitoring, measuring flooded areas, floodplain and wetlands mapping, monitoring lake and reservoir volumes and assessment of irrigated lands.

Although the application of remote sensing in hydrology did not start until 1970's a number of operational techniques exists today. Namely, satellite methods for estimation of rainfall, satellite snow mapping, and satellite imagery classification for land use/cover and other catchment characteristics used in runoff prediction. Much research work with some applications has been accomplished in other areas, including soil moisture, evapotranspiration, groundwater and water quality monitoring.

In all of the applications reviewed, remote sensing offers spatial and temporal coverage which could not be obtained by any other means. The challenge lying ahead is to develop more sophisticated mathematical models transforming remotely sensed data into hydrological parameters, and to modify existing or develop new hydrological models capable of handling essentially distributed data provided by remote sensing. The application of remote sensing techniques in hydrology and water management will grow and most likely revolutionize the discipline. One can only agree with G.A. Schultz (1988) in believing that every hydrologist starting his career after the year 2000 will be faced with remote sensing in some way or other.

5. References

Blyth, K. 1985. Remote sensing and water resources engineering. *Remote Sensing in Civil Engineering,* Kennie, T.J.M. and Matthews, M.C. editors, Surrey University Press, Glasgow and London: 289-334.

Bras, R.L. 1990. *Hydrology. An Introduction to Hydrologic Science.* Addison-Wesley, New York.

Catley-Carlson, M. 1988. The world's water: ten new commandments. *Water International,* Vol. 13, No. 4.

Engman, E.T. and Gurney, R.J. 1991. *Remote Sensing in Hydrology*. Chapman and Hall, London: 225.

Fortin, J.P., Villeneuve, J.P., Bocquillon, C., Leconte, R. and Harvey, K.D. 1990. HYDROTEL, a hydrological model designed to make use of remotely sensed and GIS data. *Proceedings,* Workshop on Application of Remote Sensing in Hydrology. Saskatoon, Saskatchewan, Editors G.W. Kite and A. Wankiewicz, Canadian Water Resources Association; 177-185.

Goodison, B.E. 1989. Determination of areal snow water equivalent on the Canadian Prairies using passive microwave, satellite data. *Proceedings,* IGARSS' 89, Vol. 3, Vancouver, B.C., 1243-1246.

Ingraham, D.V. 1980. Rainfall Estimation from Satellite Images. Vancouver, B.C., Department of Civil Engineering (Ph.D. thesis).

Kennie, T.J.M. and Matthews, M.C. 1985. *Remote Sensing in Civil Engineering.* Surrey University Press, Glasgow and London.

Klemes, V. 1981. Stochastic models of rainfall-runoff relationship. Statistical Analysis of Rainfall Runoff. *Proceedings,* Int. Symp. on Rainfall-Runoff Modeling, Mississippi State University: 139-154.

Martinec, J. 1975. Snowmelt-runoff model for streamflow forecasts. *Nordic Hydrology,* 6(3):145-154.

Rango, A., Feldman, A., George III, T.S. and Ragan, R.M. 1983. Effective use of landsat data in hydrologic models. *Water Resources Bulletin,* April: 165-174.

Rango, A. 1990. Effective use of satellite-observed snow cover data in the snowmelt-runoff model (SRM). *Proceedings,* Workshop on Application of Remote Sensing in Hydrology, Saskatoon, Saskatchewan, Editors: Kite, G.W. and Wankiewicz, A., Canadian Water Resources Assoc.: 19-31.

Schultz, G.A. 1988. Remote sensing in hydrology. *Journal of Hydrology,* 100: 239-265.

Schultz, G.A. and Barrett, E.C. 1989. Advances in Remote Sensing for Hydrology and Water Resources Management. IHP-III Project 5.1, Unesco, Paris: 102.

Soil Conservation Service. 1972. *National Engineering Handbook, Section 4, Hydrology.* U.S. Department of Agriculture, U.S. Government Printing Office, Washington, D.C. 20402.

Still, D.A. and Shih, S.F. 1985. Using landsat data to classify land use for assessing the basinwide runoff index. *Water Resources Bulletin,* December: 931-940.

REMOTE SENSING AS A TOOL IN HYDROLOGY AND WATER RESOURCES

P.K. GARG
Department of Civil Engineering
University of Roorkee
ROORKEE - 247 667 (INDIA)

ABSTRACT: The present study aims to demonstrate the capabilities of remote sensing for collection and analysis of relevant data required for hydrological and water resources investigations. In hydrological modelling the use and advantages of distributed modelling approach are described. Specific examples in hydrology and water resources are taken where remote sensing data can be used effeciently and effectively. In the end the use of Geographic Information System (GIS) is suggested as it can provide the results of analysis, spatially and temporally.

1.0 INTRODUCTION

Hydrology is the scientific study of water, its measurement on the surface of the earth and its effects. Water is one of the most prevalent substances in the earth-atmosphere system, and is most essential to the existence of life. Understanding the general laws governing the distribution of water over the earth and water balance of river basins is essential to the rational use and management of water resources. Changes in hydrology and water resources have important economic effects for water delivery and drainage to agriculture and urban areas, and for flood control (Engman and Gurney, 1990).

V. P. Singh and B. Kumar (eds.), Water Resources Planning and Management, 261–277.
© 1996 *Kluwer Academic Publishers. Printed in the Netherlands.*

India is blessed with an average rainfall of about 119.4 cm; the largest anywhere in the world for a country of a comparable size (Balkrishnan, 1987). It when considered over the geographical area of 329 M ha m amounts to nearly 400 m ha, out of which 70 M ha m is lost in the atmosphere as evaporation, 115 M ha m flows as surface runoff and the remaining 215 M ha m infiltrates into the ground (Fig. 1). The total surface water resources of the country, after considering the above, have been assessed at 180 M ha m. Due to limitations imposed by topography, climate soil conditions etc., only about 88 M ha m of surface water is considered utilisable. The present utilisation of surface water is estimated to be about 28.5 M ha m of which 95% is used for irrigation purpose and the remaining is put to other uses. Table 1 summarizes the water resources and use potential of Indian basins.

Table 1 India's Water Resources & Use Potential (from Balkrishnan, 1987)

Basin	Av. annual runoff (M ha m)	% of total runoff	Utilli-sable flow (M ha m)	% of Uti-lilsable flow	Utilisable flow as % of runoff
1. Indus	7.33	4.14	5.93	7.54	67.26
2. Ganga	45.41	25.12	22.50	28.30	41.66
3. Brahmaputra	60.07	33.41	1.23	1.88	2.08
4. West flowing rivers	27.26	15.42	7.92	10.58	25.08
5. East flowing rivers	39.74	21.91	39.80	51.70	87.25
Total	179.81	100.00	87.38	100.00	36.98

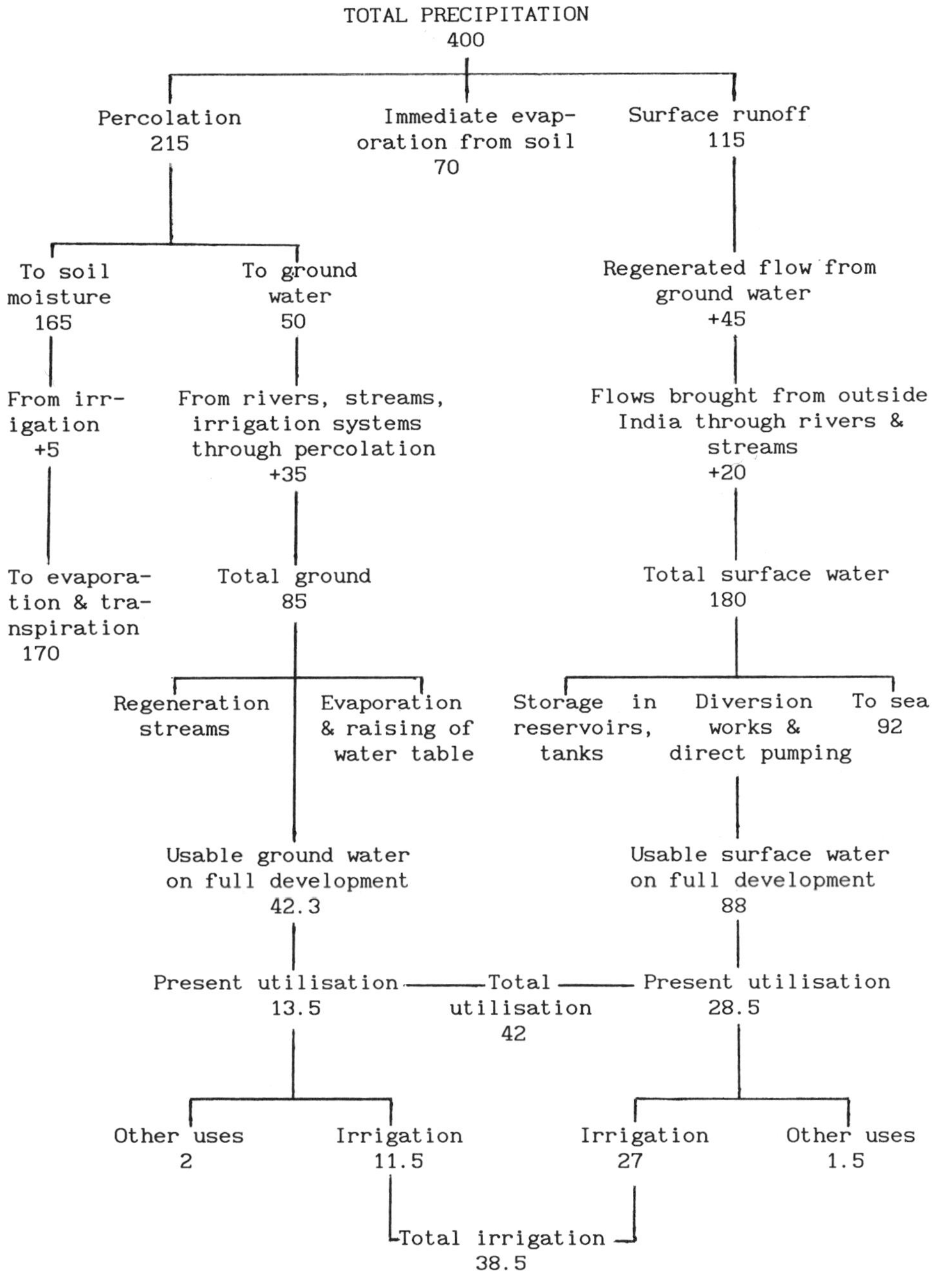

(Note: all units are in M ha m)

Fig. 1 Water Resources of India

2.0 REMOTE SENSING IN HYDROLOGY AND WATER RESOURCES

Hydrologic data required for assessment and management of water resources include precipitation, runoff, sediment yield, water storage, soil moisture, ground water yield and storage, and hydrometrological data such as evaporation, temperature, wind velocity etc (Engman and Gurney, 1990). In order to understand the complex interactions that affect the water resources, the efficient and accurate methods are required to be adopted to collect the hydrologic data. The development of remote sensing has enabled to study the spatial variability of above parameters in a quantitative way. Remote sensing data provide a unique measurement of hydrologic characteristics, and therefore offer the best opportunity for major

Table 2 Some Remote Sensing Applications in Hydrology & Water Resources (from Link, 1983)

Area	Remote Sensing Data
1. Snow cover	Visible band imagery of NOAA, LANDSAT and SPOT to map snow cover. Thermal imagery to separate clouds from snow cover.
2. Precipitation	GOES imagery to estimate intensity, duration and location by cloud indexing technique.
3. Landuse & landcover	LANDSAT, SPOT and IRS data are useful.
4. Valley & channel cross sections	Photogrammetric techniques and Airborne laser mapping systems are useful.
5. Drainage network	High resolution satellite data are useful.
6. Soil moisture	NOAA AVHRR infrared data are quite useful. LANDSAT and SPOT data are not particularly suited for soil moisture applications
7. Surface water	Infrared band satellite data of LANDSAT, SPOT and IRS data are quite useful.

advances in understanding the hydrologic balance and water features of the basin (Link, 1983). Table 2 lists some of the applications of remote sensing data in hydrology and water resources.

In hydrology and water resources, the use of hydrologic simulation models has become a common approach in nearly all analyses and studies. For example, event based models can be used to assist in the design of structures and to make decisions for selections of various watershed treatments. Whereas, continuous simulation models can be used in real-time for flood prediction, reservoir operation, runoff estimation and water supply forecasting. There are three broad categories where remote sensing data are used in hydrologic studies. In first category, simple qualitative observations/assessments are made which can be used for guiding the placement of in-situ observations, assisting in interpolation of estimates. Geometric form dimensions, patterns geographic location and distribution, area shape, length, and identification and quantification of features are the types of information derived under second category. In last category, the correlations between the remotely sensed observations and corresponding point measurements on the ground for the estimation of hydrologic parameters are developed (Balkrishnan, 1987).

2.1 Advantages of Distributed Models

Hydrologic models have evolved from simple empirical relationships to very sophisticated multi-component models (Link, 1983). The use of computers and remotely sensed data has dramatically changed the emphasis in hydrologic modelling from simplicity to sophistication incorporating hydrologic processes, meteorological conditions and watershed characteristics as inputs to the models.

Table 3 summarizes the periods of major developments in hydrologic models. The models that were initially developed, described the physical characteristics of watersheds by including average slope, flow length values, soils, landuse and vegetation cover, as input parameters that were "lumped" or averaged over the watershed area. Watersheds were later divided into sub watersheds to lump these parameters over smaller areas to more realistically represent the watershed characteristics. Finally, completely distributed type models were developed.

Table 3 Major Developments in Hydrologic Models (from Link, 1983)

Period	Model Structure
1. 1800's	Empirical Relations & Rational Formula
2. Early 1900's	Based on Unit Hydrograph
3. 1950's to 1960's	Multi-concept Water Balance Models
4. 1970's to date	Distributed Models

The capabilities of a hydrologic model are determined by its structure and quality of the input data. The required model input data are directly tied to the complexity of model structure. Advances in data collection methods (through computer) have provided increased opportunity for immediate advantage that remote sensing offers in the area of hydrologic modelling is the ability to provide distributed data on the character of the watershed such as soil moisture content, snow cover, land use/ land cover, valley and channel hydrogeometry, and the drainage network (Garg, 1991).

Distributed models are most practical when their inputs can be directly measured or developed from directly measurable parameters. Hydrologic models that fully utilize the distributed data available from remote sensing can significantly enhance the design, planning and forecasting capabilities. Such models can enhance the accuracy of runoff estimates from individual sub-basins and increase the sensitivity of models to changes in watershed response. In addition, these models can be used to predict the effects of various landuse and landcover classes on runoff and sediment yield.

The main advantages of using distributed models over lumped models are listed below. Other advantages are given in Foster (1990) & Garg (1991).

1. Increased accuracy.

2. Use of physically-based mathematical relationships.

3. Capacity to incorporate newly developed relationships.

4. Incorporate information about the spatial variability of the watershed.

5. Compatible with digital remote sensing data and Geographic Information System.

2.2 Soil Moisture Studies

Soil moisture is highly variable as the result of the inhomogeneity of soil properties, topography, land cover and non-uniformity of input from rainfall. Remote sensing data have great potential to capture the spatial and temporal variability in soil moisture (Fig. 2). The visible and near infrared region (0.4-3.0 μm), the infrared region (8-12 μm) and microwave region (1-30 cm), all are very sensitive to record variations of moisture in the soil. Schmugge (1978) summarised the properties that can be used to estimate soil moisture in different spectral wavelength regions (Engman and Gurney, 1990), as given below:

	Wavelength Region	Property Observed
1.	Reflected Solar	Soil albedo
2	Thermal Infrared	Surface temperature
3.	Active Microwave	Dielectric properties
4.	Passive Microwave	Brightness temperature

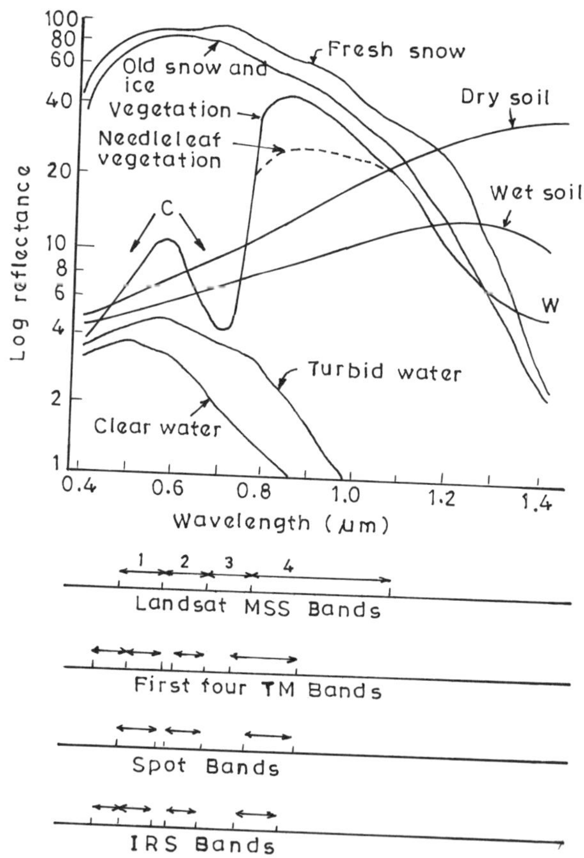

Fig.2 Reflectance Values of Water and Other Features in Various Wavelength regions (from Balkrishnan, 1987).

Reflected energy in the form of tone or colour is useful for qualitative estimates of wet or dry conditions, as wet soil will generally have a lower albedo than the dry soil (Crist and Cicone, 1984). Surface temperature is dependent upon the thermal conductivity and heat capacity of soil, and thus thermal infrared (TIR) images are useful. Microwave techniques for measuring soil moisture are based on the contrast between the dielectric properties of liquid water and dry soil.

2.3 Surface Water Mapping

Surface water inventories involve location of surface water bodies/ water contributing areas, their extent, depth and storage and distribution as well as evolution of water-environment characteristics. Water bodies absorb nearly all incident energy in both the near infrared and middle infrared channels

Standing water, in general, has low reflectance unless substantial amounts of sediment or organic materials are present. The spectral region from 0.6 to 1.1 μm is useful for mapping the areal extent of water features, and 0.7 to 1.0 μm is best for separating sediment-laden water from clear water (Fig.2). For observing variations in turbidity or delineating various classes of turbidity (<1m) or water colour in water bodies, data in 0.4 - 0.6 μm region are very useful. In moderately turbid water, average penetration depths are reduced to 1-2 m (Balkrishnan, 1987). To monitor variation in depth in water bodies, the shorter wavelengths in the visible region are preferable. For example, in clear water the maximum penetration occurs near 0.45 μm (Engman and Gurney, 1990). Ratio of visible and near infrared bands separates surface water from cloud, mountain shadows, dense urban areas and dark soils which appear on image so dark to be identified as water.

2.4 Snowcover Studies

Snowcover has been portrayed as the most difficult and complicated hydrologic parameter to measure. Snow extent, distribution, depth, water equivalent and density are all important. The wavelength regions 1.55 - 1.75 μm, and 2.10 - 2.35 μm provide capability to separate snow cover from clouds. Wavelength longer than 3 cm are used to provide an indication of snow depth (Balkrishnan, 1987).

The reflectivity of snow depends upon the properties such as the grain size and shape, water content, surface roughness, depth and presence of impurities. It decreases with age of snow in visible, infrared and near infrared regions. Rango (1985) summarized responses of some snowpack properties in various wavelength region, as given in Table 4.

Table 4 Response of Some Snowpack Properties in Various Wavelength Regions (from Rango, 1985)

Property	Visible & Near Infrared	Thermal Infrared	Microwave
1. Snow covered area	Yes	Yes	Yes
2. Depth	Only shallow	Weak	Moderate
3. Snow water equivalent	Only shallow	Weak	Strong
4. Albedo	Strong	No	No
5. Liquid water content	Weak	Weak	Strong
6. Temperature	No	Strong	Weak

Snow can readily be identified and mapped using NOAA AVHRR visible bands, LANDSAT MSS, TM bands and SPOT bands (Fig. 2). Thermal bands are perhaps the least useful for measuring snow and its properties. In near infrared band (0.8-1.1 μm) snow and ice albedo is reduced, especially when liquid water is present. Table 5 indicates the minimum size of the snow cover basin which can be identified from remote sensing data.

Table 5 Minimum Size of Basin Identified from Remote Sensing Data (from Engman and Gurney, 1990)

Data	Resolution (m)	Min. Size of basin (Km^2)
1. AIRCRAFT	3	1.0
2. LANDSAT		
TM	30	2.5 - 5.0
MSS	80	10.0 - 20.0
3. SPOT	20	2.0 - 5.0
4. NOAA		
AVHRR	1100	200 - 500

Snow cover data has been used for forecasting snowmelt. Snowmelt varies with elevation, slope, aspect and vegetation cover. Various models are available to assess the runoff from snowcover. For example, Snowmelt Runoff Model (SRM) is an empirical model, developed to simulate and forecast daily streamflow in mountain basins where snowmelt is a major component of the annual water balance (Martinec, 1975). Inputs to the model are periodic snow cover area and daily temperature and precipitation. The Precipitation Runoff Modelling System (PRMS) is a distributed parameter watershed model which treats the snowpack as a two layer system (Leavesley *et al* 1983). Heat and mass (rain or snow) are transferred across a 3 - 5 cm surface layer to the main body of the snowpack, which is maintained and modified both as a water reservoir and a heat reservoir. Seasonal runoff from snowmelt, thus obtained, is extremely important in mountainous basins to make various water management decisions.

2.5 Landuse & Landcover Mapping

Classification of landuse and landcover information has been the most wide application of remote sensing data in hydrology. For example, the quality, quantity and timing of watershed runoff are strongly dependent on landuse & landcover classes. LANDSAT MSS data have been used to identify landcover classes relevant to runoff studies. Compared to MSS, LANDSAT TM & SPOT data offer higher spatial and spectral resolution, and thus provide greater details in inventorying such classes (Garg, 1991) (Fig. 2).

Trolier and Philipson (1986) identified various landuse & landcover classes in New York State using LANDSAT TM data. Best results were obtained using Bands 3, 4 and 5 or a colour composite of these three bands. They presented the results of interpreter's ability to recognize various hydrological landuse & landcover classes on TM data, as given in Table 6. Band 4 enhanced boundaries between cultural and vegetation classes, band 3 provided detail in cultural features, and band 5 revealed subtle vegetative differences. The single most useful TM band for recognizing most classes was band 4.

Landuse information have been used to estimate SCS runoff Curve Number (CN) values for streamflow forecasting and flood studies. Jackson and Rawls (1981) evaluated the potential of using a LANDSAT

Table 6 Rating of Hydrological Landuse & Landcover Classes as Identified on LANDSAT TM Data (from trolier and Philipson, 1986)

Hydrological Landuse and Landcover Classes	T M Bands							Colour Composites	
	1	2	3	4	5	6	7	2,3 & 4	3,5 & 4
Cropland	*	*	**	**	**	x	**	**	**
Orchards/Vineyards	x	x	x	x	x	x	x	x	x
Pasture	x	x	x	x	x	x	x	x	x
Bare soil	**	**	**	x	*	*	**	**	**
Forest	**	**	**	*	*	*	**	**	**
Brush land	x	x	x	x	x	x	x	x	x
Wetlands	x	x	x	**	**	x	*	**	**
Water bodies	*	*	*	**	**	**	**	**	**
Commercial	*	*	**	**	**	x	*	**	**
Low density urban	*	*	*	x	x	x	x	x	*
Med. density urban	**	**	**	**	x	x	**	**	**
High density urban	*	**	**	**	x	x	x	**	**
Transportation	**	**	**	**	*	x	**	**	**
Tended grass	x	x	x	**	**	x	x	**	**

TM Band 1 (0.45 - 0.52 μm), Band 2 (0.52 - 0.60 μm), Band 3 (0.63-0.69 μm), Band 4 (0.70 - 0.90 μm), Band 5 (1.55 - 1.75 μm), Band 6 (10.40 - 12.50 μm) and Band 7 (2.08 - 2.35μm)

** - Best identification , * - Marginal identification
x - Not possible

data to make the Curve Number estimation process more cost effective and less tedious with a satisfactory level of accuracy. Unsupervised classification technique was used to identify the agriculture, open space, forest, disturbed land, residential, paved and commercial - industrial land cover categories of ten watersheds in Washington, ranging from 2.4 Km^2 to 87.3 Km^2 area. The landcover data were then used with the hydrologic soil type data to estimate CNs.

Zevenbergen *et al* (1988) established linear relationships between the CN values and the reflectance values derived from LANDSAT MSS data of 9 rangeland watersheds in western USA. The CN values were calculated using actual rainfall-runoff data. Vegetation indices such as Ratio, Normalised Difference Vegetation Index (NDVI), Transformed Vegetation Index (TVI) and Soil Brightness Index (SBI) were evaluated to relate with CN values, which could further be used in the direct estimation of rangeland CN values for the estimation of runoff.

3.0 USE OF GEOGRAPHIC INFORMATION SYSTEMS

The effective utilisation of large volumes of spatial data is dependent upon the existence of a geographic handling and processing system that can transform these data into useful information (Garg, 1991). A Geographic Information System (GIS) overcomes this problem as it is designed to allow users to efficiently display, analyze, manipulate, retrieve and store the large quantity of spatial data (Marble and Peuquet, 1983). GIS records all earth features with their geographic coordinates (X,Y) whichare used to tie and link all other data sets.Several GISs (e.g. **ARC/INFO, SPANS, MAP, USEMAP, ILWIS, ISRO-GIS, PAMAP,** and **IDRISI**) have been developed which are capable of performing above functions fully or partly.

GISs have been in existence for a number of years now but their capabilities have in general been rather restricted (Garg, 1991). Remote sensing data provide temporal and spatial information at low cost and in grid from that is compatible with requirements of a GIS, thus providing spatially disributed framework for the hydrologic models. A computer-based hydrologic analysis system in a GIS can easily be created by integrating the data collected from various sources such as topographic maps, field and remote sensing. The merging of remote sensing data with the Digital Elevation Model (DEM) and GIS enables types of data to be combined objectively and systematically. Once the input data are spatially co-registered, the desired output products (maps, tabular, and statistical information) can be derived by integrating them.

GIS forms an integral feature for the analysis and merging data of different characteristics and performing hypothetical basin treatments for design studies. Detailed case studies are given in Morgan and Nalepa (1982), Potter et. al. (1986), and Garg (1991). The main advantages of using GIS in hydrological and water resourses problems are (i) Grid size can be altered quickly, (ii) Intermediate resultscan be obtained at any desired step, (iii) Watershed can be displayed in 3-dimension by incorporating the elevation data, (iv) Problem areas can be located on the ground by their geographic coordinates, and (v) Results can be stored and periodically updated.

4.0 CONCLUSIONS & RECOMMENDATIONS

Remote sensing has opened new vistas for forecasting, evaluation, assessment and management of highly dynamic water resources. It provides distributed data on the characteristics of watershed such as landuse and landcover, valley and channel hydrogeometry, and the drainage network. Availability of satellite data on repetitive basis allows specific water resource problems to be monitored through various seasons and even from year to year. Remote sensing data does not completely replace conventional data but instead complements these data by permitting better interpolation for physical watershed models or operational procedures. The new models are required to be developed which use remote sensing as well as conventional data, to account for the spatial variability of basins in a more realistic way. Link (1983) provided a thorough coverage of the state of modelling and use of remote sensing data.

In future, the use of microwave and thermal data may provide additional information that may dramatically improve capabilities of models to simulate, forecast and understand hydrologic events. Use of spectral, spatical (texture) and temporal analysis will lead to much improved pixel by pixel information. Coupling watershed geometry with digital topographic data will provide the opportunity for the design of improved hydrologic models. Further development of new and improved sensors and data analysis procedures is expected to boost the application of remote sensing data to hydrology and water resources.

Advances in remote sensing will result in the capability to provide distributed data on static watershed descriptors, dynamic meterological conditions, and hydrologic process parameters for beyond the fidelity required in current models. To make optimum use of spatial and temporal characteristics of remote sensing data and integrate these with other field data, a GIS is needed to be employed. Further research should address the development and use of distributed models in GIS at both conceptual and practical levels in hydrology and water resources, particularly for those areas where remote sensing data may potentially be used.

REFERENCES

Balakrishnan, P (1987), 'Issues in Water Resources Development and Management & the Role of Remote Sensing', *ISRO Publication*, No. **TR-67-86** Bangalore.

Crist, E.P. and Cicone, R.C. (1984), 'Physically Based Transformation, of Thematic Mapper Data: VTM Tasseled Cap', *IEEE TRANS Geoscience & Remote Sensing*, **GE-22**.

Engman, E.T. and Gurney, R.J. (Eds.) (1990), *Remote Sensing in Hydrology*, Chapman and Hall, Landon.

Foster, G.R. (1990), 'Process-based Modelling of Soil Erosion by Water on Agricultural Land, In: *Soil Erosion on Agricultural Land* (J.Boardman, I.D.L. Foster and J.A. Dearing, Eds.), John Wiley & Sons, Chichester.

Garg, P.K. (1991), 'Development of a Catchment Scale Erosion Model for Semi-and Environments and its Implementation Through Remote Sensing', *Unpublished Ph.D. Thesis*, University of Bristol, U.K.

Jackson, T.J. and Rawl's, W.J. (1981), 'SCS Urban Curve Numbers from a LANDSAT Data Base', *Water Resources Bulletin*, Vol. **17**, No. 5.

Leavesly, G.H., Lumb, A.M. and Saindon, L.G. (1987), 'A Micro-computer-based Watershed Modelling and Data Management System', *Proc. of 55th Ann. Western Snow Conf.*, Vancouver, B.C.

Link, L.E. (1983), 'Compatibility of Present Hydrologic Models with Remotely Sensed Data', *Proc. of 17th Int. Symp. of Remote Sensing of Environment*, Ann Arbor, Michigan.

Marble, D.F., and Peuquet, D.J. (Eds), (1983), 'Geographic Information Systemn and Remote Sensing', In: *Manual of Remote Sensing*, ASPRS Publication, Virginia, Vol II.

Martinec, J. (1975), 'Snowmelt Runoff Model for Streamflow Forecasts', *Nordic Hydrology*, Vol **6**.

Morgan, K.M. and Nalepa, R. (1982), 'Application of Aerial Photographic and Computer Analysis to the USLE for Areawide Erosion Studies'. *Journal of Soil and Water Conservation*, Vol. **37**.

Potter, W.B., Gilliland, M.W. and Davidlong, M, (1986), 'A Geographic Information System for Prediction of Runoff and Non-point Source Pollution Model', *IAHS Publication*, No. **160**.

Rango, A. (1985), 'Assessment of Remote Sensing Input to Hydrologic Models', *Water Resouces Bulletin*, Vol. **21**.

Schmugge, T.J. (1978), 'Remote Sensing of Surface Soil Moisture', *Journal of Applied Meterology*, Vol. **17**.

Trolier, L.J. and Philipson, W.R. (1986), 'Visual Analysis of Landsat Thematic Mapper Images for Hydrologic Landuse and Cover', *Photogrametric Engineering & Remote Sensing*, Vol. **52**.

Zevenbergen, A.W., Rango, A., Ritchie, J.C., Engman, E.T. and Hawkins, R.H. (1988), 'Rangeland Runoff Curve Numbers as Determined from LANDSAT MSS Data', *International Journal of Remote Sesning*, Vol. **9**, No.3.

APPLICATION OF VARIOUS TRANSFORMATIONS FOR LANDUSE CLASSIFICATION IN DIGITAL IMAGE PROCESSING

by

Anil Kumar

National Institute of Hydrology, Jal Vigyan Bhawan, Roorkee 247667 INDIA

ABSTRACT

The utility of satellite data for multiple earth resources start beginning to unfold the information contained in these data, in the areas, such as agriculture, landuse, water resources, forestry, urban development planning and monitoring, and other allied subjects. The potential of these data for landuse studies had started in recent years and there is notable increase in the acceptance and utilization of Landsat and IRS satellite data for classification, monitoring and mapping the landuse.

With the advancement in digital image processing technique, so many new techniques is coming out and it is very difficult to re-assume the effect of any particular transformation. Therefore, two frames have been selected, one in plane and another in hills to see the effect of various transformation.

This paper presents the results of the application of various transformation for preparing the landuse classification map by digital image processing technique. The transformation which have been used in this study are ratioing, sum normalisation, principal component transformations and MMI transformations. The GIS/Image Processing System, namely ILWIS has been used to get a classified map. Confusion matrix were prepared to analyse the classification results, and this process was repeated for all the images resulting by various transformations and each time the same checking samples map was used.

Introduction

The systematic earth orbital observations began in the year 1960 with the launch of (TIROS-1), the first meteorological satellite. Latter on the first Landsat (ERTS-1) was launched in 1972 carrying in addition to other equipment, a four channel multispectral scanner scanning the earth surface in four separate spectral bands.

The utility of satellite data for multiple earth resources applications is the beginning to unfold in the areas, such as agriculture, landuse, water resources, forestry and other allied subjects. The potential of these data for vegetation oriented landuse pattern analysis has started in recent years and there is notable increase in the acceptance and utilization of Landsat and IRS data for landuse mapping. The latest Landsat sensors contain new designs like the Thematic mapper and the special cameras which allows the improvement of computational uses for image and pattern analysis.

V. P. Singh and B. Kumar (eds.), Water Resources Planning and Management, 279–293.
© 1996 *Kluwer Academic Publishers. Printed in the Netherlands.*

Objective

The main objective of this case study was to obtain the landuse information in the form of a landuse map by applying various transformations. Because all the information related to any aspect have to be extracted from the available channel. Some times it is difficult to differentiate among the various classes due to limitation of spatial, spectral and radiometric resolution of remote sensing satellites. Therefore, more images will be created by applying various transformation and the results will be analysed visually and digitally. Two frames have been selected, one in plane and another in hilly terrain. Because, It is difficult to pre assume the effect of any transformation for landuse classification.

Study Area

The himalaya mountains chain is about 2500 Kms long and 240-400 Km wide rising from the low lying Indian plains to well over 8000 m has a complex physiography. The study area is around Dehradun city, situated nearly 30^0 N latitude and 78^0 E longitude between Aravalli hills and Himalaya mountains in the north west part of India. The centre of latitude and longitude of the image is 30.04^0 N and 78.09^0 E and the image covers 185*185 km. This area is selected because it has a variety of landforms, landcover and soil types and changes in landuse area also rapid.

Landuse

The himalayan range have large variations in topography, elevation and location and this results in great contrasting landuse and climates from region to region. Not withstanding these local and regional variations in climate, the entire himalyas experiences a general weather and climatic pattern dictated by the monsoon phenomenon of Asia.

It is well known that the climatic and topographical conditions are the determining factors for landuse in a certain area, besides other factors. Climatically the temperature varies from 8^0 C in winter to 40^0 C in summer and topographically, Dehradun is situated in a valley.

Data Used

Following data have been used for the case study

----- Landsat 4 MSS with bands 4,5,6 and 7 of 14th November 1972 was used.

Details of Landsat MSS data

Band	Wavelength (m)	
4	0.5-0.6	Spatial resolution = 80m
5	0.6-0.7	Radiometric resolution= 6 bits
6	0.7-0.8	
7	0.8-0.9	

Equipment

The following equipment, system and software have been used for this study;

(i) Ramtek
The Ramtek system was used to make the skew and scene element size correction and for transferring the data from CCT to ILWIS format.

(ii) ILWIS
The processing of digital data was carried out on an IBM PC-AT using the ILWIS (Integrated Land and Water Information System) software package version 1.21.1 along with the peripherals viz: NEC Multisync II high resolution (480*640) colour monitor.

(iii)Tektronix inkjet plotter
For taking the printout of the various classified images.

Procedure

The original image was on CCT and it is converted to ILWIS format on the Ramtek System and then the files were copied onto floppy discs in the computer center, so that further work could be done on the IBM PC-AT with ILWIS package.

The main intention of this case study, is to study the effect of various transformations for the landuse classification by visual and digital interpretation technique in the proposed area. Because it was predicted that the usefulness of some transformations would be dependent on the characteristics of the area under considerations. Therefore, two windows corresponding to (750*750 element) and (300*300 element) were selected in the plain and hill areas respectively.

Firstly, histogram calculation were made for all the four bands of Landsat 4 MSS after determining the amount of haze and it was concluded that only band four required haze correction in this case.

After haze correction, destriping was done by applying the six sensor-histogram matching (destrl) program for all the four bands. These destriped and haze corrected images were used for further analysis.

All MSS bands (4,5,6 and 7) were used for the analysis inspite it is known that band 6 and 7 are highly correlated, because the cmain intention was to study the effect of various transformations in developing a landuse classification map.

A false colour composite was made by assigning red, green and blue for band 7,5 and 4 respectively for displaying and sampling. The training and checking samples map was made by sampling the window of the false colour composite. Using all the four bands a landuse classification map was obtained by running the Box classifier with a multiplication factor 2.5 and confusion matrix was prepared by running the cross and the confmat program from the

ILWIS software.

Afterwards principal component transformation was done by running the principal component program and PC_1, PC_2 and PC_3 were obtained.

Further ratioing, sum normalisation and MMI transformations were done by using the MCalc calculation program in "Spatial modelling Module" in ILWIS package.

The same training sample map was used for the classification of the images obtained by applying the various transformation.

The images obtained after the transformations were used for creating landuse classifications using the same training samples map. This procedure was repeated for all the images resulting by applying various transformations on the original bands.

Again the box classifier was used with a multiplication factor of 2.5 and confusion matrices were prepared by running the confmat program to analyse the classification results. Also this process was repeated for all the images resulting by various transformations and each time the same checking samples map was used.

Classes

Following classes were selected which exist in the (750*750) window, corresponding to plain area:

(i) Sal forest;
(ii) Mix forest;
(iii)Tea Garden;
(iv) Landuse;
(v) River1
(vi) River2
(vii)Water;
(viii)Irrigated field;
(ix) Water logged area,
(x) City;

Only three classes exist in the window (300*300), corresponding to hilly area; i.e. in himalaya mountains.

(i) Oak and Pine;
(ii) Deciduous forest;
(iii)Grassi rock;

Training and checking samples were selected by field visit and some of the classes were created to see the difference in spectral behaviour. It is proposed to study the difference in spectral behaviour of sal forest, mix forest and tea garden because they behave spectrally same in band 4,5,6 and 7. Further, water logged area, irrigated areas and city behave spectrally same in all the four bands.

Description about Various Transformation

The transformation which were used to study the change in spectral behaviour of the proposed classes are explained below briefly;

(i) Principal Component Transformation

It has been shown that all the spectral information in the four bands of a Landsat MSS image can be very well represented by two dimensional or, at the most, three dimensional data because channels are highly correlated (Mather p. 206). In the more common applications, such as landuse mapping, all such information has equal potential significance and colour composite picture production systems are frequently designed to display all the available information. Principal component analysis, a powerful method of analyzing correlated multidimensional data, can be used in such systems to facilitate the visual interpretation of a mass data having uniform a priori significance, by reducing redundancy, i,e correlation between channels (Byrne,1980). Information which appears in more that one channel is treated as being of no less potential value than any other. The procedure incidentally renders it susceptible to presentation with a three channel optical system.

(ii) Ratioing

It is most commonly used transformation in which the pixel value in one spectral channel is divided by the corresponding pixel value in second spectral channel. There are two reasons to do so. One is that certain aspects of the shape of spectral reflectance curves of different earth cover types can be brought out by ratioing. The second undesirable effect on the recorded radiance, such as the effect of variable illumination resulting from the variation in topography can be reduced (Mather,p. 189).

Ratioing of spectral bands will not have any meaningful interpretation because we are working in a domain of photon counts. Basically ratioing is a technique which provides the user a wide range of possible outcomes with a range from 0 to infinity (Bakx.p.24,L.N. 1991).

B45=(b4/b5)*40; B47 = (B4/B7)* 40; B57 = (B5/B7)*40;

Here, 4,5,6 and 7 are Landsat MSS bands and 40 is a compensation factor. The compensation factor is needed because the result of the ratio can be smaller than one.

(iii) Sum Normalisation

Sum normalisation is used to make the input data independent of the influence of external factors such as sun angle or relief. Basically, the addition or summation of input data known as representation of this influence. The ratio between the individual input bands and this sum, results in intensity independent data. If the results are displayed they appear as flat.

B4n = (B4/Int)* (90/0.58); B5n = (B5/Int)*(90/0.58);
B7n = (B7/Int)*(90/0.58);
Irt = (B4+B5+B7); and (90/0.58) is a compensation factor.

The results of normalisation is always within the interval 0 to 1, therefore it should be multiplied with a compensation factor.

(iv) MMI Transformation

It is already known that the influence of intensity on actual colour can be removed by sum normalisation. The effect of sum normalisation on the data is mapping the data onto a triangle spanned by maximum red, green and blue. For all points within this triangle the total intensity is equal to the single colour maximum intensity. Although Saturation and Hue can be used to describe every point within the triangle, a mathematically simpler method is using two orthogonal axes. An obvious choice is to have the origin of the two orthogonal axes at point where the intensity axis crosses the triangle plane. The two axes are coded M1 and M2 and M1 is going from origin to Red and M2 is orthogonal to M1 thus, parallel to the line from green to blue. m1 and m2 are sum normalised data:

m1: = 0.82*B7n - 0.41*B5n- 0.41*B4n + 127;
m2: = 0.71*B5n - 0.71*B4n +127;

and 127 is a compensation factor.

Conclusions

The main intention of this study is to see the effect of various image processing technique to improve the landuse classification and to improve the visual interpretation of the image by applying the transformations which have been explained in earlier paragraphs.

(i) By principal component transformation in the plain areas there is no improvement for the differentiation between irrigated field, water logged area and city classes. On the other hand class city is classified as class river1; table 1.2. But, the water logged area can be delineated more clearly from the surrounding classes, (figure 3(c)).
But the principal component transformation in hilly area results a flat image corresponding to PC_2 (figure 4(c)). So all the three classes can be visually differentiated. This statement gets support from the results of the confusion matrices, which maintain the overall accuracy table 2.2.

(ii) By ratioing among the band 4,5 and 7, it can be seen that there is no pixel, which can be classified as water because after ratioing the pixel value increased beyond 255. So there is no improvement in classification result table 1.3. But by the visual interpretation city and irrigated area can be clearly delineated by ratioing between band 4 and 7, (figure 2(d)).
But in hilly area there is no improvement in classification result, table 2.3 or for visual interpretation, (figure 4(d)).

(iii) By sum normalisation there is no improvement in classification result because city is classified as, irrigated field, tea garden and river 1; table 1.4. But by visual interpretation

Table 1.1 Confusion matrix with band 4,5,6 and 7 Images.

	1	2	3	4	5	6	7	8	9	10	uncl	ACC
1	20	0	0	0	0	0	0	0	1	0	0	0.95
2	0	14	0	0	0	0	0	0	0	0	4	0.78
3	0	0	21	0	0	0	0	0	0	0	5	0.81
4	0	0	0	17	0	1	0	0	0	0	0	0.94
5	0	0	0	0	29	0	0	0	0	0	7	0.81
6	0	0	0	1	0	15	0	0	0	0	1	0.88
7	0	0	0	0	0	0	17	0	0	0	2	0.89
8	0	0	0	0	0	0	0	1	6	10	2	0.05
9	0	0	0	0	0	0	0	0	3	17	0	0.15
10	0	0	0	0	0	0	0	0	0	18	1	0.95
REL	1.00	1.00	1.00	0.94	1.00	0.94	1.00	1.00	0.30	0.40		

Average Accuracy = 72.15%
Average reliability = 85.82%
Overall accuracy = 72.77%

Table 1.2 Confusion matrix with PC1, PC2, PC3 Images.

	1	2	3	4	5	6	7	8	9	10	uncl	ACC
1	20	0	0	0	0	0	0	0	0	0	1	0.95
2	0	14	0	0	0	0	0	0	0	0	4	0.78
3	2	0	20	0	0	0	0	0	0	0	4	0.77
4	0	0	0	17	0	1	0	0	0	0	0	0.94
5	0	0	0	0	31	0	0	0	0	2	3	0.86
6	0	0	0	2	0	13	0	0	0	0	2	0.76
7	0	0	0	0	0	0	17	0	0	0	2	0.89
8	1	0	0	0	0	0	0	2	16	0	0	0.11
9	0	0	0	0	0	0	0	0	20	0	0	1.00
10	0	0	0	0	0	0	0	8	11	0	0	0.00
REL	0.87	1.00	1.00	0.89	1.00	0.93	1.00	0.20	0.43	0.00		

Average Accuracy = 70.70%
Average reliability = 73.18%
Overall accuracy = 72.30%

Table 1.3 Confusion matrix with b47, b57 and b45 Images.

	1	2	3	4	5	6	7	8	9	10	uncl	ACC
1	20	0	0	0	0	0	0	0	1	0	0	0.95
2	5	10	0	0	0	0	0	0	3	0	0	0.56
3	21	0	5	0	0	0	0	0	0	0	0	0.19
4	0	0	0	16	0	2	0	0	0	0	0	0.89
5	0	0	0	0	24	4	0	2	1	2	3	0.67
6	0	0	0	0	0	16	0	1	0	0	0	0.94
7	0	0	0	0	0	0	0	0	0	0	19	0.00
8	1	0	0	0	0	2	0	1	14	1	1	0.05
9	0	0	0	0	0	0	0	1	19	0	0	0.95
10	0	0	0	1	6	3	0	2	6	1	0	0.05
REL	0.43	1.00	1.00	0.94	0.80	0.59	?	0.14	0.43	0.25		

Average Accuracy = 52.52%
Average reliability = 62.15%
Overall accuracy = 52.58%

Table 1.4 Confusion matrix with b4n, b5n and b7n Images.

	1	2	3	4	5	6	7	8	9	10	uncl	ACC
1	14	6	0	0	0	0	0	0	1	0	0	0.67
2	0	18	0	0	0	0	0	0	0	0	0	1.00
3	12	6	5	0	0	0	0	0	0	0	3	0.19
4	0	0	0	16	0	2	0	0	0	0	0	0.89
5	0	0	0	0	30	3	0	0	1	1	1	0.83
6	0	0	0	0	0	17	0	0	0	0	0	1.00
7	0	0	0	0	0	0	19	0	0	0	0	1.00
8	0	2	0	1	1	2	0	0	11	1	1	0.00
9	0	0	0	0	1	0	0	0	19	0	0	0.95
10	0	0	0	1	9	3	0	0	6	0	0	0.00
REL	0.54	0.56	1.00	0.89	0.73	0.63	1.00	?	0.50	0.00		

Average Accuracy = 65.31%
Average reliability = 65.01%
Overall accuracy = [illegible]

Table 1.5 Confusion matrix with m1 and m2.

	1	2	3	4	5	6	7	8	9	10	uncl	ACC
1	13	7	0	0	0	0	0	0	1	0	0	0.62
2	0	17	0	0	0	0	0	0	1	0	0	0.94
3	14	6	3	0	0	0	0	0	0	0	3	0.12
4	0	0	0	14	0	4	0	0	0	0	0	0.78
5	0	0	0	0	30	4	0	1	1	0	0	0.83
6	0	0	0	0	0	17	0	0	0	0	0	1.00
7	0	0	0	0	0	0	19	0	0	0	0	1.00
8	0	2	0	1	2	2	0	0	12	0	0	0.00
9	0	0	0	0	1	0	0	0	19	0	0	0.95
10	0	0	0	1	7	4	0	0	6	1	0	0.05
REL	0.48	0.53	1.00	0.88	0.75	0.55	1.00	0.00	0.47	1.00		

Average Accuracy = 62.93%
Average reliability = 66.61%
Overall accuracy = 62.44%

Table 2.1 Confusion matrix with band 4,5,6 and 7 Images.

	1	2	3	uncl	ACC
1	15	0	0	0	1.00
2	1	18	0	5	0.75
3	0	0	19	3	0.86
REL	0.94	1.00	1.00		

Average Accuracy = 87.12%
Average reliability = 97.92%
Overall accuracy = 85.25%

Table 2.2 Confusion matrix with PC1, PC2 and PC3 Images.

	1	2	3	uncl	ACC
1	14	0	0	1	0.93
2	2	21	1	0	0.88
3	0	0	17	5	0.77
REL	0.88	1.00	0.94		

Average Accuracy = 86.04%
Average reliability = 93.98%
Overall accuracy = 85.25%

Table 2.3 Confusion matrix with b47, b57 and b45 Images.

	1	2	3	uncl	ACC
1	12	3	0	0	0.80
2	0	20	0	4	0.83
3	0	0	18	4	0.82
REL	1.00	0.87	1.00		

Average Accuracy = 81.72%
Average reliability = 95.65%
Overall accuracy = 81.97%

Table 2.4 Confusion matrix with b4n, b5n and b7n Images.

	1	2	3	uncl	ACC
1	4	11	0	0	0.27
2	0	19	0	5	0.79
3	0	0	17	5	0.77
REL	1.00	0.63	1.00		

Average Accuracy = 61.04%
Average reliability = 87.78%
Overall accuracy = 65.57%

Table 2.5 Confusion matrix with m1 and m2 Images.

	1	2	3	uncl	ACC
1	4	11	0	0	0.27
2	0	23	0	1	0.96
3	0	0	17	5	0.77
REL	1.00	0.68	1.00		

Average Accuracy = 66.59%
Average reliability = 89.22%
Overall accuracy = 72.13%

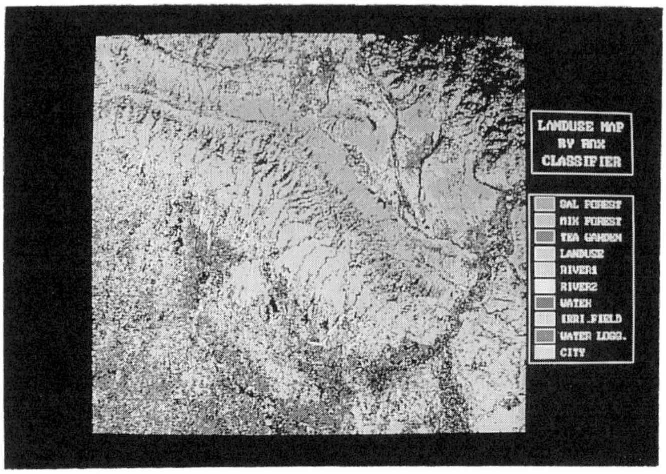

Figure 1: Landuse Map Around Dehradun

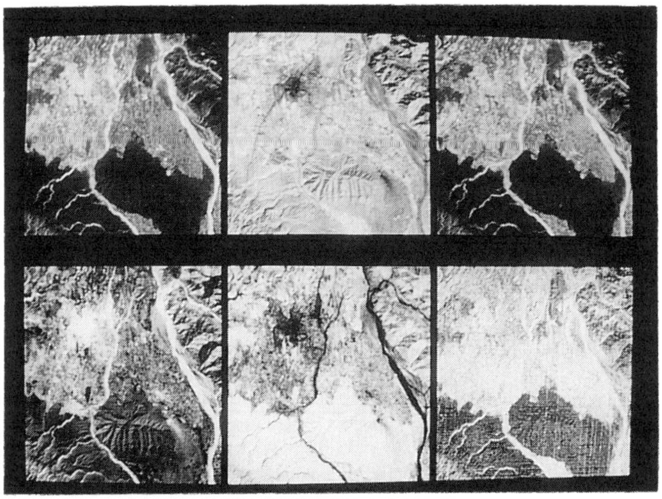

Figure 2: (a) Landsat MSS band 4 Image
(b) Landsat MSS band 7 image
(c) Principal component 1
(d) b47 Image
(e) b7n Image
(f) m2 Image, shows the difference between city and irrigated area.

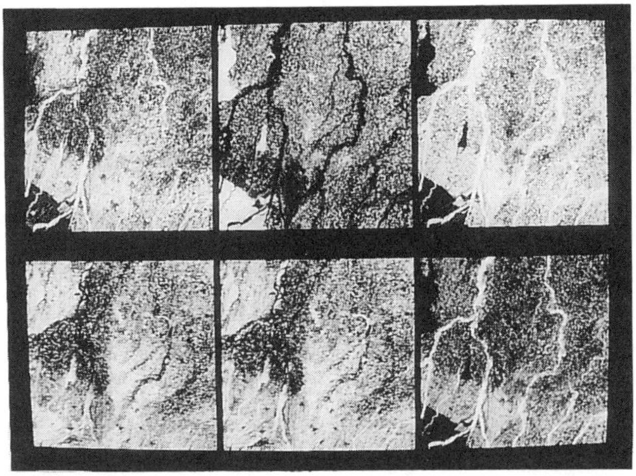

Fig.3 : (a) Landsat MSS band 4 Image (b) Landsat MSS band 7 Image (c) Principal component 1 (d) b57 Image (e) b7n image (f) m2 Image, shows the difference between River2, Landuse and water logged area.

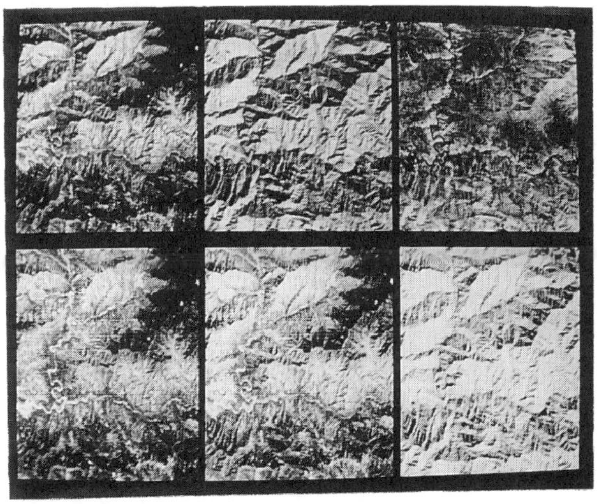

Fig.4 : (a) Landsat MSS band 4 Image (b) Landsat MSS band 7 Image (c) Principal component 1 (d) b57 Image (e) b5n Image (f) m2 Image, shows the effect of various transformation in hilly area.

Figure 5: False Colour Composite (4,5,7) Around Dehradun

city and irrigated area (figure 2(e)), river bed and surrounding soil; (figure 3(e)) can be delineated easily as compared to original image; (figure 1).
But in hilly area it results nearly a flat image, (figure 4(e)).

(iv) By MMI transformation, there is no improvement either for visual interpretation (figure 2(f)) and (figure 3(f)), or for digital classification; table 1.5 and 2.5.

(v) It can be concluded that the effectiveness of the principal component transformations depends on the type of terrain.

(vi) It is necessary to mention that the figures in the confusion matrices are not too significant because I had only a limited number of checking samples at my disposal, so there is statistical noise.

REFERENCES

(i) Bakx, J P G; "Colour coding and spectral feature extraction" Lecture notes, DIP5/IGP3, Feb. 1991.

(ii) Bryne, G R, Crapper, P F, and Mayo, K K; "Monitoring landuse cover change by principal component analysis of multitemporal Landsat data" Remote Sensing of Environment 10, 175-184.

(iii) Ellis, M C; "Statistical pattern recognition, classification and feature selection". Lecture notes, April 1991.

(iv) Mather, P M; "Computer Processing of Remotely sensed images" Wiley, Chichester 1989.

RELATIONSHIP BETWEEN FOREST COVER AND DRAINAGE DENSITY IN A TYPICAL RAINFOREST : AN INVESTIGATION USING SATELLITE DATA

M.R.Venugopal
Scientist-in-Charge
Remote Sensing Cell
CWRDM, Kozhikode
Kerala, India
Pin 673 571

E.J.James
Head,
Surface Water Division
CWRDM, Kozhikode
Kerala, India
Pin 673 571

N.V.M.Unni
Head, Forestry & Ecology Division
National Remote Sensing Agency,Hyderabad
Andhra Pradesh, India
Pin 500 037.

ABSTRACT

Vegetation cover plays a critical role in the evolution of a drainage basin to its current hydrological and geomorphological status as it strongly influences the input and output of matter and energy.Dense plant cover slows down the surface runoff and promotes infiltration thereby influencing the development of the stream networks and in turn the drainage density or the total length of the drainage channels maintained by unit area of the basin surface. In the present study, this problem is analysed in the context of Silent Valley, a typical rainforest river basin in the Western Ghats of peninsular India. Satellite data(IRS 1A ,LISS II) was the database, control-referenced with aerial data.The vegetation cover of Silent Valley was classified to locate the rainforests and estimate the afforestation coefficient of each first order stream unit.The drainage density was computed which was then correlated to the afforestation coefficient to evolve a general relationship which was verified in the case of an adjacent basin.

V. P. Singh and B. Kumar (eds.), Water Resources Planning and Management, 295–302.
© 1996 *Kluwer Academic Publishers. Printed in the Netherlands.*

1. INTRODUCTION

The evolution of drainage basin to its current status is the net resultant of various natural processes acting on its surface. These incessant forces involve flows of matter and energy moving in and out of the basin. Precipitation and insolation are the main sources of inputs of matter and energy into the basin while evapotranspiration, reflectance and emission, infilitration and runoff are the natural output routes. These processes are strongly controlled by a set of variables that include vegetation cover, soil characteristics, slope and lithology of the basin surface. Out of these , vegetation cover is of critical importance in more than one ways.Its primary effect is in transforming the kinetic energy of the falling water drops into heat by friction which cuts down the transportation capacity of water. Ground surface with sparse or scanty plant cover, for instance, with loose or non-compact soil layer will be subjected to considerable loss of particulate matter and dissolved substances.Thick vegetation on the other hand, slows down the surface runoff.It also plays a critical role in the circulation of matter and energy , through the biospherical systems of the earth.The development and sustenanace of drainage networks is a direct manifestation of this influence.Obviously the problem offers a range of perspectives for investigation with different levels of analytical compelxities .The scope of this paper is limited to the development of a simple relationship between the forest cover of a river basin with its drainage density in a typical rainforest patch of the Western Ghats of India.

2.STUDY AREA

The study area is the Silent Valley (fig. 1) which forms the catchment area of Kunthipuzha, a major tributary of the Bharathapuzha river basin in Kerala.Silent Valley, or *the Valley of Eternal Silence'*,not disturbed even by the chirping of cicados, had attracted nation-wide attention as case of a reckless natural resources exploitation strategy until it was proclaimed a National park in 1986 by the Government of India,considering its potential as a gene pool reserve.The major forest type of Silent Valley is tropical wet evergreen forests least disturbed by human inerference.It also falls in the core zone of the Nilgiri Biosphere Reserve designated under the Man and Biosphere (MAB) Programme of UNESCO.

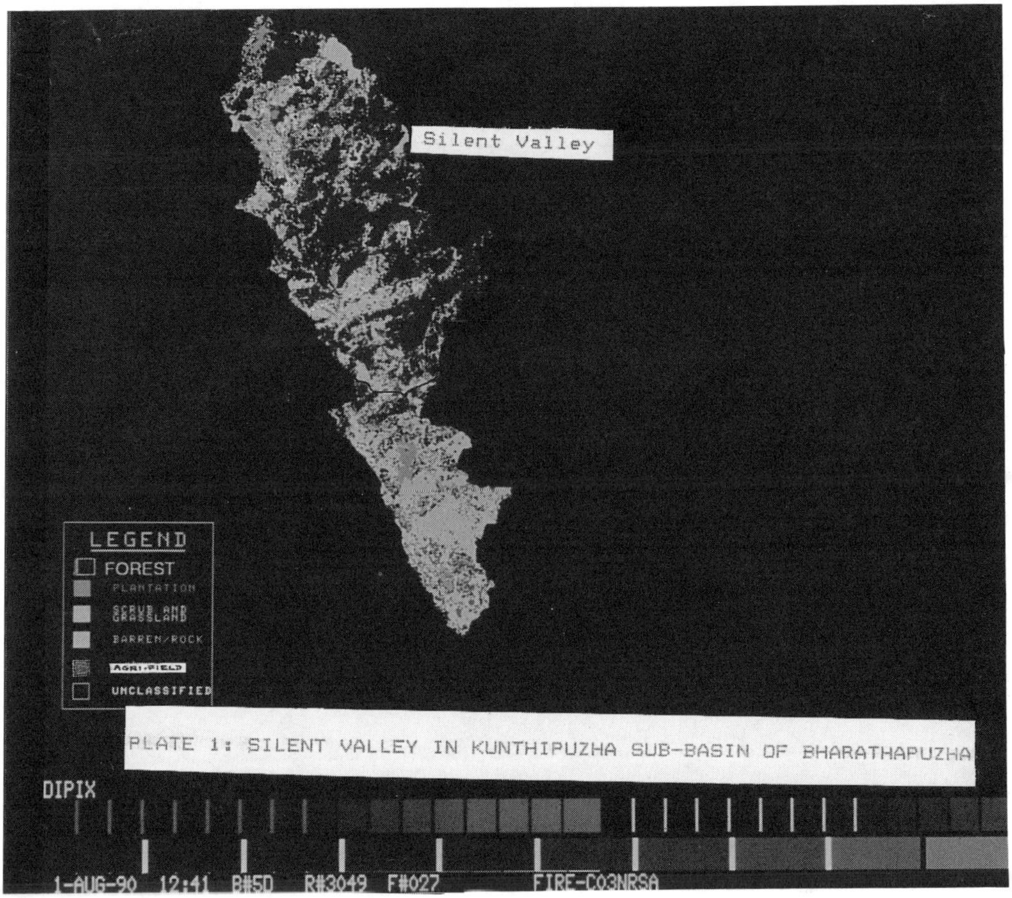

PLATE 1: SILENT VALLEY IN KUNTHIPUZHA SUB-BASIN OF BHARATHAPUZHA

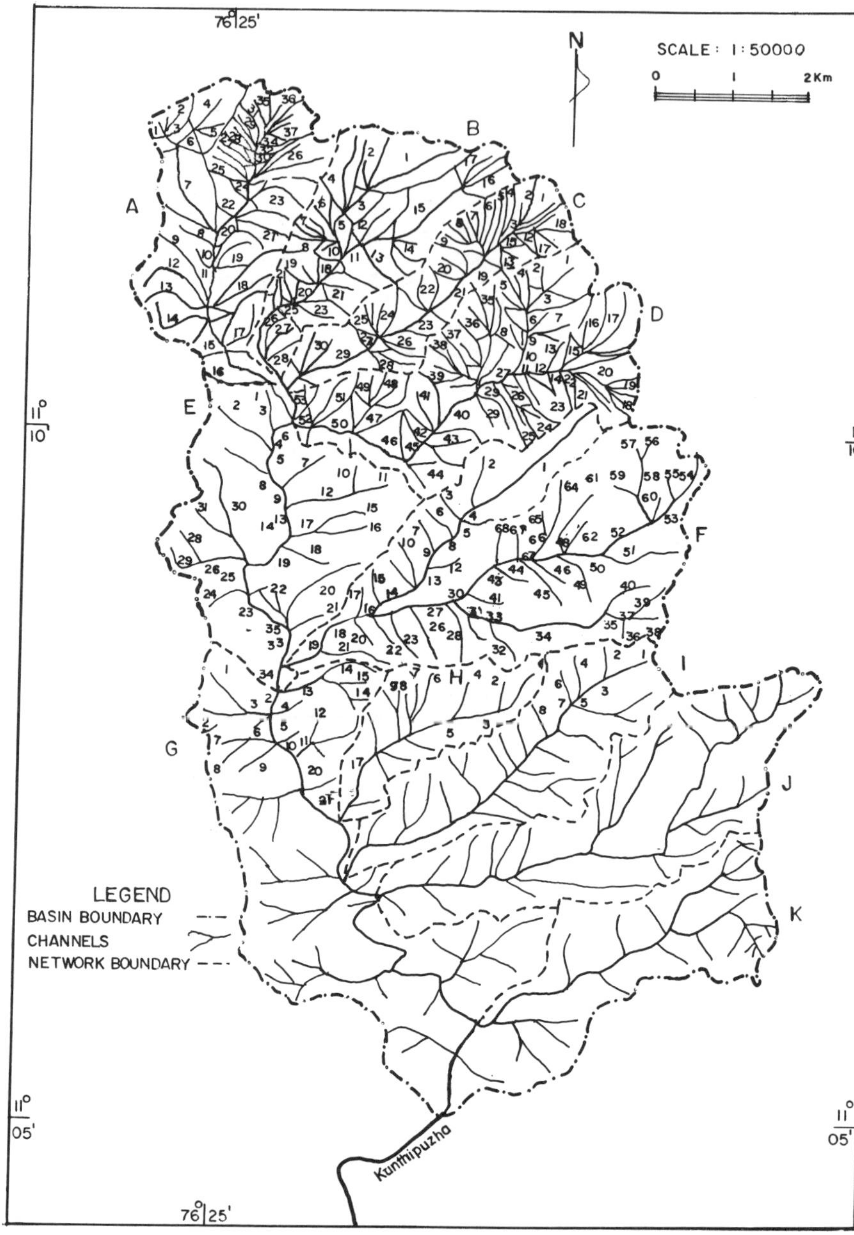

FIG. 1 NBR - SILENT VALLEY - KUNTHIPUZHA SUB BASIN - CHANNEL NETWORKS

3.DATABASE

Database of this study comprises digital data on CCTs from the Indian Remote Sensing Satellite IRS 1A (LISS II) of March 22, 1989,Landsat FCC imagery(diapositives),panchromatic B&W aerial photographs (1:60,000),SOI topomaps(1:50,000).The satellite data on the digital format were analysed using the DIPIX (ARIES) digital image processing system in a VAX 11/23 computer environment with a' Parallellepiped' classifier. Diapositives were interpreted using an optical pantograph.Ground reference data available with the Forestry & Ecology Division of National Remote Sensing Agency were also sparingly used although the major source of refernce data was aerial photographs(Unni, 1978).

4.METHODOLOGY

The landuse/cover map of Silent Valley was prepared by computer classification of LISS II data as indicated above(Plate 1). Seven cover classes identified were :

1. Dense forest
2. Plantation
3. Scrubb/grassland
4. Barren land/rock outcrop
5. Agricultural field
6. Water bodie
7. Unclassified

The few pixels which form the class '*unclassified*'have been ignored as far as this study is concerned since they form only a negligibly small portion of the area under consideration.A drainage map of Silent Valley on 1:50,000 was prepared from Landsat data by visual interpretation of diapositives and aerial photographs.The drainage system including tributaries of different orders and the finger-tip streams were identified and traced out(fig.2).Lengths of channel links and the areas associated with each link were measured. Drainage density(D_d) values were computed as the ratio of link lengths to the drainage area contributing to them.

Forest cover data for two networks identified on the upper portions of the catchment from the computer classified landcover data of the basin were demarkated and transferred on to the drainage map by overlay technique (Venugopal&James, 1990). Afforestation coefficient(f_a) was calculated for each areal element associated with each elementary channel link, as the ratio of the area under forest cover to the total area associated with the link.Regression of D_d on f_a yielded a relationship which was verified for an adjacent network .

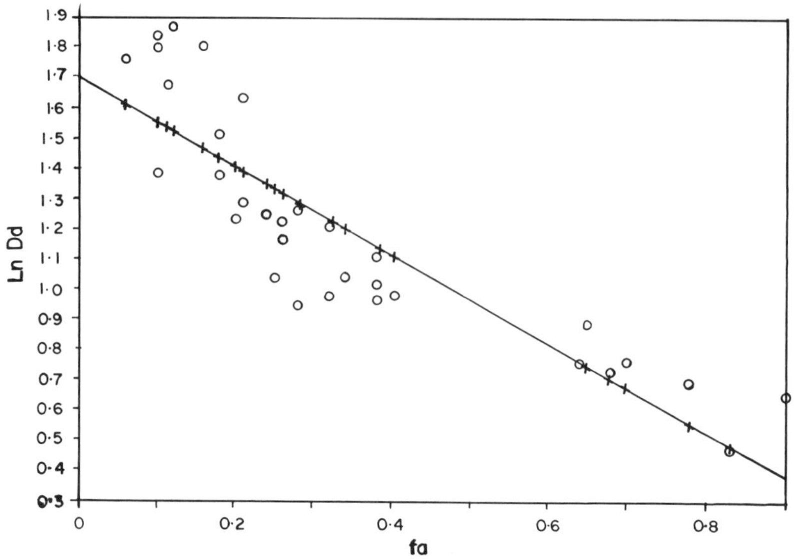

FIG 2 DEPENDENCE OF Dd ON fa

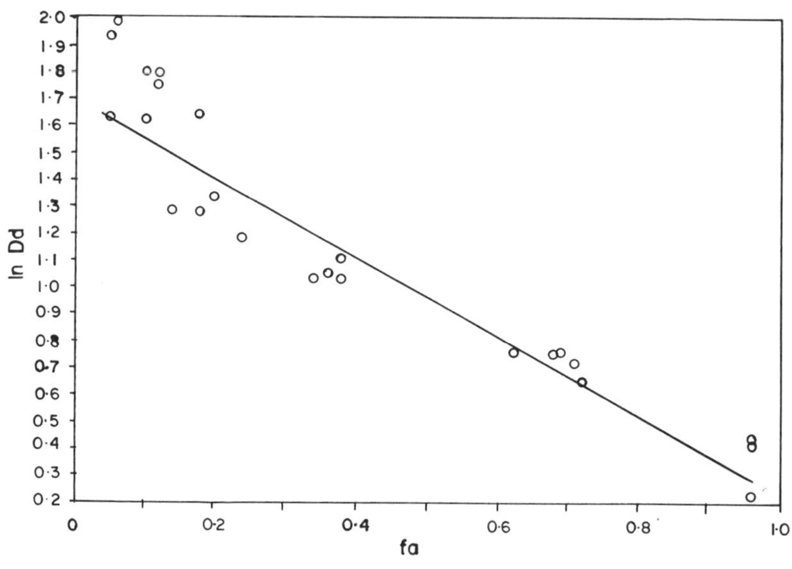

FIG 3 VERIFICATION OF Dd-Fa RELATIONSHIP

5.RESULTS & DISCUSSION

A lot of work has been done in fluvial geomorphology using drainage density as an efficient tool for probing the hydrological processes and their relationships with the river basin characteristics(Dacey,1976; Abrahams,1980). Changes in the later, anthropogenic or natural, influence the erosional processes and are reflected in the changes of the stream network which is also well-documented. However, drainge density is invariably measured as the total length of all channels of all orders in a given basin, rendering it a rather coarse parameter, limiting its applications to larger basins(Miller, 1953). The present study involves micro-catchments with mean area of less than 10 sq.km.Hence an alternative method was utilised for computation of D_d based on Shreve's concept of channel network classification.A drainage network is visualised as a collection of *links* and *nodes*.The nodes are the point of confluence of two streams or the vertex of a stream; and the link is the channel line connecting two nodes.The fundamental elements of a drainage basin are the channel link and the area associated with the link which holds a direct relation with the hydrofluvial processes working on the basin.

The values obtained for link length, basin area,forest cover and the corresponding values of drainage density,expressed as $\ln D_d$, afforestation coefficient,f_a for the network A(refer fig.1) are given in Table 1.The dependence of D_d on f_a is studied by a regression analysis by linear best fit method and the result is given as fig.2.The relationship so developed using the regression parameters may be expressed as:

$$D_d=e^{(1.700262-1.47028fa)}$$

The relationship was employed to predict the D_d values for the network B for verification.Figure 3 depicts the computed values and the observed values of D_d (Table 2) showing a fair degree of agreement between them.

6.CONCLUSION

This study has a direct bearing on the application of space technology that can be gainfully used for studying various hydro-environmental problems such as deforestation and the attending changes in hydrological parameters.Due to the repetitive coverage of satellites they can be used for monitoring the trends in the changes in the surface features of river basins, man-induced or natural.The present study is expected to serve as a baseline work on this aspect with reference to

Silent Valley; future changes can be estimated easily by satellite data of later dates.The results of the study can also be extrapolated to similar regions of tropical forests.

7.ACKNOWLEDGEMENTS

Authors are thankful to Dr P.Basak, Executive Director,CWRDM, for all the help extended towards this work.They are also grateful to the Director, NRSA, Hyderabad, for the permission granted to carry out the computer analysis of satellite data using the facilites of NRSA.

8.REFERENCES

Abrahams, A.D.,1980. `Divide angles and their relation to interior link lengths in natural channel networks', *Geograph. Anal.*,12, 2,157-171.

Dcaey,M.F.,1976.`Summary of magnitude properties of topologically distinct channel networks and network patterns'.In:Merriam,D.F.(Ed),*random processes in geology*, Springler verlag, 1976, New York, 16-38.

Miller, V.C,1953. *Quantitative study of drainage basin characteristics in geology*, Columbia University , N.Y.

Shreve,R.L., 1969.` Stream lengths and basin areas in topological random networks, *J. Geol.*,774.

Unni,N.V.M.,1978.`Computer classification and delineation of types using Landsat data in two areas of tropical forests in India', Proceedings of the 12th Internat.symp.on Rem.Sen. of Environ.. Manila, Philippines.

Venugopal, M.R., & James.E.J.,1990. `Assessment of impact of hydrologic processes on Nilgiri Biosphere reserve using remote sensing techniques', Final report of CWRDM project sponsored by DoE, Govt. of India.

UTILITY OF REMOTE SENSING TECHNIQUES IN LANDSLIDE INVESTIGATIONS : AN APPRAISAL

R. SINHA, A. KUMAR, AND P. V. SEETHAPATHI
National Institute of Hydrology,
Roorkee-247667, INDIA.

ABSTRACT The conventional methods of landslide investigations are time-consuming and are sometimes extremely difficult due to various problems such as accessibility of the area, weather conditions, field hazards etc. Remote sensing can play an important role in landslide investigations, both in the identification of landslide- prone areas and in the analysis of any specific landslide event. The major advantages of remote sensing technique are the synoptic view offered by the satellite images, studies without risking to field hazards, reproducibility of results, availability of multi-date and multi-band data, and possibility of enhancement techniques to reveal desired information. However, the satellite data (Landsat, IRS etc.) pose some limitations in landslide mapping due to constraints of scale and resolution.

Introduction

Landslides are catastrophic mass-movement processes and severely affect the hydrologic system of the region. In strict sense, landslides are defined as "rapid movements of sliding rocks, separated from the underlying stationary part of the slope by a definite plane of separation" (Zaruba and Mencl, 1982). Landslides in inhabited regions have often produced disastrous effects on the river catchments, other water bodies such as springs, lakes etc. and engineering structures. The mass-movements through landslides range from imperceptibly slow movement of the order of 1m/yr to high speed mudflows of the order of 50m/s. Specific terms used for different types of movements include creep, slump & sheet slide, debris avalanche, earthflow, rockfall etc.

The purpose of this paper is to examine the utility of remote sensing techniques vis-a-vis conventional methods for landslide investigations. Apart from visual methods of landslide mapping, various digital enhancement techniques and the need of a GIS approach have been emphasized.

Conventional Methods vis-a-vis Remote Sensing Techniques

The conventional methods of landslide investigations (see Eckel, 1958; Rib & Liang, 1978; Zaruba & Mencl, 1982) normally involve the following steps :

(i) General survey of the field area.

(ii) Detailed mapping of landslides using geological and geomorphological attributes such as lithology, structure, drainage network, landslide slopes, cracks, vegetation cover

V. P. Singh and B. Kumar (eds.), Water Resources Planning and Management, 303–308.
© 1996 *Kluwer Academic Publishers. Printed in the Netherlands.*

including state of tree growth, river terraces, and scarps.

(iii) Hydrogeological surveys including determination of depth of ground water table, range of its fluctuation, inflows of water into the slide area such as springs, wet grounds, and acquiferous fissures.

(iv) Surveying of slide area to determine the depth and shape of slide surface.

(v) Field investigation of mechanical properties of rocks such as deformational properties, shear strength, and density of sandy soils.

(vi) Geophysical surveys such as resistivity and seismic surveys to supplement the results of direct investigations.

(vii) Laboratory investigations such as mineralogical composition of rocks, soil properties, stress-strain parameters, and shear strength.

(viii) Slope stability analyses.

(ix) Integration of all these data to prepare a hazard-zoning maps.

The most important advantage of remote sensing data is the synoptic view offered by the satellite image by which studies from the point of view of monitoring as well as suggesting preventive measures for landslides can be undertaken for very large areas. Various spectral signatures on the satellite images related to tonal variations are indicative of landslide movements. The study of spatial relationships between specific geological and geomorphological features such as regional faults and lineaments, escarpments, broken vegetation, drainage patterns, distribution of lakes, swamps, and soil types can help in identifying potential landslide-prone areas. Several enhancement techniques may be applied to the digital remote sensing data to reveal the desired geomorphological information which may otherwise be not visible on the raw image or in the field. Further, the studies can be carried out in the laboratory without risking to field hazards. Also, measures aimed at erosion control and improvement of soil stability over large areas can be carried out economically through appropriate use of remote sensing data products and techniques. In addition to the above, multi-band and multi-date satellite data can provide information regarding temporal changes in landuse and landform parameters, from which inferences to potential landslide activity can be derived.

One of the major limitations of remote sensing data for landslide mapping is due to the constraints of scale and resolution. Many of the landslips occupy areas of the order of a few meters and hence would be undetectable as such on Landsat and IRS data of spatial resolution upto 30 meters or so. Therefore, Landsat (MSS and TM) and IRS (LISS I & LISS II) data have limited application for identifying most of the landslide events unless it has a wider impact on the surrounding region. These problems may, however, be overcome by using high resolution data such as SPOT and aerial photographs with a proper economisation of the funds available.

Landslide Mapping Using Remote Sensing Data

Both satellite data as well as aerial photographs should be used in conjunction with each other for landslide mapping. Following patterns on the satellite data are useful in identifying and delineating landforms associated with landslide :

(i) Tonal characteristics - localized tonal variation is a function of terrain disturbances; a tonal banding often indicates probable direction of landslide movement.

(ii) Regional differentiation between landslide and surrounding terrain - an indicator of landslide due to difference in texture of terrain appearance.

(iii) Geological features (structure) - character and distribution of lineaments in the study area, some of which may define major shear zones.

(iv) Geomorphological features - presence of scarps, ponds and lakes, drainage patterns.

(v) Spatial relationship between features - location of certain specific features aid in identification of landslide-prone areas.

Followed by regional investigations through satellite images, study of aerial photographs for specific patches should be carried out. The important features which may be recognized on the aerial photographs characterizing landslides (old or recent) are :

(i) Photocharacters - bright tone, unusually fine texture characterizing slide areas.

(ii) Vegetation cover - lack of vegetation in slide areas (recent slide), difference in vegetation pattern from surrounding areas (old slides).

(iii) Slope characteristics - sharp break in slope, mounds of possible landslide origin at their base.

(iv) Anomalous topography - scalloped escarpments suggesting landslide origins.

(v) Drainage characteristics - poor or no drainage on the slide area, anomalous constrictions in the drainage, contributing drainage to the slide area in the form of seepage and springs.

(vi) Other characters - zone of debris accumulation, presence of tension cracks on rock surfaces, bowl-shaped headwater regions of streams, cliffs or banks undercut by streams.

Digital Image Processing for Landslide Identification

The main emphasis of remote sensing investigations should be on the use of digital processing of data from a variety of sensor systems, such as Landsat TM, IRS-LISS II, and SPOT with a view to evaluate the suitability of these data for landslide detection and mapping with precision. Digital enhancement of the satellite image through filtering, ratioing, edge enhancement, contrast stretching and other transformations should help in differentiating various geomorphological features and landuse patterns in the study area. A final landuse map should be prepared using supervised/unsupervised classification of the enhanced image. Particular attention should be paid on the vegetation characteristics within the area of study. The areas of broken vegetation provide information about potential landslide areas which should be subsequently verified through field visits. Zones of extensive soil erosion, which may sometimes trigger landslips, can be mapped from satellite data.

A Digital Elevation Model (DEM) of the area can be generated using topographic maps. Slope and aspect parameters can be derived through DEM to evaluate the terrain conditions related to landslides. These parameters may subsequently be used for identifying and mapping landslide areas. Also, volumetric calculations of the expected landslide mass can be estimated and prediction about their intended location of deposition can be made. Further, investigation of the erosion potentiality of the landslide mass should be carried out under wet conditions and it's impact in and around the area should be assessed. Finally, on the basis of landcover/landuse maps, damage to human life and property can be assessed.

GIS Approach

The use of a Geographic Information System (GIS) has been strongly recommended for landslide investigations which use a variety of data sets in conjunction with each other. A GIS is designed to store, retrieve and analyze large volume of spatial data derived from a variety of sources including remote sensing data. These data are normally converted into thematic maps and are superimposed upon each other to derive the desired information in a GIS environment (Marble & Pequet, 1983).

For landslide investigations, the useful sets of data include lithology, landuse, slope, geological structure (e.g. distribution of shear zones) etc. All these data are measured/mapped on different scales and units which can be integrated using special techniques, including empirical relationships. Gupta and Joshi (1990) devised a Landslide Nominal Risk Factor (LNRF) and mapped the Ramganga catchment into three classes : (a) LNRF < 0.67 (low risk), (b) 0.67 < LNRF < 1.33 (medium risk), and (c) 1.33 < LNRF (high risk). These LNRF classes were ascribed weights of 0, 1 and 2 onto which other geoenvironmental parameters viz. lithology, landuse, and distance from major tectonic features were overlaid to finally demarcate the landslide hazard areas. Pachauri and Pant (1992) also used a rating system for landslide susceptibility mapping using geological attributes for the Algar river catchment in the Garhwal Himalaya. The important parameters used were slope, distance from fault, and dip-slope relationship to frame a database system which was imported into a GIS.

A project on identification of landslide-prone area in parts of the Garhwal Himalayas has been initiated by the authors at the National Institute of Hydrology. Apart from the tectonic and seismic disturbances, the occurrence of landslides in the region have been attributed mainly to faulty alignment of road systems in a terrain of immature geology, heavy rainfall, deforestation and slope erosion, slope toe undercutting due to fast flowing and meandering rivers, flash floods, and intensive construction activities without 'preventives'. Preliminary investigations have established a distinct lithological and structural control on landslide activities. The other important parameters being investigated are geomorphology, landuse, vegetation cover and slope aspects. All these parameters will be analysed in a GIS environment using a package ILWIS, developed by ITC, The Netherlands.

Summary and Conclusions

A typical methodology for landslide investigations using remote sensing data may therefore be summarized as follows :

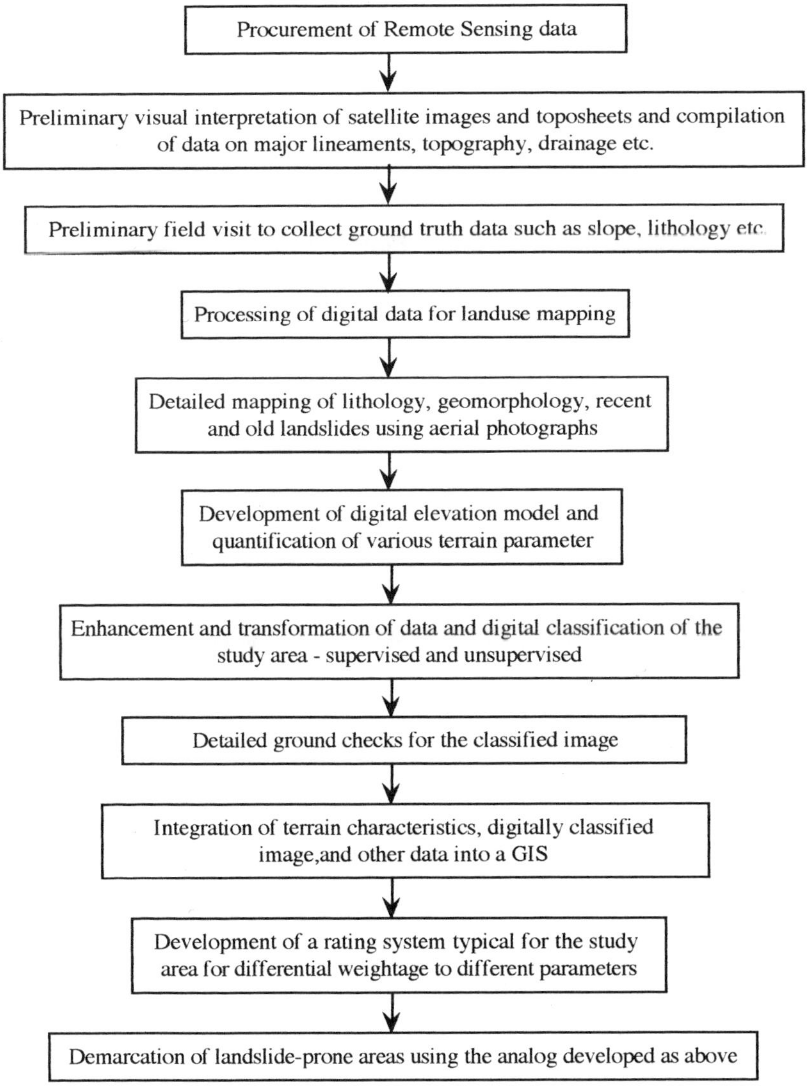

It is therefore concluded that the remote sensing data can contribute significantly to landslide investigations combined with a fair amount of field investigations.

Acknowledgements

The Ministry of Environment and Forests, Govt. of India is providing financial assistance for an ongoing project on identification of landslide-prone areas by remote sensing, being investigated by the authors at the National Institute of Hydrology, Roorkee. The Council of Scientific and Industrial Research (CSIR) is thanked for supporting one of the authors (Rajiv Sinha) through a pool officership for a related project on effect of landslides on hydrological system.

References

Eckel, E. B. (1958) ed. Landslides and Engineering Practice. Highway Research Board, Spl. Report 29, NAS-NRC Pub. 544, Washington.

Gupta, R. P. and Joshi, B. C. (1990).Landslide Hazard Zoning Using the GIS Approach- A case study from the Ramganga Catchment. Engineering Geology, 28, 119-131.

Pachauri, A. K., and Pant, M. (1992). Landslide hazard mapping based on geological attributes, Engineering Geology, 32, 81-100.

Rib, H.T. and Liang, T. (1978). Recognition and identification. In : L. R. Schuster and R. J. Krizek. (ed). Landslides analysis and control. Transportation Research Board, Nat. Academy of Sciences, Washington, p34-80.

Marble, D. F. and Piquet, D. J. (1983). Geographic Information systems and Remote Sensing. In : R. N. Colwell (editor) Manual of Remote Sensing (2nd ed), Am. Soc. Photogramm., Falls Church, p.923-958.

Zaruba, Q. and Mencl, V. (1982). Landslides and their control. Developments in Geotechnical Engg. vol.31, Elsevier, New York, 324pp.

SECTION 5:

HYDRAULIC HAZARDS AND THEIR CONTROL

HYDROLOGIC SAFETY AND REHABILITATION OF OLD DAMS

M.P. Sharma, P.E. and A. Prakash, Ph.d., P.E.

Metropolitan Water Reclamation District of Greater Chicago
111 East Erie Street
Chicago, Illinois 60611, USA

and

Dames & Moore
One Continental Tower
1701 Golf Road
Rolling Meadows, Illinois 60089, USA

ABSTRACT. Several plausible options are presented to rehabilitate, downgrade, or remove an old earthen dam so as to meet current guidelines for hydrologic safety of dams. The dam served as a source of water supply for a local establishment. This establishment has to cease operations within two years and has no further use for the dam. The dam was designed for a flood lower than 0.4 PMF. Its removal or failure is likely to result in significant flood damages to a community located within 2.5 km downstream. Results of PMF routing with and without dam failure are presented to indicate the anticipated risks. Major criteria for comparative evaluation of rehabilitation/disposition options for the dam are described.

1. INTRODUCTION

This paper describes potential rehabilitation options for an old earthen dam built in 1930 to provide water supply for an establishment which has to cease operations within two years or so. Since the construction of the dam 63 years ago, a pre-existing township in the river valley has expanded into a modern urban community approximately 2.5 km downstream from the dam. Salient features of the dam are abstracted in Table 1.

TABLE 1 - SALIENT FEATURES OF EXISTING EARTH DAM

Drainage area	19.2 sq. km.
Height of dam	39 meters
Length of dam crest	133 meters
Elevation of dam crest	1946.36 meters
Gross storage capacity	697,000 cubic meters
Storage below spillway crest	475,000 cubic meters
Spillway crest elevation	1943.31 meters
Length of spillway crest	30.78 meters

According to current hydrologic design guidelines, a dam located upstream of and so close to an urban community must be designed to safely pass the probable maximum flood (PMF). The present owner has no further use for the dam and can no longer operate and maintain it. Although the dam is too small to provide any appreciable flood control benefit during severe events like the PMF or 0.50 PMF, it has provided significant attenuation to seasonal and annual flood peaks over the past 63 years of its existence. The 100-year peak flood for the aforementioned urban community has been

V. P. Singh and B. Kumar (eds.), Water Resources Planning and Management, 311–314.
© 1996 *Kluwer Academic Publishers. Printed in the Netherlands.*

estimated by frequency analysis of data on peak flows attenuated by the dam. Flood insurance rate maps for areas downstream of the dam have been developed using this flood peak of 4.8 m^3/sec. The 100-year peak flow without the dam estimated by rainfall-runoff analysis using the 100-year 24-hour precipitation for the watershed is about 57 m^3/sec.

The community downstream of the dam has developed with the implicit understanding that most frequent floods (up to the 100-year event) will continue to be significantly attenuated by the dam. Removal of the dam at this stage would expose numerous homes and other facilities to recurring flood damages. Leaving the dam in its present condition upstream of a large community may result in extensive loss of life and property due to overtopping failure during a severe event like the PMF. Upgrading a dam which is of no use to the owner is an irretrievable investment for him. The objective of the study was to evaluate comparative risks and costs associated with different dam rehabilitation/disposition options.

2. ALTERNATIVE REHABILITATION/DISPOSITION OPTIONS

Risks and costs associated with five plausible rehabilitation/disposition options for the dam were evaluated.

2.1 No Rehabilitation

This involves establishment of a sinking fund for operation and maintenance of the dam in its present condition for the duration of its residual life of about 50 years. In this case, there is a risk of the dam being overtopped and breached during flood events exceeding 0.4 PMF. To evaluate flood risks resulting from dam failure, several breach scenarios were postulated based on historic data, empirical equations, and judgement. Relevant parameters of these scenarios are summarized in Table 2.

TABLE 2 - POSTULATED BREACH SCENARIOS

Failure Mode	Breach Characteristics					
	Case 1			Case 2		
	Bottom Width	Side Slope	Failure Time (hrs)	Bottom Width	Side Slope	Failure Time (hrs)
Fair-Weather Failure (Piping or Seismic Failure)	45.7	2V:1H	2.25	15.2	2V:1H	0.75
Overtopping Failure During PMF	87.2	2V:1H	2.25	87.2	2V:1H	1.75

Dam break flood routing computations were made using the National Weather Service DAMBRK model with 18 cross-sections of the downstream river valley (Fread, 1991). Flood routing computations for the PMF and 500-year flood without dam failure were performed using the HEC-1 model (US ACE, 1990). Estimated flood depths at the entrance to the urban community are shown in Table 3.

TABLE 3 - ESTIMATED FLOOD DEPTHS WITH AND WITHOUT DAM FAILURE

Scenario	Peak Overflows from Dam (m^3/sec)	Flood Depth at Entrance to Urban Community (m)
Fair-weather Failure Case 1	51	1.6
Fair-weather Failure Case 2	245	1.7
Overtopping Failure During PMF Case 1	637	4.3
Overtopping Failure During PMF Case 2	720	4.5
Overtopping During PMF Without Dam Failure	580	2.3
500-Year Flood Without Overtopping	108	1.5
Historic 100-Year Peak Flow With Attenuation By Dam	4.8	0.3

Evidently, dam failure is likely to result in extensive loss of life and property and catastrophic socioeconomic and political impacts. To minimize loss of life, this option will require formulation and implementation of a flood management plan supplemented by dependable flood forecasting and warning systems. Significant property losses during failure may be unavoidable.

2.2 Dam Upgrading

This involves modifications of the dam and spillway to pass the PMF without overtopping. The existing spillway is a 30.78m wide ogee-shaped side-channel capable to pass 269 m^3/sec. with no freeboard. Being a side-channel spillway, it is prone to choking due to crest submergence during high flows. To upgrade the dam, five auxiliary spillway alternatives were investigated. Results of PMF routing for each case are abstracted in Table 4.

TABLE 4 - RESULTS OF PMF ROUTING WITH AUXILIARY SPILLWAYS

Alternative	Max. Reservoir W.S. Elevation (m)	Suggested Dam Crest Elevation (m)
30.5m wide spillway, crest elevation 1944.53m	1947.10	1948.0
22.9m wide spillway, crest elevation 1944.53m	1947.61	1948.5
30.5m wide spillway, crest elevation 1943.31m	1945.93	1946.8
22.9m wide spillway, crest elevation 1943.31m	1946.38	1947.3
15.2m wide spillway, crest elevation 1943.31m	1947.25	1948.2

A 22.9m wide auxiliary spillway with a crest elevation of 1943.31m along with the existing side-channel will pass the PMF with maximum reservoir water surface nearly at the existing dam crest. As freeboard, a 0.90m high concrete parapet may be provided along the upstream edge of the dam crest. Governing concerns for this option include capital cost of rehabilitation and future maintenance and repairs.

2.3 Downgrade and Transfer

This involves lowering the existing spillway crest elevation from 1943.31 to 1938.43m with vertical sides with modification to the spillway channel to minimize choking due to crest submergence. The modified 30.78m wide spillway will pass PMF without dam overtopping. The smaller reservoir with normal pool elevation of 1938.43m will be used for debris control and enhancement of fisheries and wildlife, and can be transferred to local entities. It will have a residual storage capacity of about 259,000 cubic meters. Major concerns about this option include capital cost of modifications and finding an entity willing to take responsibility for future operation and maintenance. As an alternative, a sinking fund may be established for maintenance during residual life of the dam. Lowering the spillway crest will reduce attenuation of annual flood peaks.

2.4 Partial Removal of Dam

This involves excavation of a riprap-protected trapezoidal notch in the central portion of the dam with bottom width, side slopes, and crest elevation of 18.28m, 2H:IV, and 1932.34m, respectively. This will pass PMF with maximum water surface elevation about 8 meters below existing dam crest. The permanent pool will have storage capacity of 85,000 cubic meters and maximum depth of 6.4m. Flood peak attenuation historically provided by the dam will be significantly reduced. Therefore, formulation and implementation of a flood management plan will be required with structural and nonstructural elements to provide flood protection equivalent to that provided by the dam.

2.5 Complete Removal of Dam

This will require removal and disposal of approximately 182,000 cubic meters of embankment material and about 6.0m (49,000 cubic meters) of silt to clear the river valley to its natural shape. A comprehensive flood management plan will have to be implemented to provide flood protection to the downstream community equivalent to that provided by the dam.

3. EVALUATION AND CONCLUSION

Criteria for comparative evaluation of dam rehabilitation/disposition options are as follows:

- Subjective magnitude of political, social, and environmental consequences of failure.
- Economic consequences of failure
- Capital cost of rehabilitation/disposition
- Public acceptance of the option and willingness of local entities to take over the dam

In a free society and litigious environment, engineering and economic considerations alone cannot determine the preferred option. The matter will be resolved in future in light of the aforementioned criteria.

4. REFERENCES

- Fread, D.L., 1991, DAMBRK, The National Weather Service Dam-Break Flood Forecasting Model, Office of Hydrology, National Weather Service, Silver Spring, Md.
- US Army Corps of Engineers (USACE), 1990, Flood Hydrograph Package, HEC-1, The Hydrologic Eng. Center, Davis, Ca.

HYDROLOGICAL DESIGN OF LOW DAMS IN KANDI AREA (MIRZAPUR DAM).

Er. Ram Chand — Chief Engineer
Er. P.S. Narula — Director
Er. S.N. Goyal — Senior Design Engineer
Kandi Area Development Project, Irrigation Works,
Punjab, SCO 43/1, Sector 17-E, Chandigarh (India).

ABSTRACT : The hydrology of an area is of prime importance for design of structures like dams. The basic parameters like rate of inflow, sediment and floods are finalised from both the observed data recorded at the site over a long period of time and the synthetically prepared data. For the dams already constructed in the area, conventional spillways have been provided. In the case of Mirzapur Dam, a water cushion in the reservoir has been provided for the floods, thus containing the floods by raising the height of dam and avoiding the heavy cost of spillway structure. The maximum probable flood has been impinged at Normal Conservation Level (NCL) and routed through an emergency escape. The paper deals with the hydrological design of Mirzapur Dam in the Kandi Area.

1. INTRODUCTION

The area along foothills of Shivalik is called Kandi Area in Punjab (India). The area is intercepted by various rivulets (Choes) which create havoc in the area during monsoons causing considerable damage to crops and village habitats (1). The water flow in the rivulets is mostly during rainy days. A series of dams have been/are being constructed over the rivulets for irrigation development and flood protection in the area. The dams are being designed as per latest hydrological techniques considering the rainfall data, run-off and sediment data for fixing the height of dam. The height of dam is fixed so as to absorb the floods of the order of 25 years frequency succeeding Standard Project Flood. The conventional concrete spillway has been eliminated. An emergency escape has been provided for routing the Maximum Probable Flood. The paper deals with the hydrological design of Mirzapur Dam which is located about 25 km North-West of the city of Chandigarh (Fig.1) and is connected by a metalled road.

2. RAINFALL RUN OFF DATA

Daily rainfall data of the Mirzapur Dam catchment area (about 13.90 sq. km.) was not fully available. Incomplete data of rainfall for the period 1974-1989 and run-off for the months of January to June and October to December were available (Table-I). Since this scanty data was insufficinet for the evaluation of annual run off therefore rainfall data values from 1958 to 1988

V. P. Singh and B. Kumar (eds.), Water Resources Planning and Management, 315–325.
© 1996 *Kluwer Academic Publishers. Printed in the Netherlands.*

available for the nearby Rain Gauge Station at Chandigarh (Table- II) have been utilised to complete the missing data. This has been done by comparing the values available for both the places and evaluating an average conversion factor and utilising the same to work out the missing values at Mirzapur Dam site and used as synthetic values. Incidentally the conversion factor works out as 1.0 in this case.

3. RAINFALL/RUNOFF RELATIONSHIP

Daily discharge measurement of the rivulet (choe) has been available mainly for the monsoon months of 1974 to 1988 which has been converted into daily/ monthly runoff. A curve of Rainfall v/s Runoff was plotted for the months of July, August and September in these years (fig.2). Some daily discharge measurements were also available for other months for all these years and accordingly a separate Rainfall v/s Runoff relationship was plotted for the period (fig.2). Certain values which were suspect and also the values where Runoff is more than Rainfall were not taken into account for preparation of these curves. A most suitable curve from the Rainfall v/s Runoff relationship was arrived at and the synthetic values of monthly runoff have been worked out and adopted for design.

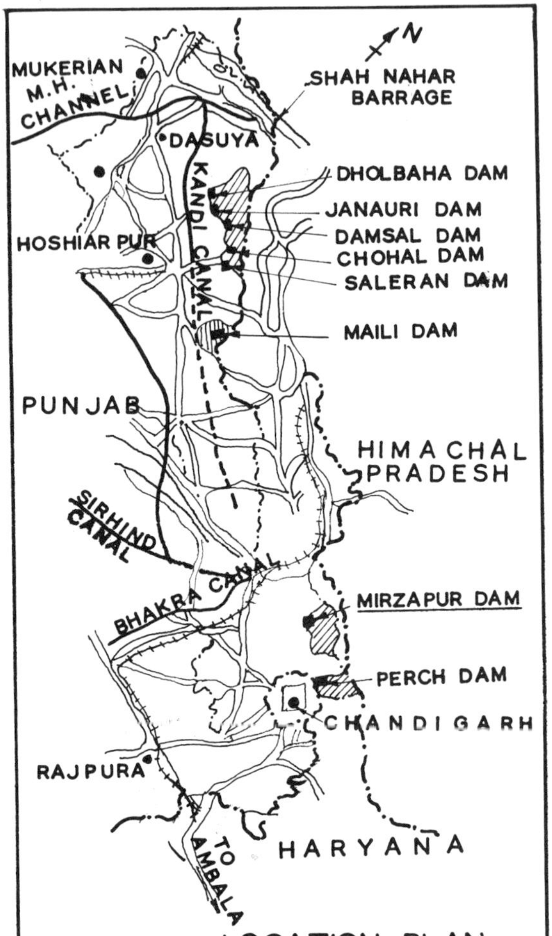

FIGURE-1 LOCATION PLAN

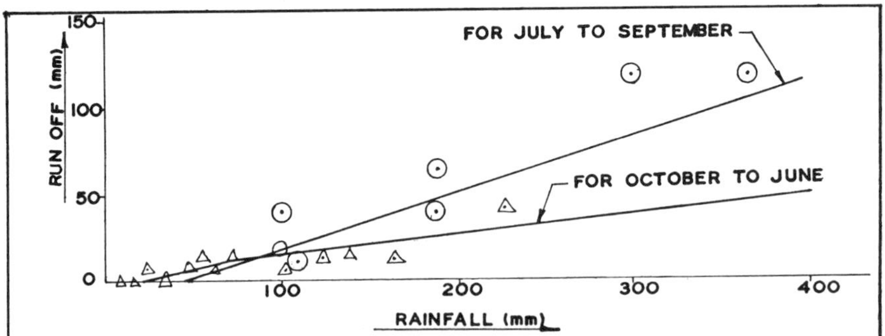

FIGURE-2 RAIN FALL V/S RUN OFF CURVE (1974-89)

TABLE-I : MONTHLY RAINFALL & RUNOFF (mm) (OBSERVED)

a) Year 1974 to 1981

Month		1974	1975	1976	1977	1978	1979	1980	1981
Jan	RF	10.0	57.5	85.0	40.0			5.0	
Feb	RF	10.0	30.0	110.0		61.5			
March	RF	9.0	30.0	7.0		50.0	12.0	10.0	
April	RF				45.0	17.5	5.0		
May	RF	10.0	12.0	64.5	110.0		11.0		
June	RF	50.0	92.0	137.0	30.0	263.5	60.5		
July	RF	139.0	191.5	433.5	560.0	324.5	247.0		
	RO		72.4	187.8	208.9	770.4	233.4	594.8	32.8
Aug	RF	283.5	286.5	276.0		499.0	123.0	122.0	
	RO	42.3	122.9	183.8	780.6	1576.5	231.7	180.0	8.3
Sept	RF	7.5	201.0	97.5	315.0	135.5	17.5		25.0
	RO		138.0		282.4	628.3	14.6	426.9	8.5
Oct	RF	10.0	5.0		90.0				
Nov	RF					46.0			25.0
Dec	RF	39.0			32.5		20.0		

b) Year 1982 to 1989

Month		1982	1983	1984	1985	1986	1987	1988	1989
Jan	RF	45.0		7.8	26.6		70.5	3.4	38.1
	RO						10.9		4.7
Feb	RF	40.0		104.7	3.6	45.4	45.6	3.5	
	RO			5.7			1.1		
March	RF	64.0		6.9		32.4	37.1	23.8	15.8
April	RF	21.0				10.1	24.5		
	RO						3.4		
May	RF					62.1	162.5	24.4	6.8
	RO					1.7	10.7	3.4	
June	RF			227.0	124.4	135.5	53.1	117.1	76.9
	RO			40.0	18.2	13.5	11.9	19.7	
July	RF	47.0	103.4	263.3	235.1	194.7	110.5	774.0	
	RO	681.8	25.9	97.1	73.4	129.9	18.5	373.8	
Aug	RF	210.0	374.8	177.1	186.5	364.8	243.7	608.1	
	RO	202.3	150.0	80.0	36.3	123.1	124.3	305.3	
Sept	RF	4.0	219.2	146.2	98.0	102.6	61.1	613.4	
	RO		80.5	15.3	40.3	2.0	24.1	354.0	
Oct	RF		25.6		72.0	25.4	76.6		
	RO					0.9	28.2		
Nov	RF					34.7			
	RO					0.3			
Dec	RF		25.6			56.7		119.1	
	RO					6.6		13.9	

NB : RF indicates Rainfall while RO indicates Run-off. Values not given in table are Not Available.

TABLE-II : MONTHLY AVERAGE RAINFALL DATA FOR CHANDIGARH (mm)

a) for Year 1973 to 1978

	1973	1974	1975	1976	1977	1978
Jan	52.9	6.7	56.8	58.3	49.4	1.0
Feb	15.6	14.1	38.4	48.5	2.0	43.7
March	15.4	5.0	34.5	11.4	5.1	72.2
April	0.0	0.0	0.0	4.2	35.9	37.0
May	12.0	14.3	23.4	46.4	70.7	0.0
June	455.5	135.0	294.2	149.4	11.5	174.8
July	395.2	242.2	164.0	524.7	340.0	253.7
Aug	274.2	239.5	365.1	267.3	289.1	463.0
Sept	196.9	14.0	119.0	28.4	254.4	265.9
Oct	122.8	5.4	8.2	1.2	17.2	8.1
Nov	0.0	0.0	0.0	0.0	0.0	33.6
Dec	34.6	39.9	0.0	2.0	63.1	7.7
Total	1575.1	716.1	1103.6	1141.8	1138.4	1360.7

b) for Year 1979 to 1984

	1979	1980	1981	1982	1983	1984
Jan	80.9	23.4	61.5	99.4	149.2	13.4
Feb	113.9	10.4	13.9	42.7	15.7	99.0
March	40.2	32.4	125.4	173.9	39.8	5.8
April	7.5	8.0	14.3	56.5	128.9	11.8
May	22.0	0.0	27.3	58.2	127.3	1.9
June	121.4	152.5	128.1	32.8	114.8	129.5
July	351.6	445.1	466.7	243.7	298.9	254.3
Aug	205.9	148.0	104.7	333.9	293.6	118.1
Sept	48.9	66.9	17.0	34.7	191.2	92.5
Oct	0.0	17.6	0.0	1.0	1.0	0.0
Nov	2.3	32.6	100.6	4.4	0.0	0.0
Dec	15.2	84.3	0.0	67.7	18.8	0.0
Total	1009.8	1021.2	1059.5	1075.8	803.5	726.3

c) for Year 1985 to 1988

	1985	1986	1987	1988
Jan	27.4	0.0	51.7	4.3
Feb	0.0	49.6	51.7	4.3
March	0.0	34.2	29.1	40.3
April	18.8	12.5	24.6	0.0
May	0.5	72.5	117.6	20.2
June	105.9	172.5	66.5	90.5
July	226.0	263.7	114.5	554.2
Aug	207.0	358.5	77.5	481.3
Sept	130.5	59.4	12.9	366.2
Oct	50.2	31.7	25.0	0.0
Nov	0.0	30.7	0.0	0.0
Dec	55.5	37.0	4.0	66.0
Total	821.8	1122.3	575.1	1627.3

4. ANNUAL RUNOFF AND WATER AVAILABILITY

Annual Rainfall (synthetic) and Annual Runoff (synthetic) for the years 1958 to 1972, partly observed and partly synthetic values for the years 1973 to 1989 have been shown (Table-III). Average Annual Run-off against Average Annual Rainfall of 1098.66mm comes to 302.77mm (420.85 HaM). Average Run-off factor comes to 0.276.

Annual Runoff values (synthetic and observed) for the years 1958 to 1989 have been arranged in ascending order (Table-IV), 75% and 66% dependable flow works out as 295.35 HaM and 321.159 HaM respectively. Reservoir losses have been taken as 12.5% and accordingly net water availability comes to 258.43 HaM and 281.01 HaM for 75% and 66% dependable flow respectively.

5. SEDIMENTATION STUDIES

Sedimentation studies have been carried out to determine the load ofsediment, to find out the level for fixing the inlet of Irrigation Outlet pipe and to find out the effective storage capacity for available water (live storage) for irrigation supply and also to control floods during various stages of the life of the reservoir by assessing average rate of sediment.

Suspended silt load observations are available for the years 1982 to 1988. From these actual observations made at site for the mean discharge and the sediment concentration (gm per litre), a co-relation has been evolved between these observations (fig.3). Synthetic values of sediment

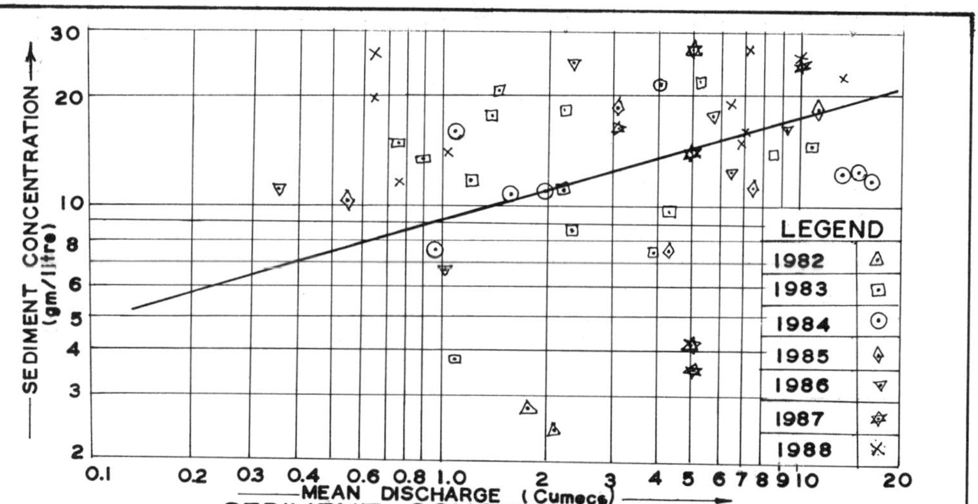

SEDIMENT CONCENTRATION V/S MEAN DISCHARGE CURVE

FIGURE-3

concentration have been worked out for the period 1974 to 1988 for which discharge measurements are available and sediment load in tonnes has been worked out. On the basis of annual run-off and corresponding sediment load, a sediment rating curve is evolved (fig.4) From the curve, average

TABLE-III : RUN-OFF FROM RAINFALL v/s RUNOFF CURVES

Year	Rainfall (mm)	Run-off (mm)	Run-off (Hect-M)
1958	1112.0	302.0	419.7
1959	1371.1	355.0	493.4
1960	984.2	277.5	385.7
1961	1192.4	256.7	356.8
1962	896.3	193.0	268.2
1963	913.3	219.0	304.4
1964	1243.1	349.0	485.1
1965	774.8	164.7	229.0
1966	1122.9	278.4	386.9
1967	1431.1	379.7	527.8
1968	931.6	192.2	267.2
1969	902.9	224.5	312.0
1970	1121.2	265.5	369.0
1971	1891.1	505.0	701.9
1972	722.5	169.7	235.9
1973	1578.0	412.0	572.6
1974	568.0	84.4	117.4
1975	905.0	354.6	492.8
1976	1210.5	449.9	625.4
1977*	1222.5	1304.0	1812.5
1978*	1397.5	3032.8	4215.6
1979*	495.0	505.4	702.5
1980*	970.0	1236.1	1718.1
1981	991.9	93.8	130.4
1982	504.1	308.2	428.4
1983	1324.3	335.4	466.2
1984	933.0	230.0	319.7
1985	801.7	184.7	256.8
1986	1064.4	283.5	394.0
1987	885.2	235.5	327.3
1988	2286.8	1070.4	1487.8
1989*	137.6	14.7	20.5
Total	29663.9	8174.9	

Average Rainfall = 1098.6 mm

Avergae Run-off = 302.7 mm

Run factor = 0.276

* Suspect & incomplete values excluded.

TABLE-IV : DISTRIBUTION OF ANNUAL RUN-OFF

Sr. No.	Year	Run-Off Hect.M	Run-Off arranged in ascending order	No.of times equalled or exceeded	Probability
1.	1958	419.7	117.4	27	100.0
2.	1959	493.4	130.4	26	96.3
3.	1960	385.7	229.0	25	92.5
4.	1961	356.8	235.9	24	88.8
5.	1962	268.2	256.8	23	85.1
6.	1963	304.4	267.2	22	81.4
7.	1964	485.1	268.2	21	77.7
8.	1965	229.0	304.4	20	74.0
9.	1966	386.9	312.0	19	70.3
10.	1967	527.8	319.7	18	66.6
11.	1968	267.2	327.3	17	62.9
12.	1969	312.0	356.8	16	59.2
13.	1970	369.0	369.0	15	55.5
14.	1971	701.9	385.7	14	51.8
15.	1972	235.9	386.9	13	48.1
16.	1973	572.6	394.0	12	44.4
17.	1974	117.4	419.7	11	40.7
18.	1975	492.8	428.4	10	37.0
19.	1976	625.4	466.2	9	33.3
20 to 23 Values for 1977 to 1980 suspect and incomplete, excluded.					
24.	1981	130.4	485.1	8	29.6
25.	1982	428.4	492.8	7	25.9
26.	1983	466.2	493.4	6	22.2
27.	1984	319.7	527.8	5	18.5
28.	1985	256.8	572.6	4	14.8
29.	1986	394.0	625.4	3	11.1
30.	1987	327.3	701.9	2	7.4
31.	1988	1487.8	1487.8	1	3.7
32.	1989	20.5			

Computation of Dependable Flow

Probability	Without reservoir Losses	With reservoir losses (12.5%)
75%	295.351 HaM	258.432 HaM
66%	321.159 HaM	281.014 HaM

TABLE V : REDUCTION IN SEDIMENT YIELD

Year of Life of Dam	%age of sediment Reduction	Sediment Yield ($HaM/km^2/year$)
0	0	0.729
5	23	0.596
10	55	0.411
15	76	0.289
20	92	0.196
25	94	0.185
40	100	0.150

annual sediment yield against average annual run-off of 420.85 HaM has been worked out as $128x10^3$ tonnes (suspended load).

As per experience on other small dams in the Kandi Area, the bed load from the choe has been adopted as 15% of suspended load. With trap efficiency of 96.5% and average density of sediment as 1402 kg/m^3, the average rate of sedimentation is of the order of 0.729 HaM/km^2/year.

The assessment of the reduction in the sediment yield over 40 years period has been made on the basis of improved vegetation in the catchment as shown in Table-V. The reservoir levels attained on deposition of silt after 15 years, 25 years and 40 years (known as zero levels) have been evaluated (2) and tabulated in Table-VI.

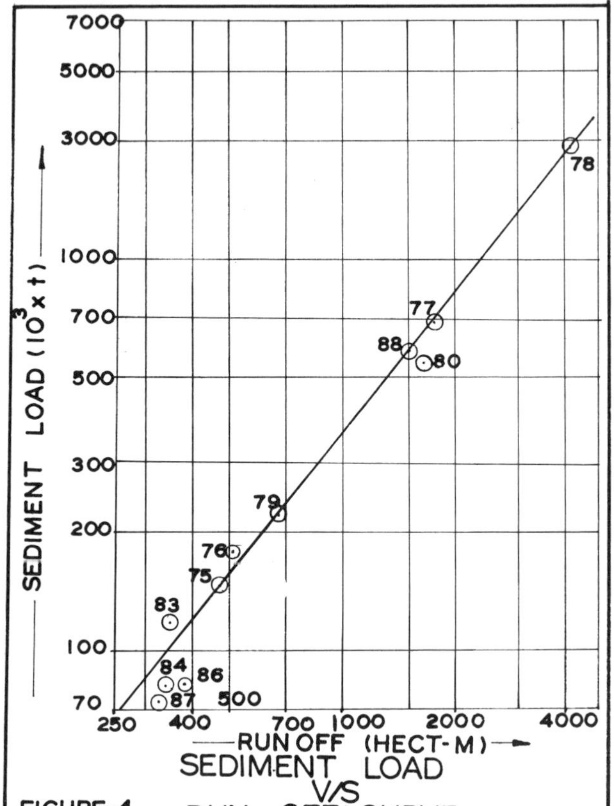

FIGURE-4

5.1 Normal Conservation Level (N.C.L.)

As per hydrological data assessment net available water (live storage) is 258.43 HaM (para-4). The live storage has been contained above the zero level (367.378m) attained after 40 years of silt deposit. The water level of reservoir works out as 375.60M. This level is known as normal onservation level (N.C.L.) (fig.6)

6. DESIGN FLOOD

For adopting the type of flood for design, the dam has to be classified either according to size by using the hydraulic head (difference in

TABLE-VI

Age of Dam (Years)	Average sediment load (Hect.M/km^2)	Total sediment load (Hect.m)	Zero Level (metre)
0	-	-	361.30
15	0.506	105.501	365.776
25	0.390	135.525	366.448
40	0.307	170.692	367.378

Maximum Water Level in the reservior and water level on the downstream of the dam if any) or the gross storage behind the dam. The storage capacity for Mirzapur Dam works out to be 4.25 million cubic meter, thereby it falls in the classification of small dam. But the hydraulic head of Mirzapur Dam is 20.60 M as such it is to be classified as intermediate dam. Hence Standard Project Flood (S.P.F.) has been adopted as design flood (3).
The S.P.F. design flood works out as below :-

a)	Synthetic Hydrograph Method	=	455.28 cumecs
b)	Improvised method (4)	=	433.45 cumecs

Maximum value of these two methods i.e. 455.28 cumecs has been adopted as design flood (fig.5).

7. IRRIGATION OUTLET

An outlet of 750mm diameter is provided for irrigation. Its inlet is to be raised in two stages, first to elevation 366.0 M and then to elevation 368.0M. These are zero levels attained on silt deposition after 25 years and 40 years respectively.

Thus the outlet remains functional during the entire life of dam i.e. 40 years. The live storage provided for irrigation is 258.43 HaM.

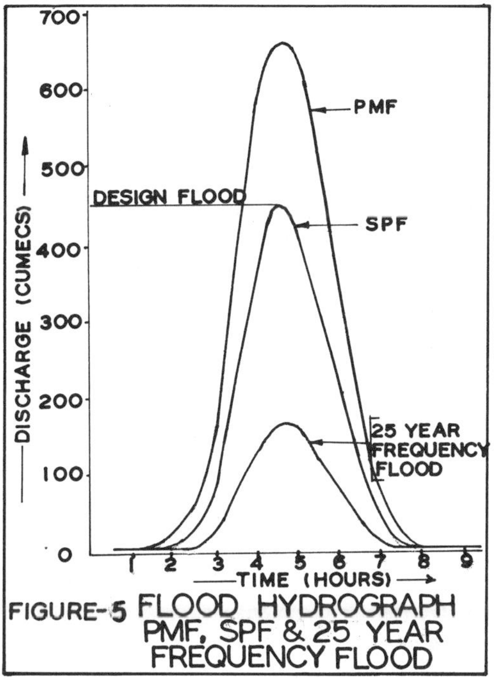

FIGURE-5 FLOOD HYDROGRAPH PMF, SPF & 25 YEAR FREQUENCY FLOOD

8. FLOOD ROUTING & HEIGHT OF DAM

Conventional spillway with crest at E1. 375.6M, bed width 20M and side slope 1.5H:1V for passing the flood waters from the reservoir was initially considered. Maximum water level attained was 378.50 M after routing the Standard Project Flood preceeded or succeeded by a flood of 25 years frequency (5). The peak to peak interval between these two floods has been taken as 3 days (2). On routing the PMF, minimum water level attained was 379.5M (5). Greater of the maximum water levels so obtained i.e. 379.5 mt. was used in determining the height of dam. Assessment of Free Board was made which is less tha 2 M (minimum) and as such 2 M Free Board was adopted (6). The top of dam comes to 381.5M.

Further, an attempt has been made to contain the flood water in the reservoir by raising the height of dam to avoid the heavy construction and annual maintenane cost of spillway structure and elimination of other strutures like Bridges/X- drainage works downstream of the proposed dam. In view of this it has been considered appropriate to contain the Standard Project Flood (S.P.F.) and 25 years frequency flood in the reservoir above normal conservation level (N.C.L.) by raising the height of dam. Thus the water level attained in the reservoir works out to 381.4 M (7). An un-regulated emergeny escape with crest at E1.381.4M, bed width as 10 M, side slope 1.5 Horizontal to 1 vertical has been provided to account for unforseen floods of higher magnitude.

The reservoir level from crest level of emergency escape channel is to be lowered to N.C.L. at slow rate as per embankment design criteria through 750mm dia irrigation outlet pipe. The studies indicate the time for lowering the water from crest of the emergency escape to NCL as about 20 days. It is assumed that the next PMF if at all occurs will be after a period of 20 days or so. The maximum probable flood has been impinged at N.C.L. (375.6M) and routed through the emergency escape. The maximum water level attained is 381.9 M, with outflow discharge of 7.36 cumecs through the emergency escape. After accounting for Free Board of 2 M the top of dam works out to E1.384.0 M and the same has been adopted (fig.6). Therefore, the total height of dam works out to 22.8 M.

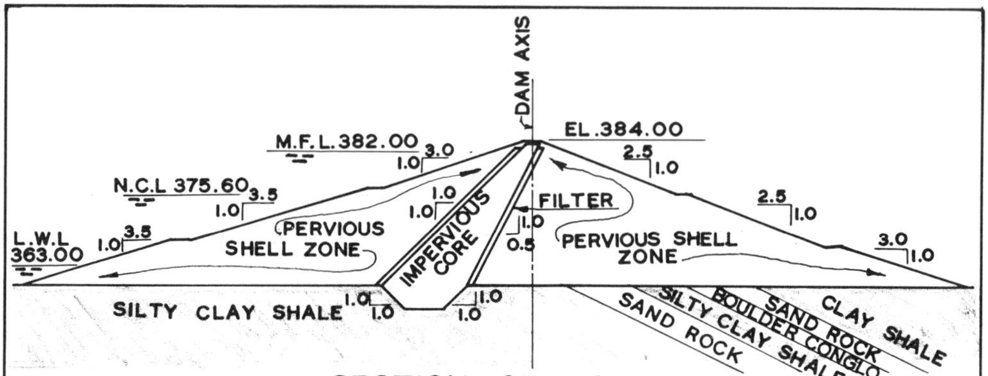

FIGURE-6 SECTION OF DAM

9. CONCLUSIONS

1. The proper assessment of hydrological data results in fixing most economical height of the dam, for flood control and irrigation requirements of the sub- mountanous area where other conventional irrigation systems are not practicable.

2. Flood absorption/flood cushion for a small/medium height dam may be provided to eliminate the construction and maintenance cost of the spillway. Consequently the structures downstream e.g. Cross- Drainage works and Bridges etc. are also eliminated.

REFERENCES :-

1. 'Budki Watershed Feasibility Report' Vol-II. Kandi Area Development Administration, Irrigation Works (Pb) SCO 43/1 Sector 17-E Chandigarh (India)

2. Indian Standard I.S:5477 (Part-I to IV)-1969:Method for fixing the capacities of Reservoir.

3. Indian Standard I.S.:11223-1985. : 'Guidelines for fixing spillway capacity.'

4. Hand Book of Hydrology Govt. of India.

5. KAD Memo No: MPD-3 'Flood Routing and Height of Dam' Kandi Area Development Administration, Irrigation WOrks (Pb) SCO:43/1 Sector 17-E Chandigarh.

6. Indian Standard I.S:10635-1983.: 'Guide Line for Freeboard Requirement in Embankment Dams'

7. KAD Memo No:MPD-3-A: 'Supplimcntary Flood Routing and Height of Dam'. Kandi Area Development Administration Irrigation Works (Pb) SCO :43/1 Sector 17-E Chandigarh (India).

ACKNOWLEDGEMENT

The authors are grateful to Er. Sanjeev Suri, Assistant Design Engineer and other officials of Kandi Canal Design Dte; Chandigarh for their valuable cooperation in preparation/compiling of this paper.

HYDROLOGIC DESIGN OF CROSS DRAINAGE WORKS ON KANDI CANAL

Er. Ram Chand — Chief Engineer
Er. R.B. Lal — Senior Design Engineer
Er. Sanjeev Suri — Assistant Design Engineer
Kandi Area Development Project, Irrigation Works,
Punjab, SCO 39/1, Sector 17-E, Chandigarh (India).

ABSTRACT : Along the alignment of the Kandi Canal, in a length of 59.5 Km, 96 Cross-drainage works are being provided. The hill torrents/rivulets (choes) cross the canal mostly near its bed or at higher levels. The paper deals with the empirical approach adopted to assess the discharge of hill torrents. The works have been designed/constructed on the basis of model studies eliminating the stilling basin by providing baffle piers and staggered blocks on horizontal floor to dissipate the energy in case of cross-drainage works having fall. Cross-drainage works of aqueduct or aqueduct with fall have been provided across the canal. Provision of syphon adqueducts has been eliminated on account of heavy silt load in these torrents during the rainy seasons.

1. INTRODUCTION

The area along Shivalik foothills is known as Kandi Area in Punjab-India. The area is having steep natural gradient and is infested with numerous hill torrents/rivulets (choes) which carry flash floods in the rainy season causing considerable damage to crops and village abadis.

To provide irrigation facilities to sub-mountainous area, a two tier irrigation system is being developed (1). In the upper reaches rock-fill/earth-fill dams are constructed and beyond the command of these dams a canal called Kandi Canal is being constructed (Fig-1). The layout of the Kandi Canal is such that it involves 96 cross-drainage works in a length of 59.5 Km. upto Hoshiarpur.

The Kandi Canal off-takes at RD. 5575 M-Left of Mukerian Hydel Channel for the irrigation of Kandi Area. The Canal is a contour channel. The irrigation is provided on the right side of the canal only as the area on the left side is at higher levels, being hill side.

2. GEOLOGY OF THE AREA

The rocks of the upper Shivalik formation are often overlain by quaternary sediments and alluvium of sub-recent to recent age. The alluvial deposits and erratic geological formation of clay bands/sand rock exist.

V. P. Singh and B. Kumar (eds.), Water Resources Planning and Management, 327–333.
© 1996 *Kluwer Academic Publishers. Printed in the Netherlands.*

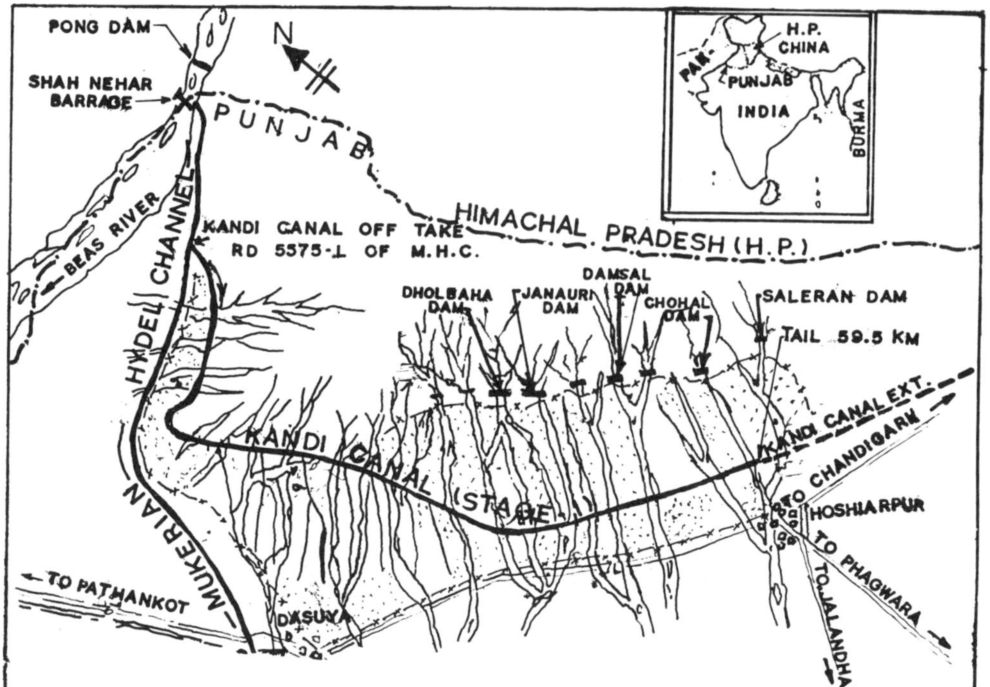

FIG.-1 LOCATION PLAN-KANDI CANAL CROSS DRAINAGE WORKS

3. CONSTRAINTS FOR CROSS-DRAINAGE WORKS

3.1 There is a large number of seasonal hill torrents/rivulets (choes) which cross the Kandi Canal. These choes are mostly wide and shallow. The sheet flow of water brings heavy load of debris/silt during rains which has a tendency to choke the waterways of cross-drainage works across the canal.

3.2 Initially it was considered to provide only superpassage crossings involving no head loss in the canal. Subsequently, the canal data revealed that most of these choes can not cross the canal in the usual normal fashion i.e. as aqueduct, syphon aqueduct or super-passage. At certain locations, the canal had to be syphoned for proper cross-drainage works.

4. METHOD FOR DESIGN

4.1 Catchment Area :

It has been worked out on the G.T.S. sheets available for the area, supplemented by field survey wherever required.

4.2 Discharge :

The practices in vogue suggest for unit hydrograph principle to work out flood peaks (2). The catchments for these hill torrents are small varying from 0.5 to 10 Sq.Km. and the flow is of flashy and turbulent nature during the rains. At other times there is hardly any flow in the rivulets. The adoptability of the unit hydrograph is not practicable. Therefore, an empirical approach has been adopted. However, the principle of unit hydrograph alongwith synthetic hydrograph has been adopted for the design of dams in the Kandi Area (3).

The design flood discharge for rivulets for fixing waterway has been worked out using Rational Formula ($Q = CIA$) and Dicken's Formula ($Q = CM^{3/4}$) (For catchment areas greater than 4.0 Sq.Km.) as per experience in the region. The value of coefficient of runoff 'C' in Rational Formula has been adopted as 0.50 to 0.75 depending upon the topography, soil condition and cultivation in the area etc. (4). Maximum intensity of rainfall has been adopted as 100mm/hour as per meteorological data of the region (2). In the Dicken's formula, value of 'C' has been adopted as 20 in view of the terrain and meteorology of the area. The value of 'C' adopted in Dicken's formula covers 50 years flood inflow peak comparable with the inflow studies carried out for different Kandi Watershed by unit hydrograph method.

As per practice for design of hydraulic works in the region, discharge has been increased by 30% for working out scour depth for the foundation and free board for the work (5).

4.3 Type of Work :

Hill torrents crossing the canal are generally wide and shallow and mostly remain dry during the year. During monsoon season, these bring debris/silt with maximum discharge. Deciding a suitable type of cross-drainage work is of utmost importance. The work provided has to suit the topography of the area and required provision of no head loss or minimum head loss at the work in the canal. For the safety of nearby village inhabitatnts, heading up in hill torrents has to be avoided/minimised.

Syphon aqueduct type works have not been preferred, keeping in view the possibility of their choking by debris/silt etc. instead the canal has been syphoned. Aqueduct with fall is provided with downstream open cut to join the N.S.L. (bed of the choe), wherever economically viable. The above type of works hve been provided even by raising the bed level of canal, restricted upto 20% depth of flow, locally.

5. SALIENT FEATURES OF DESIGN

5.1 Provision of Baffle piers/Staggered blocks for energy dissipation :

The aqueducts with fall having the upstream bed level varying in level upto 3.0 M to that of the barrels require arrangements for energy dissipation. An arrangement of baffle piers on the sloping glacis, transitions on the upstream side between approach channel and aqueduct barrel, transition on the downstream side (6) was got tested on model at Irrigation and Power Research Institute, Amritsar. It was got tested on Model for different discharge intensities so as to achieve minimum increase in velocity of flow downstream of glacis and smooth flow condition through barrel and downstream transition.

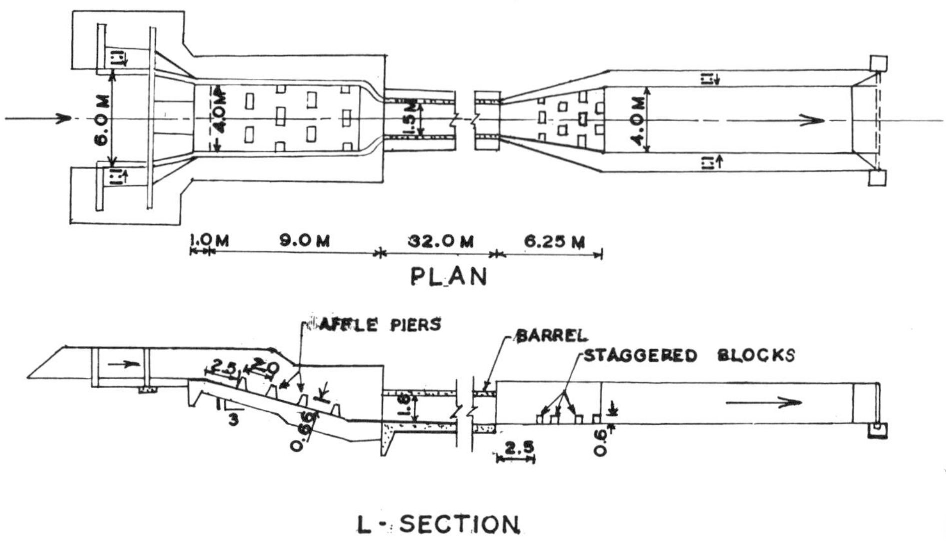

FIG-2 LAYOUT OF AQUEDUCT WITH FALL AS PER MODEL STUDIES (CANAL SECTION NOT SHOWN)

Modifications had to be carried out to achieve satisfactory results, as per arrangement of baffle piers and staggered blocks (Fig.2). It was observed that whatever energy gained at the fall is dissipated fully by the provision of baffle piers. The velocity of flow through the work was in the range of 3-4 m/sec. which was considered adequate for the movement of silt load.

Accordingly the arrangement has been incorporated in the design of cross-drainage works of aqueduct with fall. A layout of work (typical) is shown in Fig-3 alongwith design parameters. Natural flow of cross-drainage works by this arrangement has been retained and syphoning of canal has been avoided. The arrangements have proved successful during rainy seasons for the last 5-6 years in the field.

5.2 Provision of Hump in canal for aqueduct type of works :

To facilitate the design of cross-drainage works as aqueduct or aqueduct with fall, the canal bed has been raised locally upto 20% depth of flow in canal. Raising of the canal bed, reduced upstream fall and excavation for aqueduct barrel and downstream works also, thereby reducing the cost of structure. At these locations, area of flow in the canal has been kept same and suitable transitions hav been provided on the upstream and downstream sides (6).

DESIGN DATA FOR THE WORK

S.N.	PARTICULARS	DATA OF CANAL	DATA OF CHOE
1	CATCHMENT AREA	-	0.70 km^2
2	F.S. DISCHARGE / PEAK DISCHARGE	13.10 Cum.	14.67 Cum
3	F.S.L / H.F.L.	327.01 M	325.80 M
4	BED LEVEL	323.915 M	325.35 M
5	WATER SURFACE SLOPE	0.101 ‰	42.11 ‰
6	RADIUS / H.F. DEPTH	3.10 M	
7	N.S.L AT CROSSING	325.35 M	

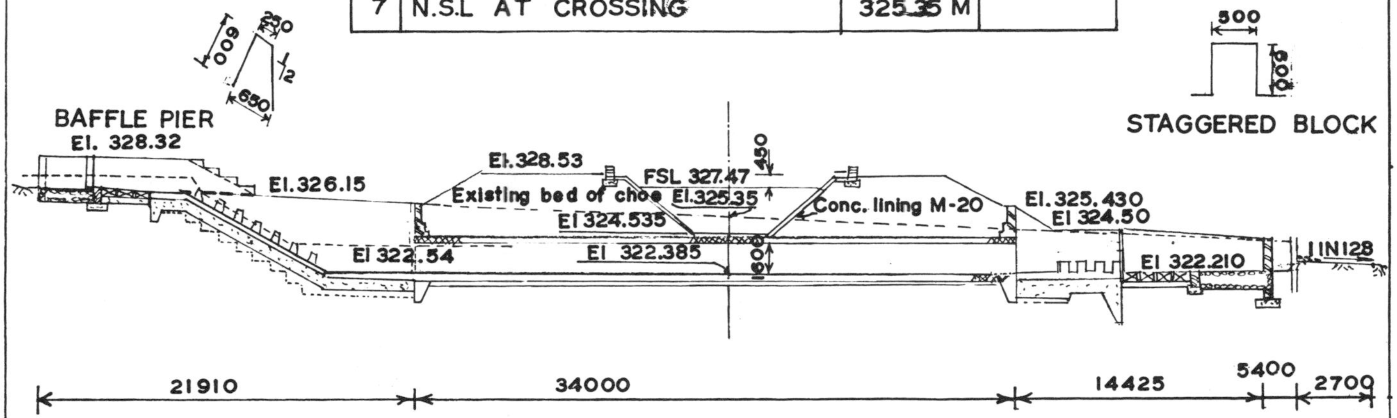

FIG.-3 LAYOUT OF CROSS DRAINAGE WORK (AQUEDUCT WITH FALL)

6. CONCLUSIONS

1. In the submountainous region and in case of contour canal, the provision of cross-drainage work for hill torrents may be planned in such a way that the flash flood passes smoothly without scouring on upstream and downstream of the work. Also to avoid heading up of water on upstream side to protect the inhabitants of the area and erosion on downstream side, protections on upstream and downstream of the work may be provided suitably.

2. Aqueduct with fall may be constructed without the provision of cistern for energy dissipation in submountainous area where silt and debris come with water during rainy season. Provision of baffle piers on the sloping glacis with staggered blocks on the horizontal floor downstream of barrel is a satisfactory alternate arrangement.

3. Syphon aqueduct type of works may be avoided in such areas as the barrels get choked during rains; instead canal may be syphoned.

4. Model studies may be carried out for cross-drainage works in the submountainous region and particularly for works of a contour canal for the safety of the canal.

5. The empirical approach for discharge assessment adopted for Kandi Area cross drainage works is quite workable/satisfatory. However, a rational approach for assessment of the discharge on the basis of rainfall data for small catchments needs to be evolved where normal guidelines like unit hydrograph, etc. do not fit in.

REFERENCES :

1. Revised Project Report of Extension and Improvement of Irrigation on Shah Nehar Canal System, Vol-I-1989. P.W.D. Irrigation Branch, Punjab, Hoshiarpur, Punjab, INDIA.

2. Estimation of design flood peak : 'A method based on unit hydrograph principle'-C.W.C. Report No:1/73. Central Water Commission, Government of India, New Delhi.

3. KAD Memo No : MPD-2A 'Hydrological Studies' Kandi Area Development Administration, Irrigation Works (Pb) SCO : 39/1 Sector 17-E, Chandigarh, INDIA.

4. Indian Road Congress Special Publication No : 13. Guidelines for the design of Small bridges & culverts, New Delhi, INDIA.

5. Indian Standard IS : 7784 (Part-I & II) : Code of Practice for design of cross-drainage works.

6. (i) Hydraulic Design of Baffle Piers for cross-drainage works on Kandi Canal. Report No. : HY/R/19/87-88 : Irrigation and Power Research Institute, Amritsar, Punjab, INDIA.

   (ii) Hydraulic Design of Hump & its transitions for drainage crossings on Kandi Canal, Report No.: HY/R/38/85-86 (Revised) : Irrigation and Power Research Institute, Amritsar-Punjab, INDIA.

ACKNOWLEDGEMENT

The authors are grateful to the officers/officials of the Kandi Area Development Administration of Irrigation Department especially Mr. H.C. Chouhan, PA to Chief Engineer/KAD, for their valuable cooperation in the preparation of the paper.

INFLUENCE OF RAINFALLS ON THE LANDSLIDES OCCURRED IN BASILICATA (ITALY) IN 1973

SDAO F.[x], SIMEONE V.[^], SPILOTRO G.[x]
[x] Department of Structural and Geotechnical Engineering and of Engineering Geology, Basilicata University, Potenza, Italy
[^] PhD Geotechnical Engineering, Bari, Italy

ABSTRACT. A lot of shallow and deep landslides were activated and/or reactivated in Basilicata region (Southern Italy) during the first months of 1973, as a consequence of a period of continuous light rains, that were not hydrologically exceptional. These landslides caused widespread serious damages throughout the whole region, affecting both towns and the infrastructural networks. Most of these landslides occured in slopes which had suffered from previous landslide movements; so in most cases they were not first failure landslides. The total amount of remobilizated landslides was so high that we can describe it as a generalised geomorphological crisis of slopes. During the previous year 1972 the total amount of rainfalls during the same period was considerably greater; however it didn't produce the same generalized instability effects, but it caused only few localised landslides. On the basis of these observations, the study carried out highlights the different distribution of the rainfalls in the two examined years. In the first one the total amount of rainfalls is due to a low number of downpours of great intensity, which favoured runoff, while during the following year rainfall events were lighter, but more frequent, favouring rainwater infiltration. Moreover in 1973 piezometric levels started from a higher level, as a consequence of the previous hydrological year.

1. INTRODUCTION

The instability of the slopes is due to several natural or anthropic factors. Rainfalls is one of the most important of them, because it operates as an active cause by increasing the total unit weight of soils and by modifying geometries of slopes by erosion, and it also operates by reducing shear strength of as a consequence of pore water pressure increasing, and by swelling and soil softening processes, and finally, in the case of cemented soils, by reducing cementation.

The effectiveness of rain depends obviously on many other factors, the most important of them are:

- morphological and structural features of the site and of the soils (i.e. the capacity of the soil surface to drain or to retain the water);
- grain size distribution and chemistry of the superficial layers (i.e. the capacity of the soil to conduct or absorb water);
- initial groundwater level and degree of saturation of the superficial layer.

However there are no doubts that the most important factor connecting to the effect of rainfall and slope instability is the time and intensity of rains with respect to water absorption capacity of the soil.

A large rainwater infiltration into the soil can be very dangerous in a lot of situations, particularly

V. P. Singh and B. Kumar (eds.), Water Resources Planning and Management, 335–343.
© 1996 *Kluwer Academic Publishers. Printed in the Netherlands.*

in the areas affected by old landslides. In this case the morphology can present typical counter-slopes and ponds which impede superficial drainage and allow infiltration of the water into the soil mass, which is affected by fractures generated during the previous movements. Moreover the detrital layers covering the landslide body modifies the permeability of the mass, allowing wide and deep circulation of water in the landslide body (Fidelibus et Al., 1990).

As a consequence of all this facts it is difficult to chose a value of a critical rain, while it seems more important to define a set of critical values of pore water pressures for each old or first failure landslide body. Recent studies (Simeone, 1992) and the present one confirm that a diffuse landslide mobilization can be induced by a not hydrologically exceptional rainfall, while the mobilization of landslides as a consequence of heavy rainfalls with an high return period is not necessarly correlated.

In this paper are examined the effects of 1972 and 1973 rainfall events on landslide mobilization in the Basilicata region. Rainfall data of these two years has been compared with that one collected over a long period of time in some representative rain-gauge stations, as well as the groundwater level in a water-bearing level measurament station.Rainfall events during 1972 winter were characterized by high return period, confirming the exceptionality of the event. Nevertheless landslide mobilitation as a consequence of these events can be considered as normal. During the subsequent winter of 1973 rainfalls were characterized by considerably smaller return period, but landslide generation or remobilization, as well as the groundwater level rise, was exceptionally high.

2. FEATURES AND DIFFUSION OF LANDSLIDES MOBILIZED BY 1973 WINTER RAINS

During the first months of 1973 and particularly during the last week of March and the first one of April, in Basilicata region and in the surrounding areas, occured rainfall events that were not particularly intense, but the rain was continuous for a period of time of more than two weeks. Following this rainfall period, all over Basilicata region, and above all in the valleys around the middle and lower section of the principal rivers, a lot of landslide phenomena of various types and dimensions, superficial and deep, were activated or remobilised. The landslides were so widespread that it would not be improper to define this event as a sort of geomorphological crisis of slopes. The extent of the area affected by landslide is confirmed by the diffuse and serious damage suffered by both large and small urban centres all over the region (Fig. 1) and by main and secondary road and railway network. No less than 40 urban centres (about 30% of the entire urban regional patrimony) were affected and suffered by damage from landslides. Some of them were isolated for several days and/or suffered from lack of water. The built-up areas of Senise, Sant'Arcangelo, Noepoli, Armento, Lauria, Maratea and San Costantino Albanese underwent notable damage. In some urban centres (Miglionico, Pisticci, Cirigliano, Gorgoglione, Stigliano, etc.) it was necessary to move people out of entire quarters as a consequence of the remobilization of significant preexisting landslide bodies (Guerricchio & Melidoro, 1988 a and b; D'Ecclesiis et al., 1991; Catenacci, 1992; Sdao & Simeone, 1993). Also the regional road and railway network underwent serious damage, in numerous points, and in some places they were destroyed by landslides.

One of the most impressive landslides [Ruderi landslide, Fig. 1 and Tab. 1 (Del Prete et al, 1977)] cut an important high-way in Basento river valley and the nearby railway too, displacing them by at least 12 metres. At Pisticci a significant landslide completely destroyed a large stretch of road.

Mass movements have affected most of the geological formations of Basilicata Region, in particular: the plio-pleistocenic soils (Blue Clays, Aliano Sands and Castronuovo Conglomerates) largely outcropping in the central-eastern portion of the region and the structurally complex formations, especially those with a high clay content and the markedly tectonized ones (Red Flysch,

Varicoloured Clays, Galestrino Flysch) that constitue the back-bone of Lucania Apennine chain. Mass movements manifest itself essentially through typologies of rotational, traslational or coumpound slide or slump- earthflow.

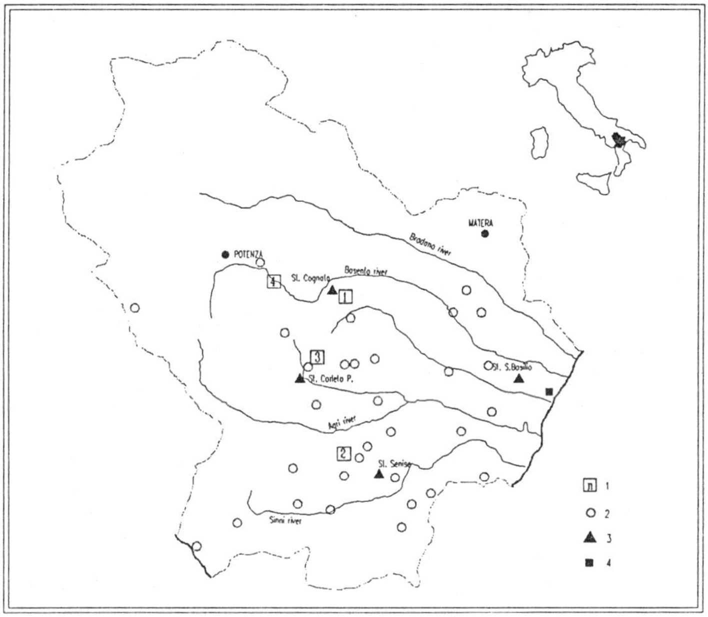

Fig. 1 - Basilicata region (Southern Italy) - (1) Main landslide remobilized after 1973 rainfalls (see tab 1); (2) Urban center damaged after 1973 rainfall events; (3) Rain-gauge station; (4) Water-bearing level measurament station.

Almost always mass movements can be considered partial or total remobilization of preexistent landslide masses, often of enormous dimensions, some of which have been quiescent for a long period of time (Del Prete et Al., 1977; Guerricchio & Melidoro, 1988; Cotecchia et Al., 1986; D'Ecclesiis et Al., 1991; Sdao & Simeone, 1993).

Most of the slopes of main fluvial valleys of the region, especially those characterized by deep layer of weathered soil, was affected by diffuse landslides (caused by soil slip, earth- and mud- flow). In some secondary hydrographic basins the area of the slopes affected by this proneness to landslide was more than 50% of the total basin surface. Also along streams there were some soil-flows involving alluvial deposits (for example in the river Basento middle valley this sort of landslides were mobilized in almost all the secondary streams). At the same time of this proneness to shallow landslides, following the rainfalls period of 1973 winter numerous large landslides occured. In most cases there were remobilization of large landslides affecting extensive areas (up to hundreds of miles of square metres) and involving layers deeper than 50 metres. In Tab. 1 are collected the main geological and geomorphological characteristics of some of the more significant large landslides which occured in the last days of March.

	LOCALITY	GEOLOGICAL FORMATION	THICKNESS AND SURFACE AREA OF LANDSLIDE	LANDSLIDE'S TYPE	REFERNCES
1	RUDERI (BASENTO RIVER)	NUMIDIAN FLYSCH (ARENACEOUS FLYSCH) AND VARICOLOURED CLAYS	> 40 m ------------ ~ 0.20 km²	TRANSLATIONAL SLIDE	DEL PRETE ET AL., 1977
2	PIANO DEI CAMPI (CASTRONUOVO S., ANDREA VILLAGE)	PLEISTOCENIC CONGLOMERATE AND SANDS	50-60 m; 0.3 km²	COMPOUND SLIDE	D'ECCLESIIS ET AL., 1991
3	BALZO DEL PRETE (CORLETO, PERTICARA VILLAGE)	GORGOGLIONE'S FLYSCH (ARENACEOUS FLYSCH)	10-20 m; 0.4 km²	SLUMP - EARTHFLOW	SDAO & SIMEONE, 1993
4	BRINDISI DI MONTAGNA (BASENTO RIVER)	VARICOLOURED CLAYS AND RED FLYSCH (CLAYEY MARLY FLYSCH)	5-12 m; 0.1 km²	SLUMP - EARTHFLOW	COTECCHIA ET AL., 1986

Tab. 1 - Some large landslide remobilized by rainfall events on 1973

Such an ample proneness to landslide greatly contributed to more or less significant morphological modifications in most of the main and secondary fluvial networks. In some cases lakes of various dimensions were produced as a consequence of the invasion and/or blocking of the river bed by the landslide mass (D'Ecclesiis et Al., 1991; Del Prete et Al., 1977).

3. HYDROLOGICAL ANALYSIS AND DATA DISCUSSION

The daily rainfall data from four different rain-gauge stations (Corleto Perticara, Senise, Cognato, San Basilio) and the water levels measured to a water-bearing levels measurament station have been considered with the aim of ascertaining the exceptionality, in terms of return period of rains occured during the first months of 1973, which induced all the landslide described above. Particular attention has been pointed out to the rains of the last week of March, because during this period there was the highest landslides mobilization (Fig.1 and Tab.1).

The choice of these rain-gauge stations has been due to the fact that they are placed in the areas mainly involved in the mobilization of landslides and was possible to collect rainfall data for a period of time sufficiently long to allow significant hydrological analysis. For each of the rain-gauge stations were considered the daily rainfall data from 1951 to 1975. The maximum values of cumulated rainfall at 1, 2, 3, 5, 10, 20, 30, 60, 90, 120, 180 days have been evaluated for each year (Cascini & Versace, 1986). These values has been regularized by using Gumbel's probabilistic distribution, which is well adapted to the interpretation of extreme events. The reliability of Gumbel's probabilistic distribution was evaluated by using Chi- quadro and Kolmogorov statistic tests, assuming a level of significance equal to 0.05. Using the statistic analysis was possible to evaluate, for each of the pluviometric stations considered, the return period of the cumulated rainfalls which preceded landslide events. The analysis clearly show that 1973 rainfalls, which preceded the landslides, are characterized by return period not particularly high (Fig. 2). So it is possible to evaluate that rainfall events that preceded landslides were not hydrologically exceptional.

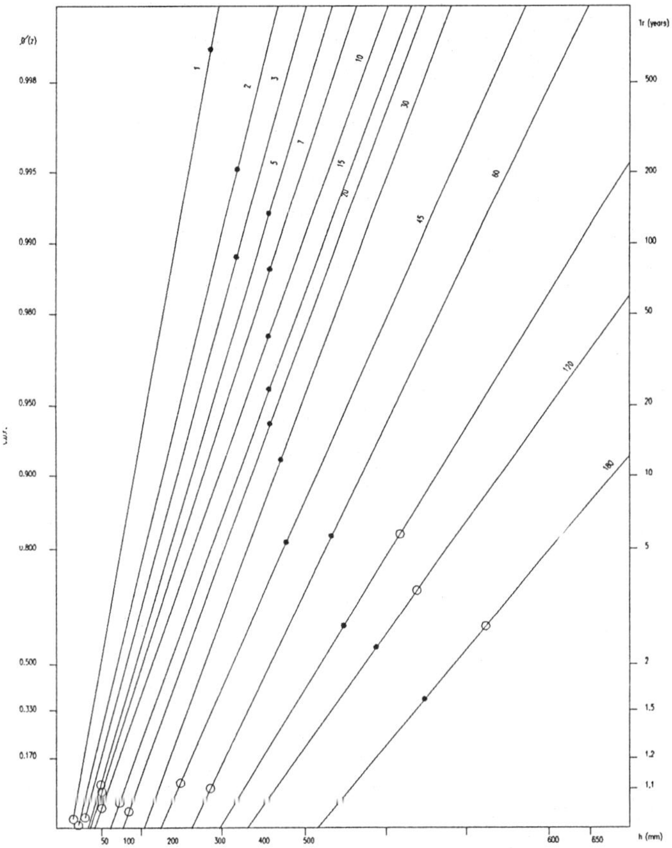

Fig. 2 - Return period of cumulative rainfalls (1-180 days) that preceded ''Ruderi Landslide'' 1973 (o) and the maximum ones of 1972 (●).

In order to clarify the influence of the precipitation on landslide movements, rainfalls which affected the entire Basilicata region in January 1972 has been considered. On that occasion, notwithstanding rainfall of notable intensity, landslide activity was not so obvious and diffuse. Only few mass movements mainly superficial were mobilized, while superficial and riverbank erosion along the main river valleys were largely diffuse. This is confirmed by the high quantities of rivers mass transport measured in some river mass transport measurement station during 1972 rainfalls period. Return period of the cumulated rainfalls over a large number of days was evaluated for each of the stations studied by using the same procedure used for 1973 rainfalls. The cumulated rainfalls of few days during the meteoric event of January 1972 have a very high return periods compared to those of the meteoric event of 1973. In some cases it was highest than a hundred of years (Fig. 2). It is therefore possible to conclude that the different reaction of slopes to hydrological events may be attributed to a different interaction of the rainfalls with the soil.

The capacity of rainwater to infiltrate into the soil and\or to runoff depends, as already previously mentioned, on a lot factors, some linked to the soil features (permeability, slope, level of saturation), others to hydrological and meteoreological factors such as: the intensity and duration of the rain, and the raindrop dimension (Simeone, 1992), and yet others to geomorphological factors (Fig. 3).

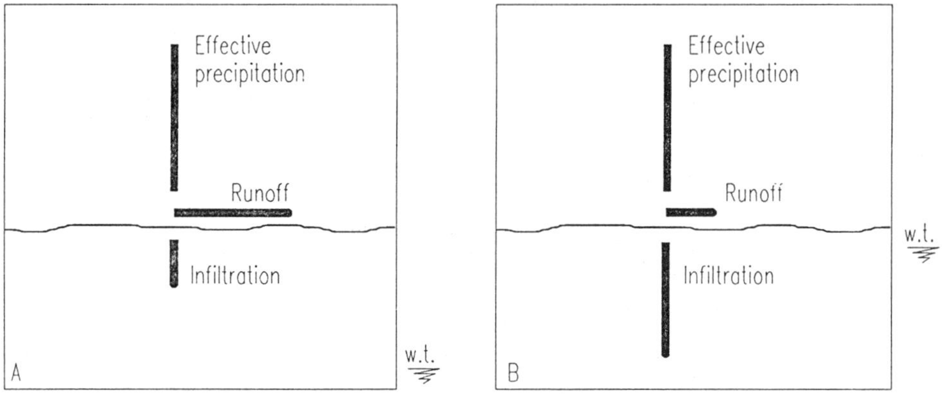

Fig. 3 - Variation of infiltration-runoff ratio depending on rain characteristic : (a) intense short rain; (b) tiny long rain.

The infiltration capacity of rainfalls in natural soils can vary from a minimum of 0.25 mm/h for soils with a high clay content up to a maximum of 25 mm/h for loose sands (Linsley and Franzini). In any case it is possible to exclude that, for silty-sandy soils, infiltration could exceed 2-3 mm/h. Therefore it is possible to conclude that rainfalls with an higher intensity does not contribute in a significant way to infiltration. In this way it is possible that rainfalls over a long period, are more dangerous, for the stability of landslides in soils of low permeability, if compared to that ones of greater intensity, which instead, are certainly more important in terms of runoff.

In the light of the above mentioned considerations, a critical rexamination of the hydrological events in Basilicata region during 1972 and 1973 could be carried out, trying to explain their different influence on the stability of the slopes. The metereological perturbation which affected all Basilicata region during 1973 was characterized by rainfall evente that, also if there were not hydrologically exceptional, were distributed over a period of time of at least one week, with rainfall events never higher than 40 mm/day (Fig. 4) and therefore probably characterized by an high effective infiltration. On the contrary, the meteorological perturbation of 1972 was characterized by torrential showers of rain, with tens millimeters of rain in few hours (Fig. 4) which therefore contributed only in a limitated way to infiltration, while they definitely induced an elevated formation of streams which produced a notable erosion. So the different behaviour of slopes, in relation to hydrological events may be attributed to the different meteorological characters of the rainfall events during the two considered periods.

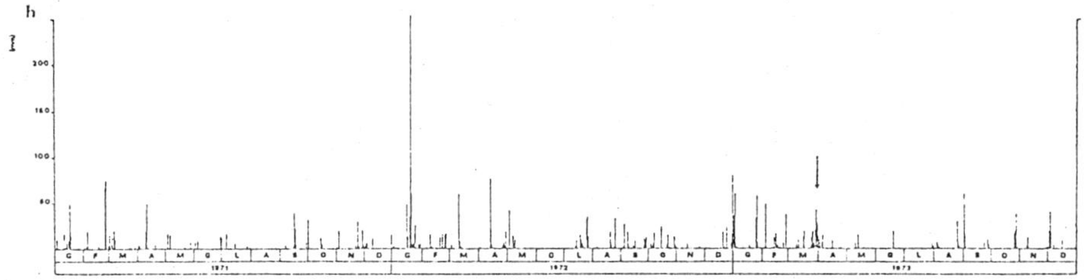

Fig. 4 - Rainfall measured at "Cognato" rain-gouge station during 1971, 1972 and 1973 years (after Simeone, 1992).

4. THE INFLUENCE OF RAINFALLS ON THE GROUNDWATER SYSTEM

The infiltration of rainwater contribute to groundwater both in saturated and non saturated zone. The stability of slopes is affected by both variations, either by changes of pore water pressure, which are negative in the not satured zone as a consequence of capillarity and become positive after full saturation, or by the increase in the level of the water table. Moreover, if the level of the water table rises, even temporarily, over the previous saturated zone, the changes in the soil effective stresses may be very important and may lead to dangerous loss of strength affecting and stability of soil slopes (Fig. 5).

As a partial confirmation of all has been said, the water levels measured at the water-bearing level measurament station near the signal box n. 51 the Taranto-Sibari railway line (Fig. 6) has been considered. At the end of March 1973 as a consequence of the rainfalls events which remobilised the landslide movements examined was registered the maximum water bearing level ever registered during the entire period of measurament taken into consideration (1951 - 1975). Water bearing data may constitute an indication, of the capacity of rain to modify pore pressure in the soil when it is not possible to have piezometric levels in the landslide bodies. The annual maximum water bearing values measured at the signal box n. 51 water-bearing level measurament station, have been regularized by using Gumbel's distribution. The maximun water bearing level registered following the rainfall of 1973 has e return period of about 50 years, while the corresponding rainfalls events measured at the nearby San Basilio rain-gauge station have a lower return period. On the contrary in 1972 after a rainfall event wihich has a return period of over a hundred years, the maximum water bearing level hed a return period of about 8 years was noticed. This provides the confirmation that on the occasion of 1973 rainfalls there was a notable infiltration of rainwater, demonstrated by an elevated value of the relationship between the variation of water bearing level and the quantity of rain.

It is necessary to consider also that rainfall phenomena of 1973 occured at the end of a rainy winter, when the piezometric levels were probably already quite high, as testified by the course of the water bearing level at the above mentioned signal box 51 water-bearing level measurament station.

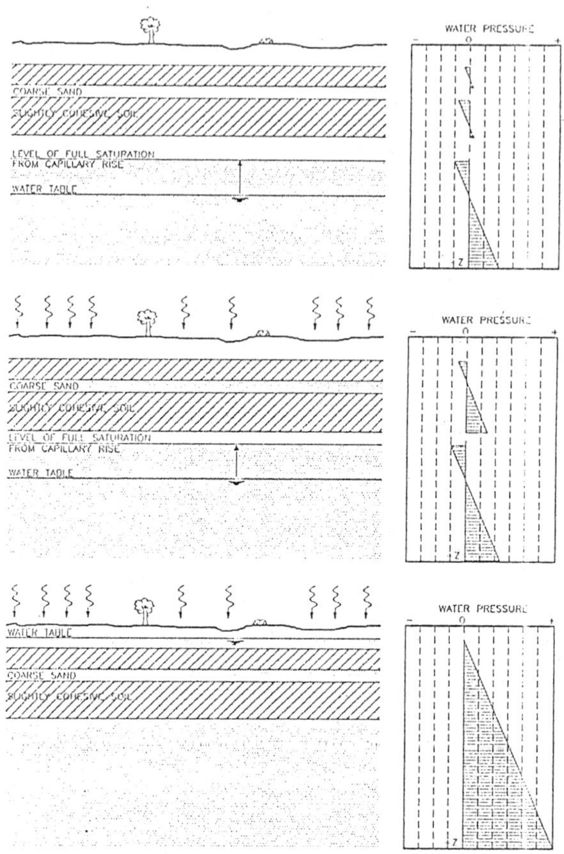

Fig. 5 - Evolution of negative pore water pressure in hydrostatic pressure after countinuous infiltration.

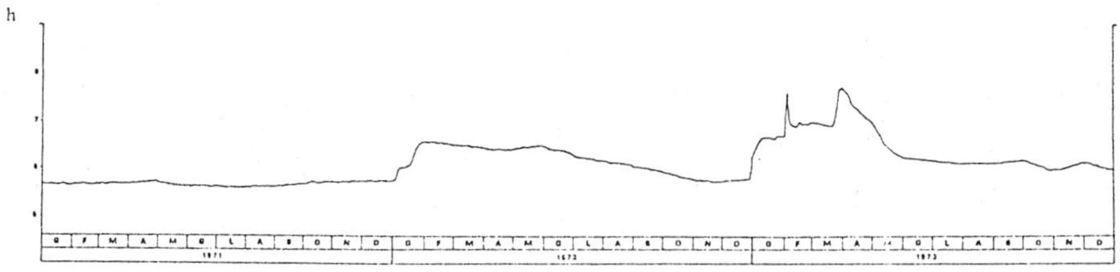

Fig. 6 - Water level measured at the signal box n. 51 water bearing level meaa:urament station during 1971, 1972 and 1973 years (after Simeone, 1992).

REFERENCES

CASCINI L. & VERSACE P. (1986) - 'Eventi pluviometrici e movimenti franosi.' *Atti XVI Convegno Nazionale di Geotecnica*, Bologna, vol. II, 171-185.

CATENACCI V. (1992) - 'Il dissesto geologico e geoambientale in Italia dal dopoguerra al 1990.' *Mem. Desc. Carta Geol. It.*, vol. 47, 301 pp, Roma.

COTECCHIA V., DEL PRETE M., FEDERICO A., FENELLI G.B., PELLEGRINO A., PICARELLI L. (1986) - 'Studio di una colata attiva in formazioni strutturalmente complesse presso Brindisi di Montagna Scalo (PZ)' *XVI Conv. Ital. di Geot.* vol. 1, 253-264, Bologna.

D' ECCLESIIS G., GRASSI D., MERENDA L., POLEMIO M. & SDAO F. (1991) - 'Evoluzione geomorfologica di un' area suburbana di Castronuovo S. Andrea (PZ) ed incidenza delle piogge su alcuni movimenti di massa.' *Geol. Appl. ed Idrogeol.*, vol. XXVI, 141-164, Bari.

DEL PRETE M., MELIDORO G., VALENTINI G. (1977) -'Influenza dell' assetto geostrutturale sulla franosita' delle formazioni pelitico- fliscioidi nella fascia orientale dell' Appennino Lucano.' *Geol. Appl. ed Idrogeol.* vol. XII, p. II, 399-414, Bari.

FIDELIBUS C., LENTI V., SDAO F. & SPILOTRO G. (1990) - 'Condizioni strutturali e al contorno nell' attivazione di frane a prevalente sviluppo longitudinale'. *Atti dei Seminari su: Runoff Processes andMorphologicalEvolution ofSlopes*. Universita' della Basilicata, C.N.R. - G.N.D.C.I., Maggio 1990.

GUERRICCHIO A. & MELIDORO G. (1988a) - 'Fenomeni franosi dell' abitato di Stigliano (Basilicata).' *C.N.R. - G.N.D.C.I., Convegno sul tema: Cartografia e Monitoraggio dei Movimenti Franosi. Sessione cartografia*, 43-63, Bologna.

GUERRICCHIO A. & MELIDORO G. (1988b) - 'Franosita' nei territori comunali di Gorgoglione e Cirigliano (Basilicata).' *C.N.R. - G.N.D.C.I., Convegno sul tema: Cartografia e Monitoraggio dei movimenti franosi. Sessione cartografia*, 45-85, Bologna.

LINSLEY R.K. & FRANZINI J.B. (1957) - *Water Resources Engineering.* Third Ed. Mc Graw-Hill Book Company.

MINISTERO LL.PP. - Servizio Idrografico . *Annali Idrologici* - Sezione di Catanzaro 1921-1987. Roma.

SDAO F. & SIMEONE V. (1993) - 'Rimobilitazioni recenti della frana di Balzo del Prete (Basilicata) ed eventi di pioggia.' In printing.

SIMEONE V. (1992) - 'Piogge e Frane in Basilicata'. *2° Conv. Naz. Giovani Ricercatori in Geologia Applicata*, Viterbo Ottobre 1992.

SECTION 6:

ENVIRONMENTAL CONSEQUENCES OF WATER-RESOURCES PROJECTS

ENVIRONMENTAL IMPACT ASSESSMENT OF ARTIFICIAL RESERVOIRS

N. ALPASLAN
Dokuz Eylul University
Faculty of Engineering
Environmental Engineering Department
Bornova, 35100 Izmir, Turkey

ABSTRACT. In developed countries, EIA has become a necessary procedure in the realization of every large scale investment and plan. Environmental impacts accruing from an artificial reservoir are numerous with varying degrees of significance. An artificial reservoir has possible impacts on: the hydrologic characteristics of the drainage area, characteristics of the atmospheric system in the form of regional climatic conditions, seismic features of the region, the ecologic, or more specifically the biologic systems of the water environment as well as those of the surrounding land, and social life within various perspectives. It is often difficult to estimate beforehand the relationships between such effects and to state whether the eventual impact of the reservoir will be negative or positive. In any case, it is important that all effects are recognized and accounted for to achieve a realistic EIA. The presented study investigates a systematic approach for EIA systems with special emphasis on effects of artificial reservoirs. Possible effects, negative or positive, of such reservoirs are discussed, and those that particularly relate to water resources engineering are covered in detail.

1. Introduction

Water has been the primary natural resource of all times, and even today, significant efforts are expended towards the development of water resources. The major means of achieving this goal is the regulation of natural flows via artificial reservoirs created by dams. Dam reservoirs serve multiple beneficial purposes such as water supply, hydroelectric energy generation, irrigation, flood control, etc. On the other hand, reservoirs are among those man-made structures which have the widest effect on ecosystems. Such effects reflect a dynamic character and occur as ecosystem modifications which cannot be recovered.

The design process of a reservoir is a complex and extensive study mastering

V. P. Singh and B. Kumar (eds.), Water Resources Planning and Management, 347–358.
© 1996 *Kluwer Academic Publishers. Printed in the Netherlands.*

technical, economic, social and administrative factors. The implementation of such design requires not only technological know-how but also significant amounts of equipment and material. The design process starts by delineation of hydrologic conditions at the dam-site. Detailed analyses follow in developing various alternative projects comprising different locations and structural properties. After the best plan is selected by feasibility studies, final design is realized by working out details of structural and operational features of the reservoir before the eventual construction starts.

Until recent years, the design of dam reservoirs has been approached in the above-mentioned sequence. Recently, however, a new dimension is added to this traditional procedure as required by significant increases in environmental problems. This new dimension has emerged in the way of ecological impact assessment to be undertaken as the final step of the above design sequence. Later, this step was generalized into the concept of environmental impact assessment (EIA). At present, when the overall feasibility of a project is to be judged, it is not sufficient to consider only the technical and economical feasibility; a thorough evaluation of environmental feasibility is equally required for the decision-making process.

Although each reservoir leads to specific environmental impacts, all reservoirs often produce similar ecological results with respect to the climate, hydrology, and the biology of the environment involved. Accordingly, it seems possible to develop a rather systematic approach to EIA of artificial reservoirs for an efficient evaluation of their impacts.

The study presented summarizes the environmental impacts of artificial reservoirs with an attempt to identify the cause-effect relationships underlying such impacts. The significance of EIA analyses is further emphasized, pointing out to the fact that a very thorough and realistic EIA procedure has actually become a requirement for feasibility analyses of artificial reservoirs.

2. A Short Review of EIA

EIA may be defined as the detailed assessment of probable effects of a particular project upon the environment. The purpose of EIA is the inclusion of environmental impact considerations in project planning, along with technical and economic considerations. An EIA must be systematic, reproducible, interdisciplinary and scientific.

In an EIA, expected impacts generally cover physico-chemical (water, air, soil, noise, radiation), biological (terrestrial and aquatic), cultural (historic, amenity, visual), and socioeconomic (population, economy, infrastructure) concepts. Such impacts may be classified as being: (a) direct or indirect; (b) primary or secondary; (c) positive or negative (d) cumulative or synergistic; (e) irretrievable or irreversible;

and (f) extreme conditions (IHEa, 1986).

When assessing impacts of any nature, a basis for comparison of different items of the same alternative or of different alternatives has to be established. Comparisons may be based on (IHE, 1986a):

a) evaluation with respect to pre-set standards;
b) ranking
c) scaling by using specific functions
d) professional judgement, and
e) assignment of significance by weighing

The final stage of EIA involves statement preparation where the following items are covered to describe (IHE, 1986a):

a) proposed action and its purpose
b) environmental setting
c) impact of proposed action
d) alternative actions and their impacts
e) unavoidable impacts
f) relation between short term use and long term productivity
g) irreversible commitment of resources
h) possible other interests that will offset adverse effects
i) relation with existing plans and policies.

3. Phases of EIA for an Artificial Reservoir

The total set of impacts from reservoir development forms an integrated system which should be treated as such in the final evaluation. On the other hand, variety of approaches are followed to analyze the individual effects of reservoir, ranging from highly formalized mathematical analyses to empirical and descriptive analyses.

In essence, the basic steps in an EIA process are "identification", "prediction", and "assessment" of impacts on the environment and of the environmental feasibility of a proposed project and alternative solutions (Klomp et al., 1980; Klomp, 1985). Likewise, the generation of impact mitigating or minimizing measures is a significant step in many procedures, as shown in Fig.1.

An EIA study, like the design process of a reservoir, requires historical data to describe the prevailing conditions of the environment prior to construction of the dam. Accordingly, not only design data but also environmental data must be collected and analyzed to assess conditions without the reservoir (Ramamoorthy and Baddaloo, 1991). The collection and analysis of environmental data serves to obtain

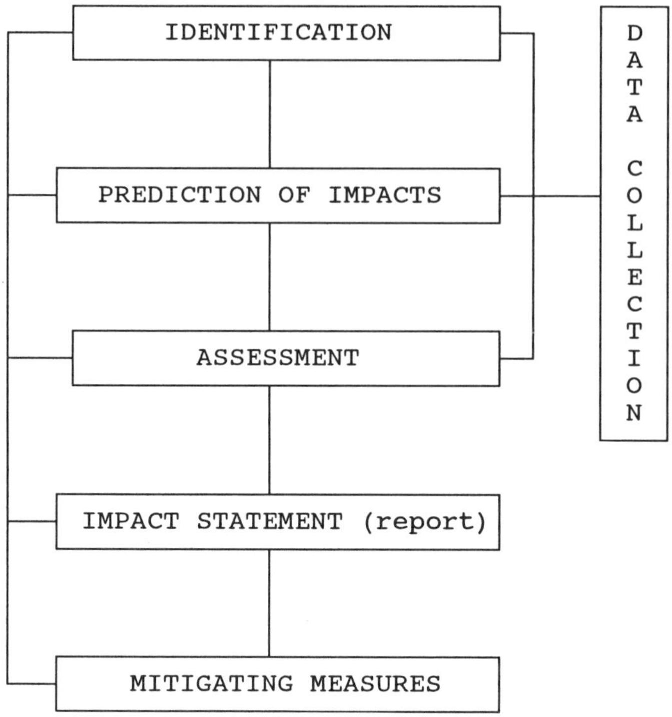

Figure 1. The main steps of an EIA study for a dam reservoir

a description of the so called "base-line conditions" which form the basis for an understanding of the system one is dealing with. By combining the information on baseline conditions and sources of impacts, it will be possible to perform the identification of potential impacts due to reservoir development. This stage of the EIA is often referred as "inventory analysis" (IHE, 1986a).

Next stage of the EIA is the prediction phase where additional data may be required to assess ecological impacts. The prediction phase can be realized by using predictive methods. Several types of predictive methods may be distinguished as (Stans, 1986):

a) experimental methods comprising:
   - illustrative models
   - physical (scale) models
   - laboratory experiments
   - field experiments

b) mathematical models in the form of:
   - empirical models, or
   - process-descriptive models

The final phase of the EIA is the stage of impact assessment by interpreting the outputs of the prediction phase. The perception of the assessment may lead to the necessity and/or desirability of impact mitigation measures and ultimately of design and management options for such measures. Finally, an integrated evaluation (synthesis) of the ecological impacts is carried out by covering alternative scenarios. On the other hand, all of the above defined phases must be realized for different stages of reservoir development, including planning, implementation, construction, operation and post-operation periods.

Once all phases of the EIA are completed, the "Environmental Impact Statement" is formulated in the form of a report. This report is subject to official and public review before it is used in the decision-making process. It may have to be revised according to reviewers's comments.

In realization of different phases of the EIA, a multi-disciplinary team of specialists including biologists, hydrologists, social scientists, and others is required to conduct the environmental impact study. While formulating the EIA, comments must be received from all agencies that have jurisdiction over the various aspects of potential impacts. The U.S. Fish & Wildlife Service, for example, must review all projects that might influence fisheries in one way or another (Linsley et al, 1992).

4. Types of Impacts

4.1. GENERAL

An artificial reservoir may have numerous impacts upon the environment. These impacts have to be classified according to their nature for a systematic approach to their identification. Accordingly, the following basic groups and subgroups of impacts may be distinguished (IHE, 1986b):

a) physico-chemical impacts comprising those in:
   - hydrology
   - hydraulics, morphology
   - water quality
b) biological impacts such as those on:
   - primary production
   - fish life
   - terrestrial vegetation

c) socio-economic impacts covering:
- domestic water use
- water transportation
- land use
- irrigation and drainage
- fishery
-recreation and tourism

d) cultural impacts including those of historical nature

More specifically, a partial list of environmental consequences of water resources projects might include the following (Linsley et al, 1992):

1) degradation of downstream channel or coastal beaches by loss of sediment trapped in a reservoir;
2) loss of unique geological, historical, archaeological, or scenic sites flooded by a reservoir;
3) flooding of spawning beds for migratory fish preventing their reproduction or destruction of spawning gravel by channel dredging or lining;
4) change in stream water temperature as a result of a reservoir leading to changes in aquatic species in the stream;
5) release of reservoir bottom water that may be high in dissolved salts or low in oxygen resulting in a change in aquatic species;
6) drainage of swamps, potholes, etc., decreasing the opportunity for survival of aquatic or amphibious animals or waterfowl;
7) change in water quality as a result of drainage from a irrigation project which may encourage growth of algae in the receiving water or lead to a change in aquatic species as salinity of the receiving body increases;
8) creation of a barrier to normal migration routes of land animals by a reservoir;
9) altering aquatic species by increased turbidity from man-induced erosion from dredging operations;
10) damage to higher species by reason of toxic materials (pesticides, toxic metals, etc.) discharged to a stream and concentrated in the food chain;
11) damage to fish by passage through pumps or turbines or overflow spillways of high dams;
12) damage to stream-bank vegetation by alteration of flow patterns in a stream.

Many more items could be added to this list, and there are probably subtle effects that have not yet been identified. A clear distinction should be made between damage that is temporary and effects that are long term and irreversible. In the following sections, the above listed impacts are discussed in further detail.

4.2. HYDROLOGIC AND MORPHOLOGIC EFFECTS

In general, the construction of a dam and its reservoir reduces the variations in river discharges and water levels. Due to reservoir regulation, the flow regime becomes more regular than the conditions which would prevail without the reservoir. Accordingly, extremes in the form of low and high flows will occur less frequently.

On the other hand, the effects of dams upon suspended solids within water are highly significant. Due to damming, the majority of suspended solid are forced to settle down within the reservoir. This effect leads to unfavorable conditions with respect to river degradation downstream of the dam. This is due to the fact that the river activates new sediments downstream as much of the transported sediments settle within the reservoir. As a result, the downstream riverbed is subjected to significant erosion and degradation. For example, the results of a mathematical model developed for analysis of the Stiegler George Power and Flood Control Project in Tanzania has revealed that the erosion in the river bed will be in the order of 4m in one year, eventually reaching 10m within 10 years. This effect further implies a reduction in water level by 7m (Klomp, 1985). In essence, these figures are closely related to properties of the river bed (i.e., characteristics of river banks, presence of nonerodible layers, presence of outcrops and armouring of the river bed); thus, they assume varying values for each river.

On the other hand, the reduction of water levels in a river affects groundwater conditions in the long range. Another effect imposed on groundwater by the reservoir occurs in the form of increased groundwater levels due to seepage from the impounded reservoir.

The effects of reservoirs upon flow regimes, sediment accumulation and transportation, and variations in groundwater levels have direct impacts on fishery and agricultural practice downstream from the dam. Accordingly, it is often the case that prevailing patterns in fishery and agriculture may have to be replaced by other types of production. Furthermore, downstream lands will be less frequently flooded in the presence of a reservoir so that the amount of fine silt and clay formerly carried by floods to agricultural lands will be reduced. As a result, the productive capability of such lands will eventually be decreased.

Less frequent flooding, lower groundwater tables, and riverbed degradation will also have serious effects on vegetation and wildlife. Degradation of riverbed and the erosion of riverbanks may lead to the formulation of relatively steep banks, which will hinder the movement of wildlife in and out of water in the river.

If a river meets the sea in the form of delta, the presence of dam or series of dams will have an impact on morphology of this delta. In other words, the morphologic properties of the delta, together with characteristics of salt intrusion and sediment transportation will change. In particular, the effects of salt intrusion will be eminent during dry seasons or throughout the impoundment of the reservoir.

Among short-term impacts, one may observe the decrease in transport of rough particles of sediment to the estuary and the delta as a result of flow regulation upstream. In the long range, a significant reduction will be observed in the amount of sediment carried to this area. Such an impact may eventually lead to morphological changes in the form of coastal erosion.

4.3. IMPACTS ON WATER QUALITY

Impact assessment related to water quality reveals a major difference between the disturbance of the old ecosystem during impoundment of the reservoir and the development of the new ecosystem during long-term operation. In essence, the construction of a reservoir causes drastic changes in water quality of both the reservoir and the downstream river system.

The major impact on water quality and the ecosystem of large reservoirs occurs as a result of changes in the oxygen budget. Particularly during the first impoundment of the reservoir and within the first year following the impoundment, the decay of vegetation submerged under water causes a high demand for oxygen. The biochemical decay of twigs, leaves, grass, and other easily degradable organic matter requires oxygen under aerobic conditions. If the decay takes place under oxygen-free (anaerobic or anoxic) conditions, it will progress rather slowly but will still continue by producing anaerobic decomposition products such as H_2S, CH_4, and NH_3 and reduced metals like Fe^{+2} and Mn^{+2}. Such reduced components will consume oxygen in case the environment changes from anaerobic to aerobic conditions (Klomp et al, 1980).

As a result of this decay process, nutrient formation occurs to lead to a new kind of vegetation in the form of extensive algal blooms and/or extensive water hyacinth. The development of low oxygen concentrations or anoxic conditions kills the prevailing aquatic life and hinders the preservation of a stable aquatic environment.

The adverse oxygen budget developing in the reservoir eventually affects downstream river conditions and ecology. However, the required oxygen levels are recovered due to reaeration as one goes further downstream from the reservoir.

It is well-known that oxygen budget is a highly complicated phenomenon that occurs as a result of various physical, biological, and chemical processes. In addition to transport phenomena occurring as inflowing water and other mixing processes, the most import processes under aerobic conditions are:

- reaeration
- biodegradation of organic carbon
- oxidation of reduced nitrogen (nitrification)
- photosynthesis by algae and water plants
- respiration by algae and other living organisms

- bottom respiration (including decay of drowned vegetation)
- chemical oxidation of reduced substances

On the other hand, the above processes may be described by mathematical models. Models have been developed for reservoirs to investigate development of thermal stratification, destratification (overturning or vertical mixing), layers such as epilimnion and hypolimnion, and temporal changes of aerobic and anaerobic (anoxic) layers within the reservoir.

Besides changes in the oxygen budget, the chemical composition of water in the reservoir also differs from the original composition of the river. Another significant change occurs as decrease in the turbidity of water due to settling of suspended solids in the reservoir. Such settling further leads to an increase of nutrients in the sediments and that of aquatic weed growth.

4.4. OTHER IMPACTS

The most significant impacts of artificial reservoirs occur as hydrologic and morphologic effects and changes in water quality as discussed in the previous sections. Other impacts have also been noted. For example, changes may be observed in temperature and the moisture content of air depending on regional topography. Such changes, which occur as result of the development of a large scale static water body in the reservoir, do not directly affect humans but have significant impacts on regional vegetation and wildlife. Another impact is associated with soil stability. Although it is not yet proven, artificial reservoirs are deemed to produce seismic effects. On the other hand, regional vegetation and wildlife are affected by impoundment of the region, erosion of the riverbed, lowering of the groundwater table, reduction in floods, changes in salt gradient, construction of various structures (e.g. dam, power plant, conveyance canals, etc.) and activities during the construction (development of construction site and access roads) (Muezzinoglu, 1991).

Changes in water quality, as discussed earlier in Section 4.3., naturally lead to changes in the aquatic ecosystem. For example, phytoplanktons in water may be increased and macroflora may develop. Such changes adversely affect reservoir life span, fishery, and the operation of turbines. As a result of increased evapotranspiration, the ecosystem experiences a stage of unbalance.

When the reservoir fully develops, passage between the right and left banks of the river becomes difficult. Accordingly, part of the land animals are forced to remain along one bank and others along the opposite bank, with no possibility of passing across the reservoir. Furthermore, the construction of a dam demolishes those species of fish which spend their lives partially in upstream regions of the river and partially in downstream sections where the river meets the sea.

Artificial reservoirs also lead to social impacts. One of the most significant impacts is the requirement of migration imposed on inhabitants of regions to be inundated by the reservoir (EPFT, 1993). Likewise, inundation of historical and natural sites of value has socio-cultural effects. For example, one of the largest springs of the world, the Dumanli Spring in Turkey with an average flow of $50m^3$/sec, was submerged within the Oymapinar dam reservoir. Again in Turkey, historical regions close to the antique city of Pergamon were inundated by the Bergama Kestel dam reservoir. A similar situation is currently being experienced in Harran plains via development of the Southeastern Anatolia Project.

On the other hand, not only the reservoir itself but also the construction site with all the equipment and worker team involved have various impacts upon the environment. Some investigations on reservoirs and diseases attributed to reservoir impacts have shown that large scale irrigation systems and microclimate produced by the presence of a reservoir increase the risk of such diseases as Malaria and Schistosomiasis (Sonmez, 1993). Another significant risk factor is constituted by probable dangerous consequences of flood occurrences if the reservoir fails to control such events. In particular, dam failures or breaks due to various reasons may even lead to catastrophic results (Goldsmith and Hilyard, 1984).

In recent EIA studies, aesthetic values have been highly appreciated. For example, the Tahtali dam and reservoir, which is planned to supply $127.7*10^6$ m^3/y water to the city of Izmir (third largest city of Turkey with a projected population of 4 million for the year 2015), is evaluated for the aesthetic conformity of its major structures (e.g.,dam, intake structure and the pumping station) with the environment. Accordingly, efforts were spent at establishing agreement between the environment and the material (kind, color, etc.) used for constructing the dam structure. Likewise, the slopes of this structure were selected to as to agree with the natural slope of the surrounding land. Similar applications were realized for the water intake structure and the water conveyance canals (DSI,1991).

5. Conclusions

In the study presented, effects of artificial reservoirs upon the environment are discussed in view of the negative impacts that may be attributed to dams and their reservoirs. As pointed out in the paper, artificial reservoirs indeed lead to adverse impacts upon environmental values. On the other hand, these systems are developed for beneficial purposes such as water supply, energy production and irrigation; thus they meet significant demands on water for various purposes. Some of these demands can be actually taken care of by other means besides artificial reservoirs. For example, energy may be produced by thermal or nuclear plants instead of hydroelectric power plants. Although it is beyond the scope of this paper, it must be emphasized at this point that, when compared to other alternatives, reservoirs have the least adverse effects upon the environment despite their negative aspects discussed in the previous sections.

Consequently, the most reasonable approach to be assumed is to identify all negative aspects prior to reservoir planning and development, to predict and to assess their possible impacts and finally to establish measures of mitigating such negative impacts. Among these mitigating measures, one may consider, according to the nature of the impact, the slow filling of the reservoir and/or additional upgrade and/or clearing of the reservoir area for water quality control, increasing of flows for salt intrusion abatement, and selecting operation rules for minimum riverbed degradation. In addition, to reduce the impacts at the construction site to a minimum, reduction of the work, evacuation and clearing of the camp after construction, permitting limited access to the area off the road, and other similar special regulations and control may be applied.

On the other hand, certain components considered in impact assessment are of a social or historical, and thereby of an intangible character. Such components must first be expressed in tangible terms to be included in benefit/cost considerations. The overall evaluation of the investment may then be worked out.

The impacts of a project on the environment may be such that its health, economic or ecological consequences are considerable, or even unacceptable. Therefore, the prediction and assessment of these impacts must be executed in an early stage of the decision making process to enable the formulation of alternative solutions and mitigating measures.

In essence, EIA of artificial reservoirs must be initiated with the design phase and continued until the operation phase. It must even be realized throughout the operational period as part of a "monitoring and assessment system" to continuously update the process of impact assessment. In other words, EIA of artificial reservoirs must be kept alive at all stages of reservoir development and operation.

7. References

Goldsmith, E. and Hilyard, N. (1984) *The Social and Environmental Effects of Large Dams,* Volume One: Overview, Wadebridge Ecological Centre, Bordaux 33 000, France.

IHE, International Hydraulics Institute, (1986)a 'Lecture notes and handouts from courses EIA', *International Hydraulics Institute (IHE), the International Course for Sanitary Engineers (ICSE).*

IHE, International Hydraulics Institute, (1986)b 'Environmental feasibility study to the Kabalebo hydropower project in Suriname S.A. - Case Study', *International Hydraulics Institute (IHE), the International Course for Sanitary Engineers (ICSE), handouts from courses EIA.*

Klomp, R., (1985) 'Identification of ecological impacts of Stiegler's Gorge Power and Flood Control Project, Tanzania', *Case study presented at conference on Environment and Development: Opportunities in Africa and the Middle East,* the Hague, the Netherlands.

Klomp, R.; Pagee, J.A.; Groot, S. (1980) 'A modelling approach to ecological impact assessment of man made lakes', Prepared *by the Delft Hydraulics Laboratory for the UNESCO/IAHR Seminar on Hydraulic Research and River Basin Development,* No: 235, Nairobi, Kenya.

Linsley, R.K.; Franzini, J.B.; Freyberg, D.L.; Tchobanoglous, G. (1992) *Water Resources Engineering.* McGraw-Hill International Editions.

Muezzinoglu, A. (1991) 'Environmental impact assessment and energy generation', In: *Environmental Impact Assessment - Examples from Applications,* (Ed. O. Uslu), Publications of The Environmental Problems Foundation of Turkey (EPFT), Ankara, Turkey, (in Turkish).

Ramamoorthy, S.; Baddaloo, E. (1991) *Evaluation of environmental data for regulatory and impact assessment,* Studies in Environmental Science 41, Elsevier Science Publishing Company Inc.

Sonmez, N. (1993) 'The large dams and environment', In: *The Southeastern Anatolia Project and Environment,* Publications of The Environmental Problems Foundation of Turkey (EPFT), Ankara, Turkey, (in Turkish).

Stans, J.C. (1986) 'Prediction and assessment of impact on the water environment', *handouts of Water Resources and Environment Division, Delft Hydraulics Laboratory,* Delft, The Netherlands.

DSI - State Hydraulic Works (1991) *Environmental Impact Assessment Report of Tahtali Drinking Water System of Izmir City.* Turkish Republic, Ministry of Construction, General Directorate of State Hydraulics Works, Ankara, Turkey (in Turkish).

EPFT - The Environmental Problems Foundation of Turkey (1993) *The Southeastern Anatolia Project and Environment.* no: 93.06.Y.0011.46, 160p., Ankara, Turkey (in Turkish).

PHYSICAL AND BIOLOGICAL IMPLICATIONS OF WATER TRANSFER: A CASE STUDY OF CALIFORNIA

Edward A McBean
Department of Civil Engineering
University of Waterloo
Waterloo, Ontario Canada N2L 3G1

Eduardo Bautista
U.S. Water Conservation Lab
Phoenix, Arizona, U.S.A. 85040

ABSTRACT: In 1991, as a measure to deal with critical water shortages induced by six years of drought, the State of California created an Emergency Water Bank. This initiative involved purchase of water from willing sellers and sale to interested buyers, establishing a marketplace for purchase and sale of water. However, the loss of water from the location of origin raised the potential for significant third-party impacts to the physical/biological regimes. Assessments of the imapcts to subsidence, water quality and wildlife habitat, as a result of the water transfers are described.

1.0 ACCEPTANCE OF WATER TRANSFERS

The history of water transfers from one river basin to another has evolved over centuries. Archaeological evidence indicates that interbasin water transfers were utilized in Babylonian times. Saggs (1962) reported that a water resource development project constructed in 2500 BC was utilized to connect the Tigris and Euphrates Rivers. Numerous additional historical examples of inter-basin transfers are described in Meador (1992). However, with increasing centralization of population into large metropolitan regions throughout the world, it is becoming readily apparent that the pressures for interbasin water transfers will continue to intensify.

The planning for water transfers requires that tradeoffs be made between alternative demands for water. Domestic and municipal needs for water have historically held priority over other uses, including environmental, recreation, irrigation, industrial and hydropower needs. However, the conflicts between evolving uses of water indicate some changes in priority will be needed.

The arguments against the developments of additional water transfers include those of legal, regulatory, contractual and third party. Although all of these are relevant for the State of California, in 1991 the State remained in the grip of one of the worst droughts in its recorded history. In developing solutions to the impasse surrounding water allocation, the State Drought Water Bank of 1991 was established which pioneered a new era in California water management. It is the examination of the functioning of this Water Bank, set up in a very short time to respond to the continuing drought, that is the

V. P. Singh and B. Kumar (eds.), Water Resources Planning and Management, 359–371.
© 1996 *Kluwer Academic Publishers. Printed in the Netherlands.*

subject of this paper. Particular attention is given to the examination of the physical and biological impacts arising from the water transfers that took place.

2.0 WATER TRANSFERS IN THE STATE OF CALIFORNIA

In order to understand the context under which the 1991 Water Bank transfers took place, it is first necessary to describe aspects about the historical land and water use within the State of California in general, and in particular, the Central Valley. Orographically, the landscape of the State is dominated by two formations, the Sierra Nevada and the Coastal Range, both extending in a North-South direction. These mountain ranges give rise to the Central Valley. The Valley is divided into two major hydrologic basins, the Sacramento River Basin to the north, and the San Joaquin River Basin to the south. The major streams from each basin meet near the center of the valley from which they then discharge into San Francisco Bay as illustrated in Figure 1. The complex of islands and sloughs formed at the confluence of these two rivers is known as the Sacramento-San Joaquin Delta (henceforth, referred to as the Delta). The southern end of the San-Joaquin Valley is separated from another important physiographic feature of the state, the Los Angeles Basin, by another mountain formation.

The climate in the State is quite varied, ranging from alpine on the slopes of the Sierra Nevadas to desert on the southeastern end of the State. Precipitation distribution is noticeably uneven in space and falls basically during the winter months. Areas in the north receive as many as 90 inches of precipitation while areas to the south get less than 10 inches. As a result, it is estimated that the annual runoff from the San Joaquin Basin is only one-third of that of the Sacramento Basin.

Temporal variability in rainfall amounts is also significant. This is manifested, for example, by the severe flood that affected the central part of the State in 1986, followed by a four year drought. It is not surprising, therefore, that southern California has historically sought to secure imported water resources to satisfy its agricultural, municipal and industrial demands.

Interbasin transfers of water have played a critical role in the demographic, agricultural and economical development of the State during the 20th Century. The history of California water development has received particular attention in recent years due to the increasing frequency of conflicts. In 1913, California became the first state in the United States to develop an interbasin transfer of water with the completion of the Los Angeles Aqueduct which carried water from Owens Valley on the eastern slopes of the Sierra Nevadas, to the Los Angeles area. Nearly 30 years later, development of the U.S. Bureau of Reclamation's (USBR) Boulder Canyon project on the Colorado River led to the construction of the Colorado River Aqueduct, a 389 km structure, which began delivering water to the Los Angeles area in 1941.

Export of water from the northern portion of the State, through the Delta region, to the southern portions of the State began in 1951 with the completion of the Delta-Mendota Canal, a unit of the U.S. Bureau of Reclamation's Central Valley Project (CVP). An important objective in the development of the CVP was to provide surface water from the Sacramento and San Joaquin Rivers to farmers in the San Joaquin Valley. Prominent features of the CVP are Shasta Dam and Reservoir, which store floodwaters in the upper

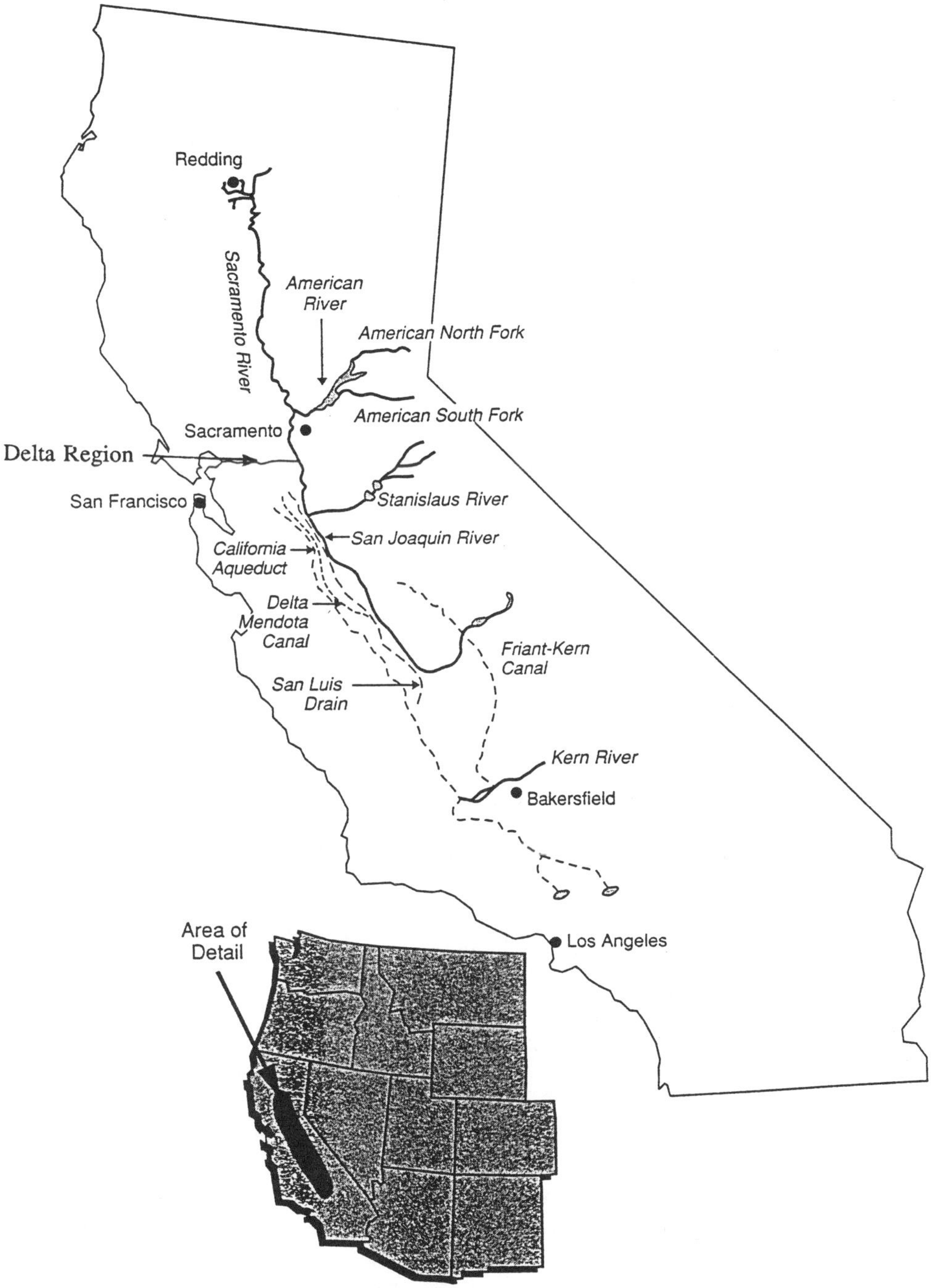

Figure 1. Main Waterways and Features, Central Valley of California

reaches of the Sacramento River Basin. Releases from Shasta flow into the Delta and are pumped at the Tracy Pumping Plant, located in the southern Delta region, into the Delta-Mendota Canal. The capacity of the CVP to deliver water for export was increased a few years later with the construction of Folsom Dam and Reservoir, which captures flows on the American River, an important tributary to the Sacramento River.

Interbasin transfer of water increased with the construction of the State Water Project (SWP), which began to deliver water to the San Joaquin Valley in 1968, and to the Los Angeles area in 1972. This project was developed to supply water to both agricultural and urban water users. The main features of this project are Oroville Dam and Reservoir which store flows of the Feather River, the main tributary to the Sacramento River, and the California Aqueduct, a 715 km long structure consisting of a series of canals, tunnels, pipelines and pumping stations, that transports water from the Banks pumping station, in the southern Delta to the southern portion of the State. As impressive as the length of this man-made conduit is the fact that the water is lifted over steep mountain ranges (over 1000 m at the Tehachapi Mountains, between the San Joaquin Valley and the Los Angeles basin) to its final destination.

It follows from the above, that crucial to the operation of both the CVP and the SWP is the export of water through the Delta. Nearly thirty percent of the State's total water supplies, including municipal supplies to about 20 million people in southern California, are exported through the Delta. These inflows to the Delta travel by various routes (see Figure 1) to the pumps in the southern Delta. When freshwater flows out of the Delta and into Suisun Bay are low, water in the western Delta becomes brackish beause they mix with salty ocean water entering as tidal inflow. Further, during periods of low flow and high pumping into the aqueducts, water is drawn upstream into the San Joaquin River. The resulting reverse flow disorients anadromous fish species such as striped bass, salmon and steelhead, which travel through the Delta from their spawning grounds to the Pacific Ocean. As a result, during operational periods that cause adverse water quality conditions, more water than is needed for export to consumers, must be released from reservoirs within the Sacramento Basin to repel intruding sea water and to maintain required minimum flows into the western Delta channels.

The Delta's importance, however, is not limited to water alone. Many different entities have vital concerns in the Delta, including agriculture, cities and industry, fish and wildlife, transportation, and recreation. The Delta, consisting of 700,000 acres, was once a tule marsh. Reclamation of the Delta began in the 1850s such that by 1930, virtually all the marsh had vanished, to be replaced by farms growing a large variety of crops. A total of 515,000 acres of land are now farmed in the Delta. In addition to farmlands, there are an estimated 66000 acres of wetland habitat and 53,000 acres of open waterways in the Delta which support a diverse biological community, including species of economic importance. With a surface of 1,153 square miles and encompassing 1,100 miles of coastline the Delta offers much in the way of recreation. As the third largest use, behind agriculture and natural gas exploration, respectively, recreation in the Delta attract 12 million users annually. The prominence of this region is recognized in the fact that water and land uses are subject to special regulatory actions, and are the foci of extensive research and conservation efforts.

It follows directly from the above that operation of both the CVP and SWP is complicated by the need to assure minimum water quality conditions in the Delta. The

U.S. Bureau of Reclamation and the California Department of Water Resouces (DWR) share the responsibility of meeting specific water quality objectives set by the State Water Resources Control Board through coordinated releases from their reservoirs and export operations through the Delta.

As apparent from the examples given above, the precedent and practice of water transfer is fully entrenched within the State of California. Directly relevant to this research are the operational features of the CVP and SWP and the transfer of water through the Delta.

3.0 THE 1991 STATE DROUGHT WATER BANK

Following four consecutive drought years and in contemplation of a fifth year of below average precipitation, Pete Wilson, Governor of the State of California, declared in February 1991 a Drought Emergency Plan. The principal feature of the plan was the formation of an Emergency Water Bank Program. The role of the Water Bank, to be coordinated by the State's Department of Water Resources, was to facilitate one year sales of water by individual water rights holders to satisfy emergency municipal, industrial and agricultural demands. Additionally, the Bank was to provide water to protect fish and wildlife resources and to guarantee carrryover storage for 1992.

An indication of the need for action is apparent form an examination of the entries in Table 1. For comparison purposes, the table includes the corresponding statistics for 1977, one of the driest years on record. The economic and environmental impacts of the drought were already large and were apparently going to continue to grow with the continuation of the drought. The two major water projects in the State, the CVP and the SWP, announced severe cutbacks in their deliveries to junior contractors in order to meet their obligations to senior water rights holders. For example, SWP projected deliveries to urban contractors at 10 percent of their entitlement while agriculture was projected to receive no water at all. In response, the Water Bank was created.

The operation of the Water Bank was described in detail in the DWR's publication entitled "The 1991 Drought Water Bank" (DWR, 1992). In summary, DWR offered to purchase water from farmers and other voluntary sellers using the following mechanisms:
i) Land fallowing: farmers were paid for an amount of water equivalent to the consumptive use of their crops. Consumptive use was estimated from historical data and eligible acres were determined based on lands that were cultivated or enrolled in federal set-aside programs in 1990.
ii) Groundwater exchange and direct pumping: with exchange contracts, CVP agricultural contractors were able to sell the smaller of their entitlement (reduced by 25% in 1991) or their average annual diversions based on the three years of maximum withdrawals, and to substitute this water with groundwater. With pumping contracts, groundwater was pumped directly into the Sacramento River system and, thus, did not involve any surface water.
iii) Surface storage: this involved the purchase of storage from local projects.

Sellers of water were paid 125 \$/acre-ft by the Water Bank. This price was established through negotiations between sellers, buyers and the Water Bank. Urban demand largely determined the final price of water. This water was then resold for 175

$/acre-ft. The sale price was deemed effective at the SWP Delta pumping plants and, thus, additional costs were incurred by the purchasers for pumping and conveyance to the final point of delivery. The result was that, for example, in agricultural regions, water was resold for as much as 265 $/acre-ft and was exclusively used to keep high-valued perennial crops alive.

Table 1. Summary of Statewide Water-Year Data (Oct.1 to Oct.1)

	1977	1986	1987	1988	1989	1990	1991	1992
Precip	45	128	61	82	86	69	76	66
Runoff	20	140	48	47	72	45	43	40
Reser.Storage	35	119	84	66	74	60	61	59
SRI (MAF)	5	26	9	9	15	9	8	-
Year type	C	W	C	C	D	C	C	C

* as of 12/1/91
C = Critical
D = Dry
W = Wet
Reference: Department of Water Resources, State of California, "California's Continuing Drought, 1987-1991", December 1991.

Of the total Water Bank purchases, approximately fifty percent of a total of 821,000 acre-feet came from land fallowing, thirty-three percent from groundwater (exchange and pumping) and the rest from reservoir storage.

The Emergency Water Bank was widely commended by many as an effective means of responding to a serious water emergency. The Bank demonstrated that a central agency could act as an intermediary in reallocating scarce water resources: the Bank contracted with upstream water rights holders to relinquish their diversions and delivered an equivalent amount to users downstream. However, the Water Bank was also the source of controversy. Although there were precedents to the long-term interbasin transfer of water in the State of California, interbasin water marketing similar to the 1991 Water Bank had not been implemented in the past, mostly due to legal constraints. Short-term transfers by means of water marketing, have been advocated for many years by economists on the basis of economic efficiency on the premise that it enables the reapportionment of water from lower-valued uses (agriculture) to higher-valued applications (municipal/industrial). Other arguments posed in favor of free-market transfers are the high financial and environmental cost of building new infrastructure. Nonetheless, there are legal, economic and social implications to the transfer of water.

Water rights in the State of California have been established largely on the basis of "beneficial use" doctrine. Current laws do not specifically identify water marketing as a "beneficial use" of water. Even if water marketing was allowed by law, enforcing contractual obligations (amounts surrendered by the seller and diverted by the buyer) would be complicated.

From an economics point-of-view, sales of agricultural water represent for the region of origin, reduced income and tax revenues, greater unemployment in the agricultural and agricultural services sectors, and diminished economic prosperity. In addition, there are also physical, biological and environmental implications of short-term water transfers by means of water marketing. Regional water and land uses and the environmental quality of the receiving waterbodies would be directly affected by this kind of transaction. The focus of the technical analyses completed by the authors, was the determination of the implications to the region of origin, specifically in relation to the physical and biological implications, as a result of the Water Bank transfers.

4.0 WATER BANK TRANSFERS AS AN INCREMENT

4.1 Range of Features

The Emergency Drought Water Bank transfers represent only an incremental change to an ongoing system of water transfers to other parts of the State of California. Nevertheless, the assessment of the impacts of these transfers was an important set of questions. The isolation of the impacts as a result of the Water Bank from the numerous other ongoings was a substantial challenge. Analyes were undertaken to separate out the effects of the 1991 Water Bank transfers from the effects of other forces such as the ongoing drought, the economic recession, land fallowing under federal farmland set-aside programs, and the ongoing transfers through the California water delivery system.

The implementation of the Water Bank transfers directly affected crop acreage distribution (due to land fallowing) and water supplies in the study area (e.g. due to irrigation from groundwater pumping). The hypothesis is that changes in land and water use were the primary factors determining the magnitude and importance of the overall effects on the physical/biological system. Secondary effects included subsidence, water quality, and wildlife habitat.

Technical details of the assessment are more fully described in Bautista and McBean (1993).

4.2 Cropping Acreages

The amount of irrigated land has remained relatively stable over the last decade within Yolo and Solano Counties, two of the primary counties of origin to the Water Bank transfers. Figure 2 indicates the crop distribution in Yolo County, as an example. There have been variations in the dryland base such that not all available farmland is planted for cropping every year. Effects such as the unknown potential length of the drought and the crop marketing conditions influence the yearly cropping patterns. Thus, the changes in crop acreage from year-to-year are only partly a result of the implementation of the Water Bank.

Numerous complicating factors arose in attempts to isolate out the impacts of the Water Bank transfers. For example, the growers were not required to idle the farmland but only to not irrigate the land. In addition, because of features such as the continuing drought and agricultural set-aside programs, for some crops the decrease in acreage far exceeded the amounts of land involved in the transfers e.g. the acreage of rice decreased by 11,000 acres, a change in acreage that far surpassed the 6400 acres that were fallowed by the Water Bank transfers. The acreage of some crops actually increased in 1991, presumably due to market conditions (e.g. tomatoes and alfalfa).

The changes in cropping acreage are relevant since typically only a portion of the applied water is beneficially used by a crop. The excess volume is lost as deep percolation beyond the crop root zone and as surface runoff (tailwater). If the source of the applied water is rainfall or imported water, then the percolated fraction recharges the regional aquifer. In farming areas provided with continuous networks of irrigation and drainage channels, portions of the tailwater losses can be recovered by downstream users. Thus, the fallowing of extensive quantities that would otherwise be irrigated using imported water, reduces the recharge of groundwater and the availability of tailwater. On the other hand, Water Bank transfers arising from groundwater exchange and pumping directly deplete the regional groundwater resource.

The impacts of the Water Bank transfers were assessed by computing the demands and supplies of water using a water balance model. As an example, Table 2 summarizes the difference in irrigated surface acreage and groundwater use between 1989 and 1991. Results show that irrigated acreage decreased relative to 1989 levels in the Delta Unit whereas irrigated acreage in the Lower Cache drainage area unit remained essentially constant but a substantial reduction in total applied water was computed. Groundwater use increased in the Lower Cache drainage unit. The values in the last row represent the increased loss of groundwater storage due to increased pumpage and to decreased recharge. Given that a potential overdraft condition already existed in 1989, the contribution of the Water Bank transfers to the total loss of groundwater storage is profound.

Table 2. Difference in Quantity of Agricultural Groundwater Used and Impact on Ground Water Storage in 1989 and 1991 (for Yolo County)

	County Total	Lower Cache Unit	Delta Unit
Irrigated Surface	(15.3)	1.4	(22.5
Groundwater Pumpage	89.0 **20.0**	70.0 **20.0**	(2.0) **0.0**
Groundwater Recharge from Applied Surface Water	(40.0) **30.0**	(30.0) **20.0**	(11.0) **10.0**
Total Losses from Pumpage and Lower Recharge	130.0 **48.0**	100.0 **40.0**	9.0 **8.0**

Note: All values in 1000's of Acre-feet. The values in 'bold' indicate the amount attributed to the water transfers

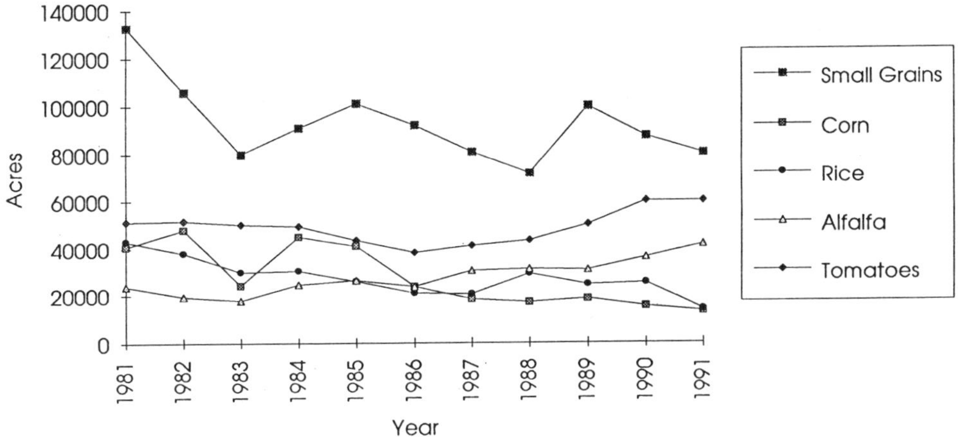

Figure 2. Annual Acreages for Different Crops, for Yolo County

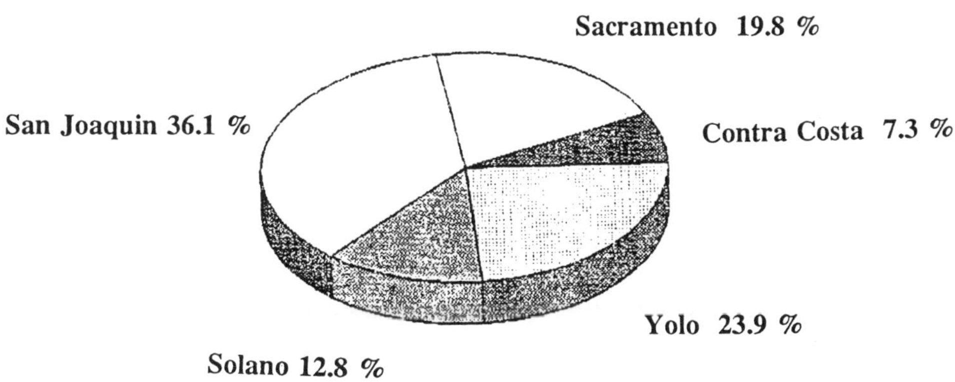

Figure 3. Water Bank Fallowings by County (acres)

Total of 144,000 acres

4.3 Subsidence

The continuation of land subsidence is a definite cause for concern. The result of subsidence is a lowering of flood control levees and changes in grades in irrigation canals and ditches. With the Water Bank transfers, there are two particularly important features:
(i) due to the fallowing of lands, there is less groundwater replenishment;
(ii) due to reduced surface water availability, increased groundwater pumpage results.
The result is increased subsidence levels from the regions of groundwater withdrawal. Unfortunately, field instrumentation was not in place at the time of the Water Bank transfers and the estimation of subsidence is complicated. Hydrology Consultants (1992) indicated that as an extreme situation during 1991, lands near Woodland, California may have subsided as much as 0.4 feet.

4.4 Water Quality

Water quality changes may have occurred due to changes in the water directions and waste discharges associated with changes in agricultural practices. The implications are that a smaller volume of surface water diverted for irrigation and a smaller volume of agricultural return flows discharged back into the river system will have occurred.

A series of different metrics were considered in evaluation (see Bautista and McBean (1993) for details) but only trihalomethanes (THMs) will be discussed here. THMs are by-products of the water disinfection process. The relevance of regulating the maximum levels of this contaminant in treated water is that the THMs are suspected carcinogens. Carbon precursors are very diverse and not all of them produce THMs; the focus is on total trihalomethane formation potential (TTHMFP).

Given the significance of agricultural drainage to the TTHMFP levels, it is possible that extensive land fallowing caused by the Water Bank transfers could have resulted in lower TTHMFP levels. A total of 144,000 acres of crops were fallowed (see Figure 3). The drainage volumes from the irrigated lands peak twice a year, once during the winter as a result of precipitation and leaching practices, and the second time at the peak of the irrigation season. The contributions of irrigation to the TTHMFP levels was examined by estimating the amount of return flows that would have been produced by land fallowed by the Water Bankt ransfers. The resulting analyses indicate that without the land fallowings, higher TTHMFP levels would have occurred. However, the monitoring results (e.g. see Figure 4) no net benefit was observed; the negative influence of the ongoing drought overshadowed the potential water quality improvements as a result of the Water Bank.

4.5 Impacts on Waterfowl Populations

Historically, presently less than 300,000 acres of wetland remain in the Central Valley of California, 70 percent of which are in private ownership. Ninety-five percent of the

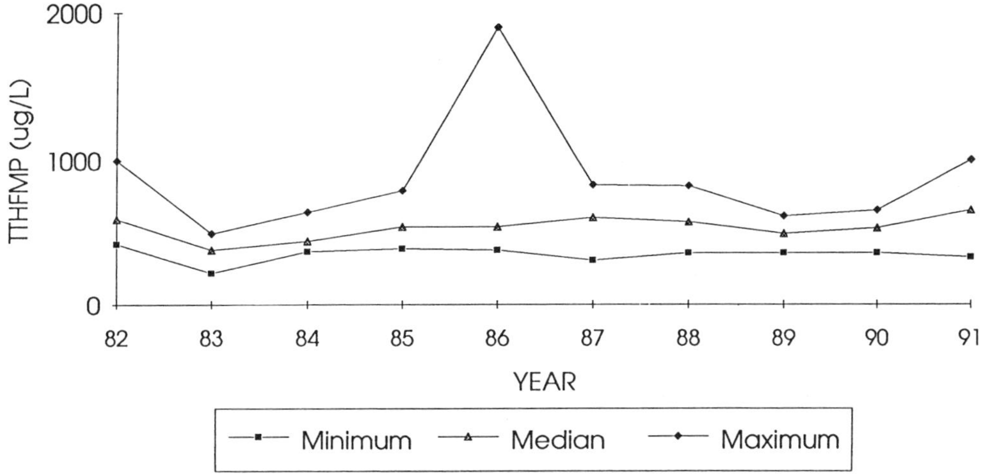

Figure 4. TTHMFP at Banks Pumping Station, 1982-1991

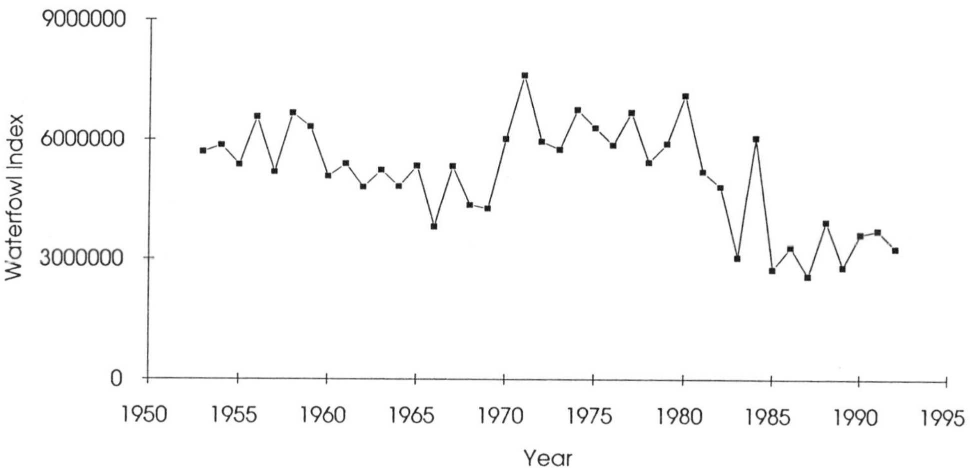

Figure 5. Mid-Winter Waterfowl Inventory, California 1952-1991

available wetlands do not have natural supplies and require artificial flooding and management. The nesting density, and whether a bird will use a specific area for nesting, depends on numerous factors including the species of waterfowl, vegetation type and density, proximity to wetlands, and abundance and species composition of predators. Reduction in suitable habitat will lead to a decrease in the food available. Overcrowding of remaining habitats, another potential impact of the Water Bank transfers, can result in increased disease transmission. All of these effects are increased by land fallowing.

The value of habitats, be they natural or artificial, depends on the time of year. The types of habitats that must be considered in evaluating the impact on waterfowl populations includes agricultural fields, uplands, tidal and non-tidal marshes, channels and open water, and lakes.

The Water Bank transfers took place where managed wetlands exist. The mid-winter waterfowl inventory for all of California shows a declining trend for the past several decades. Some fluctuations have occurred but a general decline is apparent from Figure 5. Extensive land fallowing will exacerbate the rate of decline. The loss of 6500 acres of rice fields due to fallowing represents a loss of waterfowl opportunity. However, coordination by Department of Water Resources, State Department of Fish and Game and private landowners allowed the managed wetlands to be flooded earlier than usual. Field operations also were adjusted to minimize impacts on waterfowl nesting.

The Water Bank contracts stipulated that all fallowed fields must be kept clean, in order to reduce water loss through transpiration. however, after it was discovered that mowing or disking the fields that waterfowl nests were destroyed, the Department of Water Resources amended its policies. A subsequent policy allowed fields to go undisturbed in order to protect nesting waterfowl.

5.0 SUMMARY ASSESSMENT OF IMPACT

Since it appears unlikely that the State of California will embark on major new water development schemes, some reallocation of existing supplies is inevitable and will result in increases in the pressure for water transfers. Thus, there was the potential for the transfers of water as a result of the Water Bank to have had a number of impacts. The potential impacts in the physical and biological regimes identified in the study included assessments of impacts with respect to groundwater pumpage, land subsidence, surface water quality, and waterfowl.

A very important difficulty in the assessment arose because the Water Bank transfers represented only an increment of approximately 5 percent to an ongoing transfer of water; thus, this necessitated isolation of the effects of Water Bank transfers in response to a small change in an ongoing transfer program.

The most significant negative environmental impact of water transfers was increased depletion of groundwater resources. Other environmental impacts, whether detrimental or beneficial, are difficult to distinguish from the effects of the ongoing drought and changing agricultural market conditions.

References

Bautista, E., and McBean, E., "Effects on Physical and Biological Resources - Actual, Perceived and Anticipated", ed. H. Vaux, H. Carter, and A. Scheuring, Sharing Scarcity: Gainers and Losers in California Water Transfer, University of California, Davis, 1993.

Hydrology Consultants Inc., "Preliminary Analysis of Potential Land Subsidence/Impacts of Water Transfers from Yolo County, California", prepared for Yolo County Flood Control and Water Connection District, Davis, California, September 1992.

Meador, M.R., "Interbasin Water Transfer: Ecological Concerns", Fisheries, Vol. 17, pp.17-22, 1992

Saggs, H., The Greatness that was Babylon, Hathorn Books Inc., N.Y., 1962

HYDROLOGICAL IMPACTS OF THE PATHAKHALI KONAI BEEL PROJECT

M Mozzammel Hoque

Institute of Flood Control and Drainage Research
Bangladesh University of Engineering and Technology
Dhaka

ABSTRACT

The construction of Flood Control Drainage and Irrigation (FCDI) projects is, in fact, an intervention against the nature. The changes in landscape resulting from the construction of flood control embankment affect the environment. The impacts of the Pathakhali-Konai project on the water resources system of the area have been presented in this paper. The specific impacts on surface water, groundwater, erosion and deposition, and flooding and drainage conditions have been evaluated.

The annual maximum discharge of the Bangsi river, the only surface water source of the project, shows virtually no change during pre-project period; during post project period the results show a significant increasing trend. In general, a rising trend in groundwater during monsoon has been observed where during winter an insignificant variation has been found. It has been revealed that the project has accelerated the sand deposition resulting in increased drainage congestion and decreased in land productivity. The study reveals that there is an increased erosion during the post project period in the river Bangsi at the vicinity of the project area. The study also reveals that the embankment has successfully protected the area from the spring flood, but the protection from monsoon floods and the drainage congestion have not been improved significantly by the embankment.

V. P. Singh and B. Kumar (eds.), Water Resources Planning and Management, 373–386.
© 1996 *Kluwer Academic Publishers. Printed in the Netherlands.*

INTRODUCTION

The topography, rainfall, and the river regime of Bangladesh are ideal for agriculture. But due to its location Bangladesh is subject to frequent floods which seriously affect the agricultural development. The Flood Control, Drainage, and Irrigation (FCDI) project has been considered as one of the alternatives to achieve a break through in agricultural development. Since its inception, upto 1990 the Bangladesh Water Development Board (BWDB), an organization for planning and execution of water development activities in Bangladesh, has completed 437 FCDI projects. The implemented projects have shown positive impacts on over all economy of the country, however, in many cases the adverse impacts have been reported.

The Pathakhali Konai is a small scale Flood Control, Drainage and Irrigation Project and known as the Early Implementation Project (EIP). The project is so called because, the planning, design, and implementation of this type of projects is usually done in two to three years time and the benefit from the project to the people is quick and high compared to the large Scale FCDI projects. Tho project cost is small and the management is simpler due to application of appropriate Technology (Shahjahan, 1991). The project was initiated in 1967 and implemented in 1979 and further extended in 1983. The underlying objectives was to construct the flood control embankment and drainage structure to increase food production as rapidly as possible.

The Pathakhali Konai, project has been undertaken in the present study to evaluate its impacts on water resources systems, inside and outside the project area. This project has been selected, because the project is located very close to Dhaka, the capital of Bangladesh and the implementation of the project was completed more than 10 years ago and 10 years may considered long time to show any possible impacts of the project.

PROJECT DESCRIPTION

Project Location, Boundary and Features

The Pathakhali Konai project is located at Mirzapur in Tangail district at longitude 90° 8'N and latitude 24°6'E, about 60 km North and North West of Dhaka, the capital of Bangaladesh. The project is suited along the eastern side of the Bangsi River from Tarappur to Hatubhangha in Mirzapur (see Figure 1) is protected

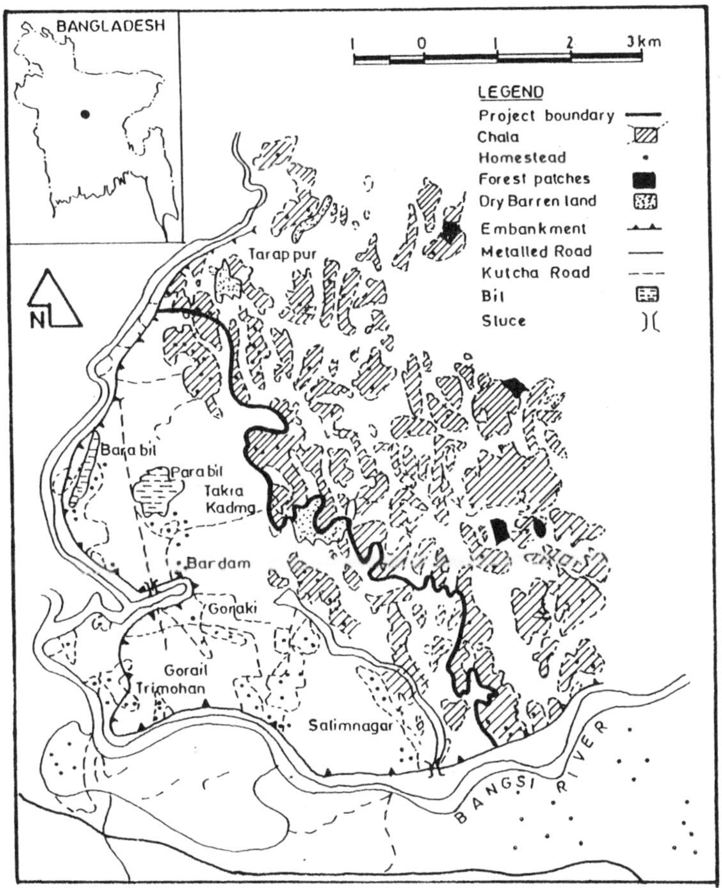

Figure 1: Pathakhali Konai Beel Project Area

with a river side embankment of total length of 15.5 km with two sluice gates, one of them is installed at upstream at Bardam for water intake and the other one is at downstream at Salimnagar for drainage purpose. The Eastern and the Northern boundary of the region is marked by a slightly raised ground and forests of the Modhupur Tract. The completed project is supposed to serve an estimated area of 2500 ha. The medium and marginal farmers' groups are predominant in the area. These groups are most receptive to new farming systems, inputs, and technology.

Climate and Hydrology

The area has a pronounced tropical monsoon climate. The mean temperatures range from about 85°F in April to about 65°F in January. The mean annual temperature is about 78°F. The mean daily temperature fluctuation in the dry season is about 24°F and in the rainy season about 9°F. The mean relative humidity rages from about 65% in March to about 90% in June and August. About 95% of the annual rainfall occurs in 7 months from April to October.

Lying on the left bank of the river Bangshi, the area is essentially a flood plain of the river. One small flood-season channel, namely, the Pathakhali Khal, issuing out of the main river Bangshi and rejoining it downstream, divides the region into two halves. A few marshes exist in this region. Flood waters from the Bangshi used to enter the area through the Pathakhali Khal and inundate it almost annually.

Soil and Topography

The Pathakhali Konai project area belongs to the older Jamuna meander flood plain. The textural variation of these area are due to sedimentation rather than to soil formation. The materials of the soil has 42 to 55 % clay and 10 to 30 % sand regardless of the state of weathering. The clay percentage decreases through the subsoil to the topsoil, with proportionate increase in the silt and sand percentages. Topsoils, generally have 12 to 35 % sand. There is no evidence of clay accumulation

METHODOLOGY OF STUDY

To assess the impact of the Pathakhali Konai embankment on water resources system of the area, an analysis of surface water, groundwater, drainage conditions, erosion and deposition, and irrigation water availability has been made. The study is based on the secondary hydrological data, field visits, and questionnaire survey. The river water level and discharge and groundwater data have been collected from the Hydrology Directorate of Bangladesh Water Development Board. The data collected from the secondary source have been supplemented, when required, by collecting data during the field visits. The data have been analyzed statistically and graphically (Hoque, 1992). The graphical results have been presented in this papers.

RESULTS AND DISCUSSIONS

Surface Water System

The river Bangsi is the main surface water source in the project area. To evaluate the sustainable water use and environmental impacts of the embankment, it is necessary to make an evaluation of availability of surface water and long term changes in the river flow. For making an assessments of discharge and water level of the river Bangsi, data from Mirzapur station have been analyzed graphically. The results of analysis of the annual highest stage show a decreasing trend upto 1979 but this trend is not observed during 1979 to 1990. The trend is not significant when the total period of analysis is considered. The results of the yearly lowest water level show no change either upto 1979 or for the period from 1979 to 1990, when considered separately. But, if the total period of analysis is considered, a significant decreasing trend is observed. The results of the analysis are shown in Figures 2a and 2b.

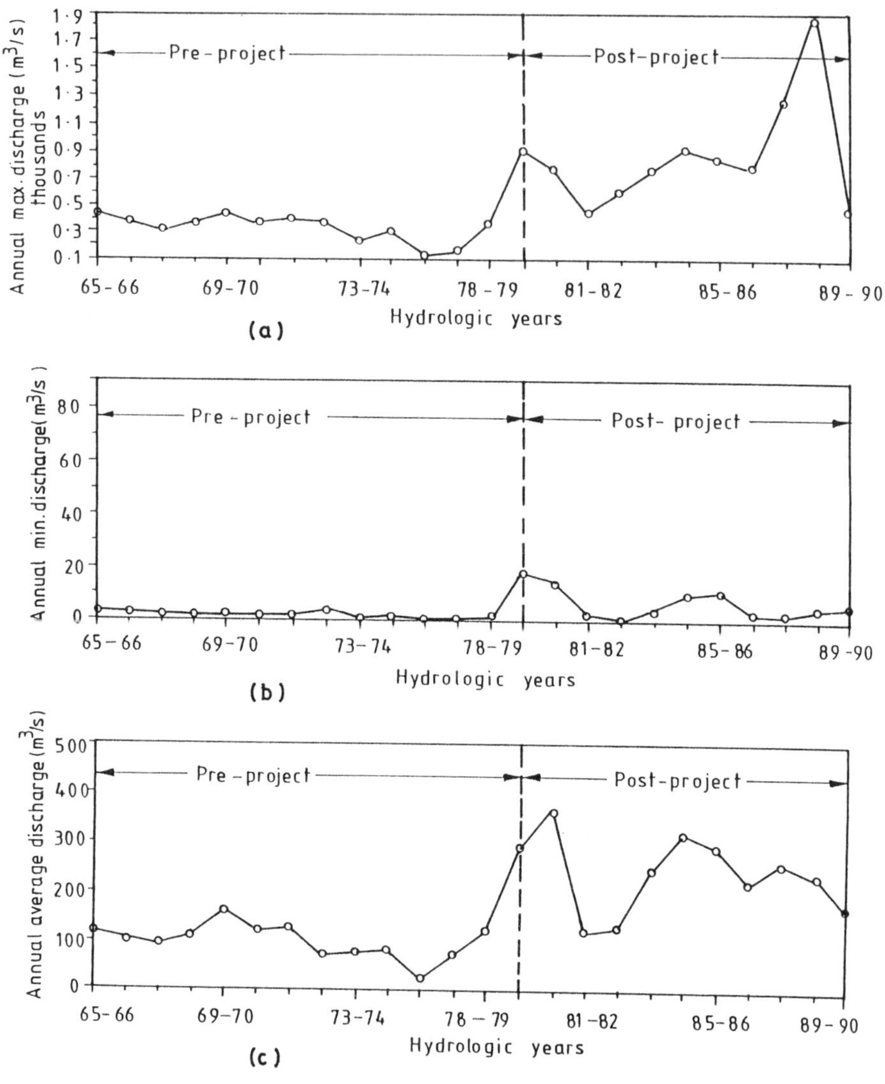

Figure 2: Variation of Bangsi River Discharge at Mirzapur Station: (a) Annual Maximum Discharge, (b) Annual Minimum Discharge (c) Annual Average Discharge.

The results show virtually no trend upto 1979 in annual maximum discharge of the Bangsi River at Mirzapur, but from 1979 to 1990 a significant increasing trend is observed. When the results of total period of analysis are considered, a significant increasing trend is observed (see Figure 2c).

For annual minimum discharge of the Bangsi river at Mirzapur the results show again no trend, neither increasing nor decreasing upto 1979. No trend is observed from 1979 to 1990 and for the whole period of analysis (see Figure 3a).

The results of annual average discharge show that upto 1979 there is a decreasing trend which may be statistically considered insignificant. The trend observed from 1979 to 1990 is increasing but again the trend is statistically insignificant. However, when the results of the total period is considered, the trend is significant(see Figure 3b).

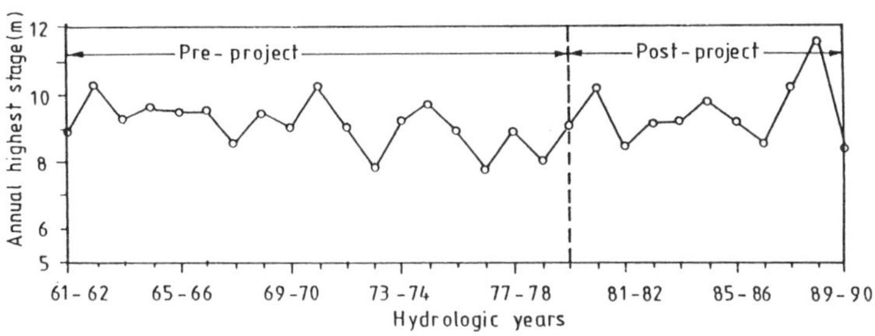

(a)

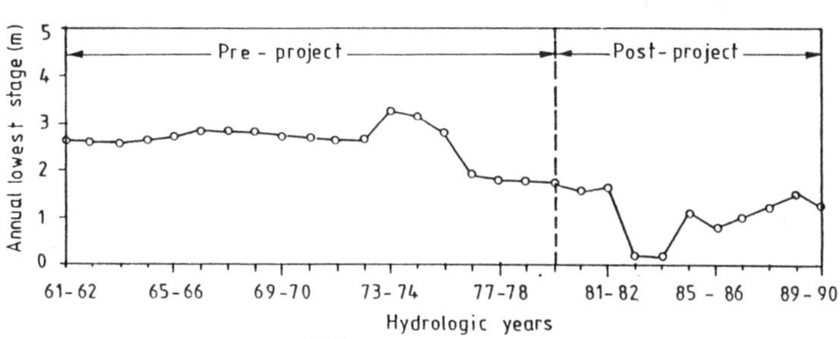

(b)

Figure 3: Variation of Bangsi River Water Level at Mirzapur Station: (a) Annual Highest Water Level, (b) Annual Lowest Water Level.

Groundwater System

During dry season groundwater is used in the project area for irrigation purpose. The water is withdrawn generally by Deep Tube wells (DTW) and Shallow Tube wells (STW). Currently 30 DTW and 85 STW are being used for this purpose.

The project area does not have any groundwater observation well. As a result, the observation wells installed outside at the vicinity of the project area have been considered to see sustainability of long term use of groundwater as a resource. The location of the selected observation wells are shown in Figure 4. The weekly observed yearly minimum, maximum, and mean water level data from 1954 to 1990 for nine selected wells have been considered for a trend analysis.

The analysis for observation well TA-01 shows rising trend in water table from 1971 to 1991 during the monsoon period but in dry months the variation is less significant. The results from the observation wells TA-02 and DA-57 do not show any change where the results of DA-92 show significant increasing trend in only yearly lowest water level. The results from observation well TA-22 show rising trend in water table during monsoon periods through the period of analysis, but in dry period the water level variation is insignificant with exception in the year 1988. The results for TA-01 are shown in Figures 5a to 5b.

Bank Spilling and Drainage Congestion

The project area was vulnerable to flood, sand deposition, and erosion in the pre-embankment situation. About 40 - 60 % of the crops would damage almost every year by flood. The average elevation of the project area is less than the elevation of the flood water level of the river Bangsi. The major part of the project area is inundated by flood depth of 1.8 to 3.6 m and the average terrain level of the Konai beel is 5.40 m. The drainage to Bangsi river on gravity is impossible for a period of 4-5 months each year. At lower levels of the river, the project area drains out properly except the beels.

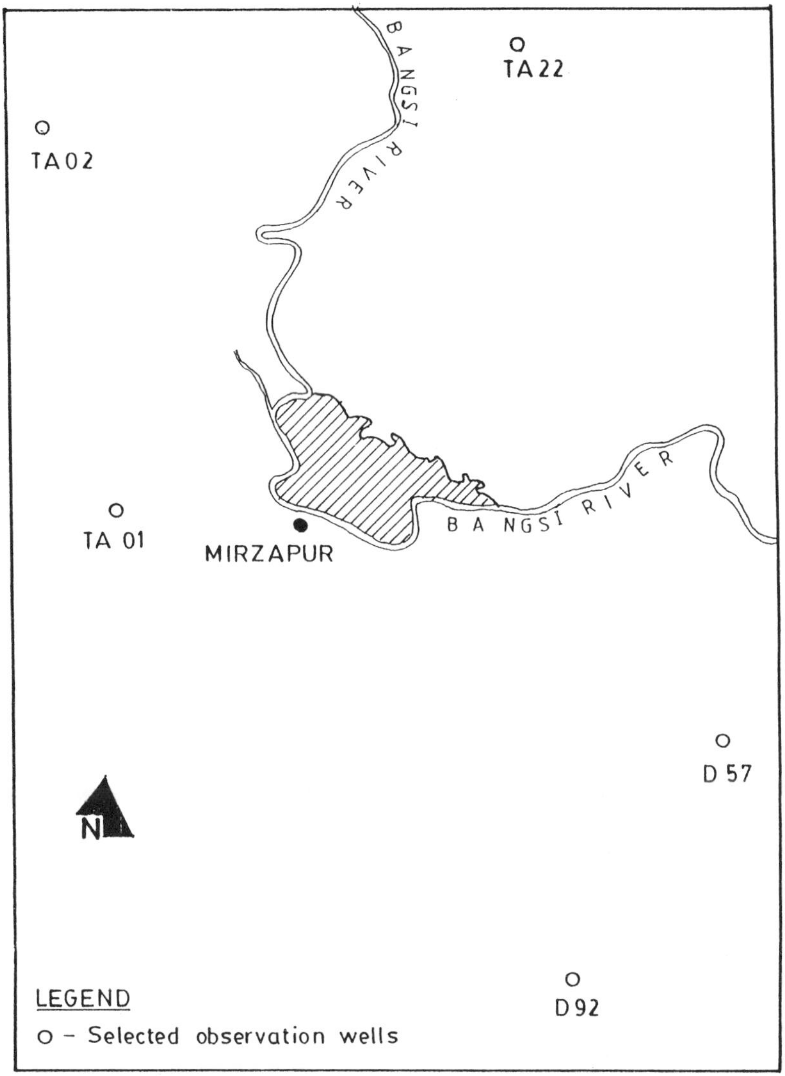

Figure 4: Location of Groundwater Observation Wells.

About 90 % of the project area is flooded each year. Flooding inside the project area is caused by local heavy rainfall, floods from the North-East undulating high land area, and backwater flow through the breaching of embankment which is caused due to the embankment failure. Flood from the North-East hilly region causes serious damages, because, there is no suitable passage of the flood water to Bangsi river.

The amount of land flooded each year after the construction of embankment has not changed significantly from the pre-embankment condition.

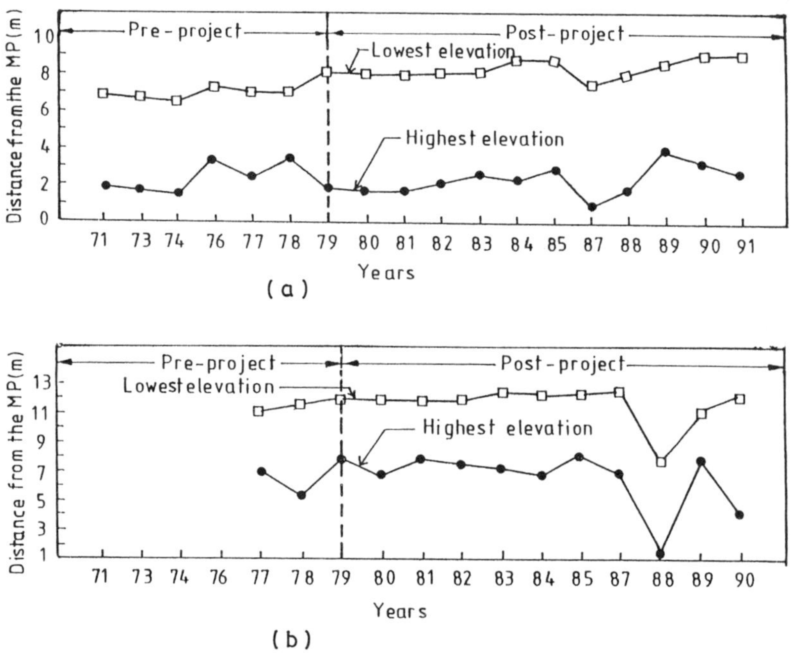

Figure 5: Variation of Groundwater Table in the Project Area: (a) Observation Well TA-01, (b) Observation Well TA-22.

The natural drainage system of the project area has been disturbed by the construction of flood control embankment. Drainage facilities have been included in the project to drain out the excess rainfall from the project area. Insufficient drainage in the pre-project area restricted the availability of land for cultivation. The drainage system is not improved as expected after the construction of embankment. The sluice gate at Bardam is in proper condition but that at Salimnagar is not in operation. The drainage canal at the inlet of the sluice gate has been silted with sand deposition making the regulator in effective.

The questionnaire survey indicates that the flood water due to internal rainfall does not drain properly. An estimate based on responses reveals that about 25 % of the crop is normally damaged due to drainage congestion. The cultivation of Boro and Ropa paddy are suffered due to the drainage congestion.

Erosion, Deposition and Sedimentation

The normal water flow path was interrupted by embankment construction and as a result, deposition of sand in some places has been accelerated. During dry season, the water level at Mirzapur Ferry Ghat and 0.5 km upstream and 2 km down stream is about 5 to 7m. But other places at the vicinity of the project the river remains almost dry. During past embankment period, the river Bangsi in some places are being severely scoured. At Feringgipara (Trimohan) the left bank and at Mirzapur Ferry Ghat (see Figure 6a and 6b), the right bank and the river bed are being scoured. Near Jugirkofa, the left bank and the river bed scouring are also seen. Flood associated with sand deposition was a genuine threat in the past to the farmers to grow crop. The embankment has successfully prevented the sand from entering into the crop fields. The effect of flash flood in the pre-embankment situation is well controlled by the embankment and the sluice gates.

The local farmers raised complaints that sand deposited since the construction of the embankment between the Bangsi river and the embankment. During 1988 abnormal flood, the embankment failed

(a)

(b)

Figure 6: Picture Showing Bank Erosion: (a) Feringgipara (Trimohan), (b) Mirzapur Ferry Ghat.

near Trimohan and flood water entered into the project area, causing sand deposition. Trimohan, Jagirkufa, Goraki-Garaki were badly affected. However, the farmers of these area had removed the deposited sand to get back the fertility of their lands (see Figure 7).

Figure 7: Picture Showing Sand Deposition on Agricultural Land.

Water Availability

The number of irrigation units and irrigation intensity are increasing very rapidly during the post project period. At the beginning of irrigation season there is no problem of water availability in the whole project area. Gradually with pumping, the discharge rate is reduced. It is observed during the field visit and questionnaire survey that STWs are greatly affected in the peak irrigation season and virtually the hand tube wells fail to draw water. Normally, the surface water dose not face problem except one LLP at Garaki.

It was found that the pumping units are not installed properly. As a result, the DTWs and the LLPs are in good condition, though their discharge is reduced to some extent.

CONCLUSIONS

The Comparison between the pre and post project conditions has shown significant impact of the embankment on the hydrological conditions of the project area. The study has been carried out with the secondary data and with the information collected from questionnaire survey. Therefore, the results presented may be considered only indicative. However, to evaluate the impact quantitatively, systematic monitoring should be carried out on river discharge, groundwater, siltation, erosion, and water quality.

REFERENCE

1. Hoque, M.M. and others (1992), Preliminary Environmental Examination of the Pathakhali-Konai Project, Draft Report, Institute of Flood Control and Drainage Research, Bangladesh University of Engineering & Technology and the Royal Institute of Technology, Stockholm, Sweden.

2. Shahjahan, M.(1991), Project Description of Site Selected for Intensive Environmental Study, Proceedings of the Workshop on Methodology for Sustainable Water Resources Management and Development in Bangladesh, January 16-17, 1991 held at Bangladesh University of Engineering and Technology, Dhaka, WR Report No.B1, 1991 Royal Institute of Technology, Sweden.